CAMBRIDGE LIBRARY COLLECTION

Books of enduring scholarly value

Botany and Horticulture

Until the nineteenth century, the investigation of natural phenomena, plants and animals was considered either the preserve of elite scholars or a pastime for the leisured upper classes. As increasing academic rigour and systematisation was brought to the study of 'natural history', its subdisciplines were adopted into university curricula, and learned societies (such as the Royal Horticultural Society, founded in 1804) were established to support research in these areas. A related development was strong enthusiasm for exotic garden plants, which resulted in plant collecting expeditions to every corner of the globe, sometimes with tragic consequences. This series includes accounts of some of those expeditions, detailed reference works on the flora of different regions, and practical advice for amateur and professional gardeners.

A Treatise on the Theory and Practice of Landscape Gardening

This 1841 work by the American landscape designer and writer Andrew Jackson Downing (1815–52), reissued here in its 1849 fourth edition, was the first such book published in the United States. Downing, the son of a nurseryman, saw that a 'taste for rural improvements of every description is advancing silently, but with great rapidity in this country', and he aims to provide the prosperous east-coast dweller with a guide to beautifying his surroundings. The emphasis is on landscape and overall effects rather than the minutiae of gardening, with chapters on plantations, specimen trees, and the construction of walks, water features, and other architectural elements. Downing went on to edit The Horticulturist magazine and to work on significant landscape projects, including the grounds of the White House and the surroundings of the Smithsonian Institution, before he was tragically killed, aged only 36, in an explosion on a river steamer.

Cambridge University Press has long been a pioneer in the reissuing of out-of-print titles from its own backlist, producing digital reprints of books that are still sought after by scholars and students but could not be reprinted economically using traditional technology. The Cambridge Library Collection extends this activity to a wider range of books which are still of importance to researchers and professionals, either for the source material they contain, or as landmarks in the history of their academic discipline.

Drawing from the world-renowned collections in the Cambridge University Library and other partner libraries, and guided by the advice of experts in each subject area, Cambridge University Press is using state-of-the-art scanning machines in its own Printing House to capture the content of each book selected for inclusion. The files are processed to give a consistently clear, crisp image, and the books finished to the high quality standard for which the Press is recognised around the world. The latest print-on-demand technology ensures that the books will remain available indefinitely, and that orders for single or multiple copies can quickly be supplied.

The Cambridge Library Collection brings back to life books of enduring scholarly value (including out-of-copyright works originally issued by other publishers) across a wide range of disciplines in the humanities and social sciences and in science and technology.

A Treatise on the Theory and Practice of Landscape Gardening

With a View to the Improvement of Country Residences, Comprising Historical Notices and General Principles of the Art

ANDREW JACKSON DOWNING

CAMBRIDGE
UNIVERSITY PRESS

University Printing House, Cambridge, CB2 8BS, United Kingdom

Cambridge University Press is part of the University of Cambridge.
It furthers the University's mission by disseminating knowledge in the pursuit of
education, learning and research at the highest international levels of excellence.

www.cambridge.org
Information on this title: www.cambridge.org/9781108083294

© in this compilation Cambridge University Press 2019

This edition first published 1849
This digitally printed version 2019

ISBN 978-1-108-08329-4 Paperback

This book reproduces the text of the original edition. The content and language reflect
the beliefs, practices and terminology of their time, and have not been updated.

Cambridge University Press wishes to make clear that the book, unless originally published
by Cambridge, is not being republished by, in association or collaboration with,
or with the endorsement or approval of, the original publisher or its successors in title.

VIEW IN THE GROUNDS AT BLITHEWOOD, DUTCHESS Cº N.Y.
THE RESIDENCE OF ROBERT DONALDSON ESQ.

A

TREATISE

ON

THE THEORY AND PRACTICE

OF

LANDSCAPE GARDENING,

WITH A VIEW TO

THE IMPROVEMENT OF COUNTRY RESIDENCES.

COMPRISING

HISTORICAL NOTICES AND GENERAL PRINCIPLES OF THE ART,
DIRECTIONS FOR LAYING OUT GROUNDS AND ARRANGING PLANTATIONS,
THE DESCRIPTION AND CULTIVATION OF HARDY TREES,
DECORATIVE ACCOMPANIMENTS TO THE HOUSE AND GROUNDS,
THE FORMATION OF PIECES OF ARTIFICIAL WATER, FLOWER GARDENS, ETC.

WITH REMARKS ON

RURAL ARCHITECTURE.

BY A. J. DOWNING,
AUTHOR OF DESIGNS FOR COTTAGE RESIDENCES, ETC.

"Insult not Nature with absurd expense,
Nor spoil her simple charms by vain pretence;
Weigh well the subject, be with caution bold,
Profuse of genius, not profuse of gold."

LONDON:
LONGMAN, BROWN, GREEN, AND LONGMANS.
1849.

TO

JOHN QUINCY ADAMS, LL.D.,

EX-PRESIDENT OF THE UNITED STATES;

THE LOVER OF RURAL PURSUITS,

AS WELL AS

THE DISTINGUISHED PATRIOT, STATESMAN,

AND SAGE;

THIS VOLUME,

BY PERMISSION,

IS RESPECTFULLY AND AFFECTIONATELY

DEDICATED,

BY HIS FRIEND,

THE AUTHOR.

PREFACE

TO THE FOURTH EDITION.

IT is even more gratifying to the author of this work to know, from actual observation, that the public taste in Rural Embellishment has, within a few years past, made the most rapid progress in this country, than to feel assured by the call for a fourth edition, that his own imperfect labors for the accomplishment of that end have been most kindly appreciated.

In the present edition considerable alterations and amendments have been made in some portions—especially in that section relating to the nature of the Beautiful and the Picturesque. The difference among critics regarding natural expression and its reproduction in Landscape Gardening, has led him more carefully to examine this part of the subject, in order, if possible, to present it in the clearest and most definite manner.

The whole work has also been revised, and more copiously illustrated, and is now offered in a more complete form than in any previous edition.

A. J. D.

Newburgh, New York, Jan. 1849.

PREFACE.

A TASTE for rural improvements of every description is advancing silently, but with great rapidity in this country. While yet in the far west the pioneer constructs his rude hut of logs for a dwelling, and sweeps away with his axe the lofty forest trees that encumber the ground, in the older portions of the Union, bordering the Atlantic, we are surrounded by all the luxuries and refinements that belong to an old and long cultivated country. Within the last ten years, especially, the evidences of the growing wealth and prosperity of our citizens have become apparent in the great increase of elegant cottage and villa residences on the banks of our noble rivers, along our rich valleys, and wherever nature seems to invite us by her rich and varied charms.

In all the expenditure of means in these improvements, amounting in the aggregate to an immense sum, professional talent is seldom employed in Architecture or Landscape Gardening, but almost every man fancies himself an amateur, and endeavors to plan and arrange his own residence. With but little practical knowledge, and few correct principles for his guidance, it is not surprising that we witness much incongruity and great waste of time and money. Even those who are familiar with foreign works on the subject in question labor under many obstacles in practice, which grow out of the difference in our soil and climate, or our social and political position.

These views have so often presented themselves to me of late, and have been so frequently urged by persons desiring advice, that I have ventured to prepare the present volume, in the hope of supplying, in some degree, the desideratum so much felt at present. While we have treatises, in abundance, on the various departments of the arts and sciences, there has not appeared even a single essay on the elegant art of Landscape Gardening. Hundreds of individuals who wish to ornament their grounds and embellish their places, are at a loss how to proceed, from the want of some *leading principles*, with the knowledge of which they would find it comparatively easy to produce delightful and satisfactory results.

In the following pages I have attempted to trace out such principles, and to suggest practicable methods of embellishing our Rural Residences, on a scale commensurate to the views and means of our proprietors. While I have availed myself of the works of European authors, and especially those of Britain, where Landscape Gardening was first raised to the rank of a fine art, I have also endeavored to adapt my suggestions especially to this country and to the peculiar wants of its inhabitants.

As a people descended from the English stock, we inherit much of the ardent love of rural life and its pursuits which belongs to that nation; but our peculiar position, in a new world that required a population full of enterprise and energy to subdue and improve its vast territory, has, until lately, left but little time to cultivate a taste for Rural Embellishment. But in the older states, as wealth has accumulated, the country become populous, and society more fixed in its character, a return to those simple and fascinating enjoyments to be found in country life and rural pursuits, is witnessed on every side. And to this innate feeling, out of which grows a strong attachment to natal soil, we must look for a counterpoise to the great tendency towards constant change, and the restless spirit of emigration, which form part of our national character; and which, though to a certain extent highly necessary to our national prosperity, are, on the other hand, opposed to social and domestic happiness. "In the midst of the continual movement which agitates a democratic community," says the most philosophical writer who has yet discussed our institutions, "the tie which unites one generation to another is relaxed or broken; every man readily loses the trace of the ideas of his forefathers, or takes no care about them."

The love of country is inseparably connected with the *love of home*. Whatever, therefore, leads man to assemble the comforts and elegancies of life around his habitation, tends to increase local attachments, and render domestic life more delightful; thus not only augmenting his own enjoyment, but strengthening his patriotism, and making him a better citizen. And there is no employment or recreation which affords the mind greater or more permanent satisfaction, than that of cultivating the earth and adorning our own property. "God Almighty first planted a garden; and, indeed, it is the purest of human pleasures," says Lord Bacon. And as the first man was shut out from the *garden*, in the cultivation of which no alloy was mixed with his happiness, the desire to return to it seems to be implanted by nature, more or less strongly, in every heart.

In Landscape Gardening the country gentleman of leisure finds a resource of the most agreeable nature. While there is no more rational pleasure than that derived from its practice by him, who

"Plucks life's roses in his quiet fields,"

the enjoyment drawn from it (unlike many other amusements) is unembittered by the after recollection of pain or injury inflicted on others, or the loss of moral rectitude. In rendering his home more beautiful, he not only contributes to the happiness of his own family, but improves the taste, and adds loveliness to the country at large. There is, perhaps, something exclusive in the taste for some of the fine arts. A collection of pictures, for example, is comparatively shut up from the world, in the private gallery. But the sylvan and floral collections,— the groves and gardens, which surround the country residence of the man of taste,—are confined by no barriers narrower than the blue heaven above and around them. The taste and the treasures, gradually, but certainly, creep beyond the nominal boundaries of the estate, and re-appear in the pot of flowers in the window, or the luxuriant, blossoming vines which clamber over the porch of the humblest cottage by the way side.

In the present volume I have sought, by rendering familiar to the reader most of the beautiful sylvan materials of the art, and by describing their peculiar effects in Landscape Gardening, to encourage a taste among general readers. And I have also endeavored to place before the amateur such directions and guiding principles as, it is hoped, will assist him materially in laying out his grounds and arranging the general scenery of his residence.

The lively interest of late manifested in Rural Architecture, and its close connexion with Landscape Gardening, have induced me to devote a portion of this work to the consideration of buildings in rural scenery.

I take pleasure in acknowledging my obligations and returning thanks to my valued correspondent, J. C. Loudon, Esq., F. L. S., etc., of London, the most distinguished gardening author of the age, for the illustrations and description of the English Suburban Cottage in the Appendix; to the several gentlemen in this country who have kindly furnished me with plans or drawings of their residences ; and to A. J. Davis, Esq., of New York, and J. Notman, Esq., of Philadelphia, architects, for architectural drawings and descriptions.

CONTENTS.

SECTION I.

HISTORICAL SKETCHES.

Objects of the art, page 18. The ancient and modern styles, p. 21. Their peculiarities, p. 23. Origin of the modern and natural style, p. 31. Influence of the English poets and writers, p. 32. Examples of the art abroad, p. 38. Landscape Gardening in North America, and examples now existing, p. 40.

SECTION II.

BEAUTIES OF LANDSCAPE GARDENING.

Capacities of the art, p. 61. The beauties of the ancient style, p. 62. The Beautiful and the Picturesque; their distinctive characteristics, with illustrations drawn from nature and painting, p. 63. Nature and principles of Landscape Gardening as an imitative art, p. 66. The Production of Beautiful Landscape, 67. Of Picturesque do., 68. Simple beauty of the art, p. 78. The principles of Unity, Harmony, and Variety, p. 80.

SECTION III.

WOOD AND PLANTATIONS.

The beauty of trees in rural embellishments, p. 85. Pleasure resulting from their cultivation, p. 88. Plantations in the ancient style; their formality, p. 89. In the modern style, p. 94. Grouping trees, p. 95. Arrangement and grouping in the Graceful school, p. 101. In the Picturesque school, p. 102.

xii   CONTENTS.

Illustrations in planting villa, ferme ornée, and cottage grounds, p. 113. General classification of trees as to forms, with leading characteristics of each class, p. 123.

SECTION IV.

DECIDUOUS ORNAMENTAL TREES.

The history and description of all the finest hardy deciduous trees. Remarks on their effects in Landscape Gardening, individually, and in composition; their cultivation, etc. The oak, p. 139. The elm, p. 152. The plane or buttonwood, p. 158. The ash, p. 162. The lime or linden, p. 167. The beech, p. 171. The poplar, p. 175. The horse chestnut, p. 181. The birch, p. 184. The alder, p. 189. The maple, p. 191. The locust, p. 196. The three-thorned acacia, p. 200. The Judas tree, p. 202. The chestnut, p. 204. The Osage orange, p. 209. The mulberry, p. 211. The paper-mulberry, p. 214. The sweet gum, p. 215. The walnut, p. 218. The hickory, p. 222. The mountain ash, p. 226. The ailantus, p. 230. The Kentucky coffee, p. 232. The willow, p. 234. The sassafras, p. 241. The catalpa, p. 242. The persimon, p. 244. The peperidge, p. 246. The thorn, p. 248. The magnolia, p. 250. The tulip-tree, p. 255. The dogwood, p. 259. The ginko, p. 261. The American cypress, p. 264. The larch, p. 268. The Virgilia, p. 276. The Paulownia, p. 278.

SECTION V.

EVERGREEN ORNAMENTAL TREES.

The history and description of all the finest hardy evergreen trees. Remarks on their effects in Landscape Gardening, individually and in composition. Their cultivation, etc. The pines, p. 280. The firs, p. 290. The cedar of Lebanon, and Deodar cedar, p. 296. The red cedar, p. 300. The arbor vitæ, p. 301. The holly, p. 304. The yew, p. 306.

SECTION VI.

VINES AND CLIMBING PLANTS.

Value of this kind of vegetation;—fine natural effects, p. 312. The European ivy, p. 316. The Virginia creeper, p. 316. The wild grape-vine, p. 317. The bittersweet,—the trumpet creeper, p. 317. The pipe vine, p.

CONTENTS.   xiii

318. The clematis,—the wistaria, p. 319. The honeysuckles and woodbines, p. 320. The climbing roses, p. 322. The jasmine and periploca, p. 323. Remarks on the proper mode of introducing vines, p. 324. Beautiful effects of climbing plants in connexion with buildings, p. 325.

SECTION VII.

TREATMENT OF GROUND—FORMATION OF WALKS.

Nature of operations on ground, p. 327. Treatment of flowing and of irregular surfaces to heighten their expression, p. 328,—of flats or level surfaces, p. 331. Rocks, as materials in landscape, p. 334. Laying out roads and walks; the approach, p. 336. Rules by Repton, p. 339. The drive and minor walks, p. 341. The introduction of fences, p. 343. Verdant hedges, p. 344.

SECTION VIII.

TREATMENT OF WATER.

Beautiful effects of this element in nature, p. 347. In what cases it is desirable to attempt the formation of artificial pieces of water, p. 348. Regular forms unpleasing, p. 350. Directions for the formation of ponds or lakes in the irregular manner, p. 351. Study of natural lakes, 352. Islands, p. 358. Planting the margin, p. 360. Treatment of natural brooks and rivulets, p. 363. Cascades and water-falls, 364. Legitimate sphere of the art in this department, p. 366.

SECTION IX.

LANDSCAPE OR RURAL ARCHITECTURE.

Difference between a city and country house, p. 369. The characteristic features of a country house, p. 370. Examination of the leading principles in Rural Architecture, p. 371. The harmonious union of buildings and scenery, 377. The different styles, p. 380. The Grecian style, its merits and associations, p. 381. Its defects for domestic purposes, p. 382. The Roman style. The Italian style, p. 385 ;—its peculiar features, and examples in this country, p. 388. Associations of the Italian style, p. 390. Swiss style, p. 392. The pointed or Gothic style,—leading features, p. 693. Castellated buildings, p. 396. The Tudor mansion, p. 398. Examples here, p. 400. The Elizabethan style, p. 401. The old English cottage,—its features, p. 402. Associations of the pointed style, p. 405. Examples in this country, p. 409. Individual tastes, p. 411. Entrance lodges, p. 412.

SECTION X.

EMBELLISHMENTS; ARCHITECTURAL, RUSTIC, AND FLORAL.

Value of a proper connexion between the house and grounds, p. 419. Beauty of the architectural terrace, and its application to villas and cottages, p. 420. Use of vases of different descriptions, p. 424. Sundials, p. 427. Architectural flower-garden, p. 428. Irregular flower-garden, p. 429. French flower-garden, p. 430. English flower-garden, p. 430. Mingled flower-garden, p. 437. General remarks on this subject, p. 437. Selections of showy plants, flowering in succession, p. 438. Arrangement of the shrubbery, and selection of choice shrubs, p. 442. The conservatory and green-house, 448. Open and covered seats, p. 454. Pavilions, p. 456. Rustic seats, p. 456. Prospect towers, p. 459. Bridges, p. 460. Rockwork, p. 461. Fountains of various descriptions, p. 466. Judicious introduction of decorations, p. 472.

APPENDIX.

I. Notes on transplanting trees, p. 475. Reasons for frequent failures in removing large trees, p. 476. Directions for performing this operation, p. 478. Selection of subjects, p. 479. Preparing trees for removal, p. 481. Transplanting evergreens, p. 482.

II. Description of an English suburban residence, *Cheshunt Cottage*, p. 484. With views and plans showing the arrangement of the house and grounds, p. 485. And mode of managing the whole premises, p. 487.

III. Note on the treatment of Lawns, p. 525.

IV. Note on professional quackery, p. 527.

V. Note on roads and walks, p. 530.

ESSAY ON LANDSCAPE GARDENING.

SECTION I.

HISTORICAL SKETCHES.

Objects of the Art. The ancient and modern styles. Their peculiarities. Sketch of the ancient style, and the rise and progress of the modern style. Influence of the English poets and writers Examples of the art abroad. Landscape Gardening in North America, and examples now existing.

" L'un à nos yeux présente
D'un dessein régulier l'ordonnance imposante,
Prête aux champs des beautés qu'ils ne connaissaient pas,
D'une pompe étrangère embellit leur appas,
Donne aux arbres des lois, aux ondes des entraves,
Et, despote orgueilleux, brille entouré d'esclaves;
Son air est moins riant et plus majestueux
L'autre, de la nature amant respectueux,
L'orne sans la farder, traite avec indulgence
Ses caprices charmants, sa noble négligence,
Sa marche irrégulière, et fait naître avec art
Des beautés du désordre, et même du hasard."

DELILLE.

U R first, most endearing, and most sacred associations," says the amiable Mrs. Hofland, "are connected with gardens; our most simple and most refined perceptions of beauty are combined with them." And we may add to this, that Landscape Gardening, which is an artistical combination of the beautiful in nature and art—an union of natural expression and harmonious cultivation—is capable of affording us the highest and most intellectual enjoyment to be found in any cares or pleasures belonging to the soil.

The development of the Beautiful is the end and aim of Landscape Gardening, as it is of all other fine arts. The ancients sought to attain this by a studied and elegant regularity of design in their gardens; the moderns, by the creation or improvement of grounds which, though of limited extent, exhibit a highly graceful or picturesque epitome of natural beauty. Landscape Gardening differs from gardening in its common sense, in embracing the whole scene immediately about a country house, which it softens and refines, or renders more spirited and striking by the aid of art. In it we seek to embody our *ideal* of a rural home; not through plots of fruit trees, and beds of choice flowers, though these have their place, but by collecting and combining beautiful forms in trees, surfaces of ground, buildings, and walks, in the landscape surrounding us. It is, in short, the Beautiful, embodied in a home scene. And we attain it by the removal or concealment of everything uncouth and discordant, and by the introduction and preservation of forms pleasing in their expression, their outlines, and their fitness for the abode of man. In the orchard, we hope to gratify the palate; in the flower garden, the eye and the smell; but in the landscape garden we appeal to that sense of the Beautiful and the Perfect, which is one of the highest attributes of our nature.

This embellishment of nature, which we call Landscape

Gardening, springs naturally from a love of country life, an attachment to a certain spot, and a desire to render that place attractive—a feeling which seems more or less strongly fixed in the minds of all men. But we should convey a false impression, were we to state that it may be applied with equal success to residences of every class and size, in the country. Lawn and trees, being its two essential elements, some of the beauties of Landscape Gardening may, indeed, be shown wherever a rood of grass surface, and half a dozen trees are within our reach; we may, even with such scanty space, have tasteful grouping, varied surface, and agreeably curved walks; but our art, to appear to advantage, requires some extent of surface—its lines should lose themselves indefinitely, and unite agreeably and gradually with those of the surrounding country.

In the case of large landed estates, its capabilities may be displayed to their full extent, as from fifty to five hundred acres may be devoted to a park or pleasure grounds. Most of its beauty, and all its charms, may, however, be enjoyed in ten or twenty acres, fortunately situated, and well treated; and Landscape Gardening, in America, combined and working in harmony as it is with our fine scenery, is already beginning to give us results scarcely less beautiful than those produced by its finest efforts abroad. The lovely villa residences of our noble river and lake margins, when well treated—even in a few acres of tasteful fore-ground,—seem so entirely to appropriate the whole adjacent landscape, and to mingle so sweetly in their outlines with the woods, the valleys, and shores around them, that the effects are often truly enchanting.

But if Landscape Gardening, in its proper sense, cannot be applied to the embellishment of the smallest cottage residences in the country, its principles may be studied with advantage, even by him who has only three trees to plant for ornament; and we hope no one will think his grounds too small, to feel willing to add something to the general amount of beauty in the country. If the possessor of the cottage acre would embellish in accordance with propriety, he must not, as we have sometimes seen, render the whole ridiculous by aiming at ambitious and costly embellishments; but he will rather seek to delight us by the good taste evinced in the *tasteful simplicity* of the whole arrangement. And if the proprietors of our country villas, in their improvements, are more likely to run into any one error than another, we fear it will be that of too great a desire for display—too many vases, temples, and seats,—and too little purity and simplicity of general effect.

The inquiring reader will perhaps be glad to have a glance at the history and progress of the art of tasteful gardening; a recurrence to which, as well as to the history of the fine arts, will afford abundant proof that, in the first stage or infancy of all these arts, while the perception of their ultimate capabilities is yet crude and imperfect, mankind has, in every instance, been completely satisfied with the mere exhibition of *design* or *art*. Thus in Sculpture, the first statues were only attempts to imitate rudely the *form* of a human figure, or in painting, to represent that of a tree: the skill of the artist, in effecting an imitation successfully, being sufficient to excite the astonishment and admiration of those who had not yet made such advances as to enable them to appreciate the superior beauty of *expression*.

Landscape Gardening is, indeed, only a modern word, first coined, we believe, by Shenstone, since the art has been based upon natural beauty; but as an extensively embellished scene, filled with rare trees, fountains, and statues, may, however artificial, be termed a landscape garden, the classical gardens are fairly included in a retrospective view.

All late authors agree in these two distinct and widely different modes of the art; 1st, the Ancient, Formal, or Geometric Style; 2d, the Modern, Natural, or Irregular Style.

THE ANCIENT STYLE. A predominance of regular forms and right lines is the characteristic feature of the ancient style of gardening. The value of art, of power, and of wealth, were at once easily and strongly shown by an artificial arrangement of all the materials; an arrangement the more striking, as it differed most widely from nature. And in an age when costly and stately architecture was most abundant, as in the times of the Roman empire, it is natural to suppose, that the symmetry and studied elegance of the palace, or the villa, would be transferred and continued in the surrounding gardens.

Nothing fills so grand a place in the history of the gardening of antiquity, as the great hanging gardens of Babylon. A series of terraces supported by stone pillars, rising one above the other three hundred feet in height, and planted with rows of all manner of stately trees, shrubs and flowers, interspersed with seats, and watered and supplied with fountains from the Euphrates; all this was indeed a princely effort of the great king, to recall to his Median queen the beauties of her native country. The "Paradises" of the Persians seem not only to have had straight walks bordered with blossoming trees, and overhung with exquisite lines of roses and other odoriferous shrubs, but to have been interspersed with occasional thickets, and varied with fountains, prospect towers, and aviaries for singing birds.

The Athenians borrowed their taste in gardens from Persia. The lime tree and the box lined their walks, and bore patiently the shears of symmetry; and a passion for fragrant flowers seems to have been greatly indulged by them. Their most celebrated philosophers made the sylvan, or landscape gardens of their time, their favorite schools. And the gardens of Epicurus and Plato appear to have been symmetrical groves of the olive, plane, and elm, enriched by elegant statues, monuments, and temples, the beauty of which, for their peculiar purpose, has never been surpassed by any example of more modern times. Among the Romans, ornamental gardening seems to have been not a little studied. The villas of the Emperors Nero and Adrian were enriched with everything magnificent and pleasing in their grounds; and the classically famous villas of Cicero at Arpinum, and of Pliny at Tusculum, with Cæsar's

> "Private arbors, and new planted orchards,
> On this side Tiber,"

are among the most celebrated specimens of the taste, among the ancients. Pliny's garden, of which a pretty minute account remains,—filled with cypresses and bay trees, planted to form a coursing place or hippodrome, adorned with *vis-à-vis* figures of animals cut in box trees, and decorated with fountains and marble alcoves, shaded by vines—seems, indeed, to have been the true classical type of all the later efforts of modern continental nations in their geometric gardens.

Of the latter, the Italians have been most successful in

their ornamental grounds. Their beautiful marbles seem to have been supplied by Art in too great profusion to be confined even to the colonnades of their villas, and broad enriched terraces, vases, and statues, everywhere enliven, and contrast with, the verdure of the foliage; trees and plants being often less abundant than the sculptural ornaments which they serve to set off to advantage. An island —Isola Bella—in one of their little lakes, has often been quoted as the most highly wrought type of the Italian taste; "a barren rock," says a spirited writer, "rising in the midst of a lake, and producing but a few poor lichens, which has been converted into a pyramid of terraces supported on arches, and ornamented with bays and orange trees of amazing size and beauty." The Villa Borghese, at Rome, is one of the most celebrated later examples, with its pleasure grounds three miles in circumference, filled with symmetrical walks, and abounding with an endless profusion of sculpture.

The old French gardens differ little from those of Italy, if we except that, with the same formality, they have more of theatrical display—frequently substituting gilt trellises and wooden statues for the exquisite marble balustrades and sculptured ornaments of the Italians. But we must not forget the crowning glory of the Geometric style, the gardens of Louis XIV. at Versailles. A prince whose grand idea of a royal garden was not compassed under two hundred acres devoted to that purpose, and who, when shown the bills of cost in their formation, amounting to two hundred millions of francs, quietly threw them into the fire, could scarcely fail, whatever the style of art adopted, in producing a scene of great splendor. He was fortunate, too, in his gardener, Le Notre, whose ideas, scarcely less superb than those of his master, kept pace so closely with his fancies, that he received the honor of knighthood, and was made general director of all the buildings and gardens of the time.

"The gardens of Versailles," says a tasteful English reviewer, "may indeed be taken as the great exemplar of this style; and magnificent indeed they are, if expense and extent and variety suffice to make up magnificence. To draw petty figures in dwarf-box and elaborate patterns in parti-colored sand, might well be dispensed with where the formal style was carried out on so grand a scale as this, but otherwise the designs of Le Notre differ little from that of his predecessors in the geometric style, save in their monstrous extent. The great wonder of Versailles was the well known labyrinth, not such a maze as is really the source of so much idle amusement at Hampton Court, but a mere ravel of interminable walks, closely fenced in with high hedges, in which thirty-nine of Æsop's fables were represented by painted copper figures of birds and beasts, each group connected with a separate fountain, and all spouting water out of their mouths! Every tree was planted with geometrical exactness, and parterre answered to parterre across half a mile of gravel. 'Such symmetry,' says Lord Byron, 'is not for solitude;' and certainly, the gardens of Versailles were not planted with any such intent. The Parisians do not throng there for the contemplation to be found in the 'trim gardens' of Milton. There is indeed a melancholy, but not a pleasing one, in wandering alone, through those many acres of formal hornbeam, when we feel that it requires the 'galliard and clinquant' air of a scene of Watteau; its crowds and love-making—its hoops and minuets—a ringing laugh and merry tamborine

—to make us recognise the real genius of the place. Taking Versailles on the gigantic type of the French school it need scarcely be said that it embraces broad gravelled terraces, long alleys of yew and hornbeam, vast orangeries, groves planted in the quincunx style, and water-works embellished with, and conducted through every variety of sculptured ornament. It takes the middle line between the two other geometric schools—admitting more sculpture and other works of art than the Italian, but not overpowered with the same number of 'huge masses of littleness' as the Dutch. There is more of promenade, less of parterre; more gravel than turf; more of the deciduous than the evergreen tree. The practical water-wit of drenching the spectators was in high vogue in the ancient French gardens; and Evelyn, in his account of the Duke de Richelieu's villa, describes with some relish how 'on going, two extravagant musketeers shot at us with a stream of water from their musket barrels.' Contrivances for dousing the visitors—'especially the ladies'—which once filled so large a space in the catalogue of every show place, seem to militate a little against the national character of gallantry; but the very fact that everything was done to surprise the spectator and stranger, evinces how different was their idea of a garden from the home and familiar pleasures which an Englishman looks to in his."

It is scarcely necessary for us to say, that this new splendor of the French in their gardens was more or less copied, at the time, all over Europe. "*Ainsi font les Français—voilà ce que j'ai vu en France*," was the law of fashion in the gardening taste from which there was no higher court of appeal. But, in copying, every nation seems to have mingled with the "grand style" some elementary notions of its own, expressive of national character or locality. The most marked of these imitators were the Dutch, whose style of ornamental gardening seems sufficiently unique to be worthy of being considered a separate school.

And how shall we characterize the Dutch school, which even to this day, in the Low Countries, has scarcely given way to the continental admiration for the *"jardin Anglais;"* this double distilled compound of labored symmetry, regularity, and stiffness, which seems to convey to the quiet owners so much pleasure, and to the tasteful traveller and critic so much despair! A stagnant and muddy canal, with a bridge thrown over it, and often connected with a circular fishpond; a grass slope and a mound of green turf, on which is a pleasure or banqueting house with gilt ornaments; numberless clipped trees, and every variety of trellis-work lively with green paint; in the foreground beds of gay bulbs and florists' flowers, interspersed with huge orange trees in tubs, and in the distance smooth green meadows—such are the unvarying features of the Hollander's garden or grounds.* The true Dutchman looks upon his garden as a quiet place to smoke and be "content" in; if he lazily saunters through, it is rather to enjoy the gay pencillings of some new bed of tulips than to enjoy the elegance and harmony of its design, the variety of scenery, or the freshness and beauty of the foliage. At the same time, he is neither exclusive nor secret with the stores of enjoyment which he has within its bounds; and very many of the private villas near Rotterdam, and in other parts of Holland, have mottoes like those inscribed

---

* In the neighborhood of Antwerp, not a long time since, was the villa of M. Smetz, where, among many things that were pretty, was the odd conceit of a lawn on which were a shepherd, his flock of sheep, and his dog cut in stone, and always looking "pastoral and country-like."

over the gateways—"Tranquil and Content," "My desire is satisfied"—(*genegentheiel is volden*)—"Friendship and sociability," and numerous others of a similar import.

As modern landscape gardening owes its existence almost entirely to the English, we must take a rapid glance at the early condition of the art in Great Britain, and its subsequent development to the present time.

It would appear to be an undeniable fact in the history of ornamental gardening that, from the time of William the Conqueror down to the latter part of the reign of Queen, Anne, and the beginning of that of George I., nothing was considered garden scenery except it was in the formal and geometric style.

The royal gardens of Henry VIII., at Nonsuch Palace, laid out in the beginning of the sixteenth century, may perhaps be taken as a type of the highest ideal of a garden at that period. Heutzner, in speaking of this place, after describing it as abounding in every species of costly magnificence, adds,—

> "This, which no equal has in art or fame,
> Britons deservedly do Nonsuch name."

Loudon remarks that "these gardens are stated, in a survey taken in the year 1650, above a century after Henry's death, to have been cut and divided into several alleys, compartments, and rounds, set about with thorn hedges. On the north side was a kitchen garden, very commodious, and surrounded with a wall fourteen feet high. On the west was a wilderness severed from the little park by a hedge, the whole containing ten acres. In the privy gardens were fountains and basins of marble, one of which is 'set round with six lilac trees, which trees bear no fruit, but only a pleasant smell.' In the kitchen garden were seventy-two fruit trees and one lime tree. Lastly, before the palace, was a neat handsome bowling-green, surrounded with a balustrade of freestone." Another writer, describing Nonsuch when in perfection, says, "In the pleasure and artificial gardens are many columns and pyramids of marble, two fountains that spout water, one round and the other like a pyramid, upon which are perched small birds that stream water out of their bills. There is besides another pyramid of marble full of concealed pipes, which spirt upon all who come within their reach."

In the reign of Elizabeth "trim gardens" seem to have been in high favor. Hatfield was one of the great estates of that period, and its gardens were described as "surrounded by a piece of water, with boats rowing through *alleys of well cut trees*, and labarynths made with great labor. There are jets d'eau, and a summer house, with many pleasant and fair fish ponds." The *Gardener's Labarynth*, a work intended to direct the taste of that day (1571), gives plates of "knotts and mazes curiously handled for the beautifying of gardens."

During the reign of James I. many fine country seats were either created or improved. Both the descriptions and the engravings of gardens of that period agree in placing before us grounds surrounded by high walls, divided into regular squares, compartments, or parterres, and ornamented with all kinds of trained and clipped trees, interspersed with statues—and, in the finest examples, not omitting that delightful puzzle of the time a "labarynth."

Lord Bacon attempted to reform the national taste during this reign, but apparently with little immediate success. He wished still to retain shorn trees and hedges, but proposed winter or evergreen gardens, and rude or neglected spots, as specimens of wild nature. "As for the making of knots or figures," says he, "with divers colored earths, they be but toys. I do not like images cut out in juniper or other garden stuff: they are for children."*

One gets a condensed idea of the taste of this and the previous century or two by a work published at Oxford by Commenius during the Commonwealth. "Gardening," says he, "is practised for food's sake in a kitchen garden and orchard, or for pleasure's sake in a green grass-plot and an arbor." In his details of the ornamental garden he adds, "the pleacher (*topiarius*) prepares a green plat of the more choice flowers and rarer plants, and adorns the garden with pleach-work; that is, with pleasant walks and bowers, &c., to conclude with water-works." He also informs us, respecting the parks, that "the huntsman hunteth wild harts, whilst he either allureth them into pitfalls, or killeth them, and what he gets alive he puts into a park."

In the reign of Charles II. the fame of Versailles, the most superb of all geometric gardens, created a sensation in England. Le Notre was of course immediately sent for by this monarch. He planted St. James and Greenwich parks, and thus aided by royal patronage, inspired the nobility with a desire for some of the more splendid formations of the French school of design. Chatsworth, the magnificent seat of the Duke of Devonshire, was laid out in a grandly formal manner, and the Earl of Essex and Lord Capel were among the foremost to emulate the glories of Versailles in their country places—the former nobleman

* Encyclopædia of Gardening.

sending his gardener (Rose) to France, in order to make himself thoroughly acquainted with all the beauties of that Royal garden.

The period of William and Mary's reign was remarkable for no great deviation from this style, except perhaps in substituting partially the Dutch formalities—such as iron trellis-work, clipped yews, and a greater profusion of verdant sculpture. Embroidered parterres and vegetable sculpture are said indeed to have arrived at their highest perfection in this period, or towards the year 1700; and we may get a good notion of the subjects most in vogue, by an extract from Pope's keen satire, written as late as 1713 (in the early part of Anne's reign), when it was beginning to get into disrepute.

> INVENTORY OF A VIRTUOSO GARDENER. Adam and Eve in yew; Adam, a little shattered by the fall of the tree of knowledge in the great storm; Eve and the serpent, very flourishing. Noah's ark in Holly; the ribs a little damaged for want of water.
> The tower of Babel, not yet finished.
> St. George, in box; his arm scarce long enough, but will be in a condition to stick the dragon by next April.
> Edward the Black Prince, in cypress.
> A pair of giants stunted, to be sold cheap.
> An old maid of honor, in wormwood.
> A topping Ben Jonson, in laurel.
> Divers eminent modern poets, in bays; somewhat blighted.
> A quick set hog, shot up into a porcupine, by being forgot a week in rainy weather.
> A lavender pig, with sage growing in his belly.

Whatever may have been the absurdities of the ancient style, it is not to be denied that in connexion with highly decorated architecture, its effect, when in the best taste—as the Italian—is not only splendid and striking, but highly suitable and appropriate. Sir Walter Scott, in an essay on landscape embellishment, says, "if we approve of Palladian architecture, the vases and balustrades of Vitruvius, the enriched entablatures and superb stairs of the Italian school of gardening, we must not, on this account, be construed as vindicating the paltry imitations of the Dutch, who clipped yews into monsters of every species, and relieved them with painted wooden figures. The distinction between the Italian and Dutch is obvious. A stone hewn into a gracefully ornamented vase or urn, has a value which it did not before possess: a yew hedge clipped into a fortification, is only defaced. The one is a production of art, the other a distortion of nature."

It must not be forgotten that, during all this period, or nearly six centuries, *parks* were common in England. Henry I. (1100 to 1135) had a park at Woodstock, and four centuries later, or during the reign of Henry VII., Holinshed informs us, that large parks or inclosed forest portions, several miles in circumference, were so common, that their number in Kent and Essex alone amounted to upwards of a hundred.

Although these parks were more devoted to the preservation of game and the pleasures of the chase than to any other purpose, their existence was, we conceive, not wholly owing to this cause; but we look upon them as indicating that love of nature and that desire to retain beautiful portions of it as part of a residence, which form the groundwork of the taste for the modern or landscape gardening, since the latter is only an epitome of nature with the charms judiciously heightened by art.

THE MODERN STYLE. Down to the time of Addison, in the beginning of the eighteenth century, the formal style reigned triumphant. The gardener, the architect, and the sculptor—all lovers of regularity and symmetry, had retained complete mastery of its arrangements. And it is worthy of more than a passing remark, that when the change in taste did take place, it emanated from the poet, the painter, and the tasteful scholar, rather than from the practical man.

In the poetical imagination, indeed, the ideal type of a modern landscape garden seems always to have been more or less shadowed forth. The Vaucluse of Petrarch, Tasso's garden of Armida, the vale of Tempe of Ælian, were all exquisite conceptions of the modern style. And Milton, surrounded as he was by the splendid formalities of the gardens of his time, copied from no existing models, but feeling that EDEN must have been free and majestic in its outlines, he drew from his inner sense of the beautiful, and from nature as he saw her developed in the works of the Creator. There, the crisped brooks,—

> " With mazy error under pendant shades
> Ran nectar, visiting each plant, and fed
> Flowers worthy of Paradise, *which not nice Art
> In beds and curious knots*, but Nature boon
> Pour'd forth profuse, on hill and dale and plain,
> Both where the morning sun first warmly smote
> The open field, and where the unpierced shade
> Imbrown'd the noontide bowers; *thus was this place
> A happy rural seat of various view.*"

But it required more than poetical types to change the long rooted fashion. The lever of satire needed to be applied, and the golden links that bind Nature and Art, more clearly revealed, before the old system could be made to waver.

The glory and merit of the total revolution which about this time took place in the public taste, belong, it is generally conceded, mainly to Addison and Pope. In 1712 appeared Addison's papers on Imagination, considered with reference to the works of Nature and Art. With a delicate and masterly hand, at a time when he possessed, through the "Spectator," the ear of all refined and tasteful England, he lifted the veil between the garden and natural charms, and showed how beautiful were their relations—how soon the imagination wearies with the stiffness of the former, and how much grace may be caught from a freer imitation of the swelling wood and hill.

The next year Pope, who was both a poet and painter, opened his quiver of satire in the celebrated article on verdant sculpture in the Guardian, where he ridiculed with no sparing hand the sheared alleys, formal groves, and

> "Statues growing that noble place in,
> All heathen goddesses most rare,
> Homer, Plutarch, and Nebuchadnezzar,
> Standing naked in the open air!'"

Pope was a refined and skilful amateur, and his garden at Twickenham became a celebrated miniature type of the natural school. In his Epistle to Lord Burlington, he developed sound principles for the new art;—the study of nature; the genius of the place; *and never to lose sight of good sense;* the latter, a rule which the whimsical follies of that day in gardening, seemed, doubtless, to render especially necessary, but which the discordant abortions of ambitious, would-be men of taste, prove is one soonest violated in every succeeding age.

The change in the popular feeling thus created, soon gave rise to innovations in the practical art. Bridgeman, the fashionable garden artist of the time, struck, as Horace

Walpole thinks, by Pope's criticisms, banished verdant sculpture from his plans, and introduced bits of forest scenery in the gardens at Richmond. And Loudon and Wise, the two noted nurserymen of the day, laid out Kensington gardens anew in a manner so much more natural as to elicit the warm commendations of Addison in the Spectator. It is not too much to say that Kent was the leader of this class. Originally a painter, and the friend of Lord Burlington, he next devoted himself to the subject, and was, undoubtedly, the first professional landscape gardener in the modern style. Previous artists had confined their efforts within the rigid walls of the garden, but Kent, who saw in all nature a garden-landscape, demolished the walls, introduced the *ha-ha*, and by blending the park and the garden, substituted for the primness of the old inclosure, the freedom of the *pleasure-ground*. His taste seems to have been partly formed by Pope, and the Twickenham garden was the prototype of those of Carlton House, Kent's *chef d'œuvre*. And, notwithstanding his faults, "his temples, obelisks, and gazabos of every description in the park, all stuck about in their respective high places," notwithstanding that his passion for natural effects led him into the absurdity of sometimes planting an old dead tree to make the illusion more perfect, we have no hesitation in according to Kent the merit of first fully establishing, in practice, the reform in taste which Addison and Pope had so completely developed in theory.

Among the landmarks of the progress of the taste, we must not refuse a passing notice of what seems to have been an unique and beautiful specimen of the new feeling for embellished nature—Leasowes, the "sentimental farm" of Shenstone. From contemporary accounts, it appears to have been originally a grazing farm, from which, by tasteful arrangement and planting, and pretty walks, seats, roothouse, urns, and appropriate inscriptions, the poet created a scene of much pastoral and poetical beauty.

The modern style was now running high in popular favor in England, but the next professor of the art, Brown, who seems to have been a mannerist not without some sympathy with nature, but not capable of grasping her more varied and expressive beauties, "Capability" Brown, as he was nicknamed, saw in every new place *great* capabilities, but unfortunately his own mind seems to have furnished but one model—a round lake, a smooth bare lawn, a clump of trees and a boundary belt—which he expanded, with few variations, to suit the compass of an estate of a thousand acres, or a cottage with a few roods. His works were often on a grand scale, and he boasted that the Thames would never forgive him for the rival he had created in the artificial lake at Blenheim. "The places he altered," says Loudon, "are beyond all reckoning. Improvement was the fashion of the time; and there was scarcely a country gentleman who did not, on some occasion or other, consult the gardening idol of the day." Mason, the poet, praises this artist, and Horace Walpole apologizes for not praising him. Daines Barrington says, "Kent hath been succeeded by Brown, who hath undoubtedly great merit in laying out pleasure grounds; but I conceive, that, in some of his plans, I see traces rather of the kitchen gardener of old Stowe, than of Poussin, or Claude Lorraine."

This mannerism gave rise finally, to the celebrated work *On the Picturesque* by Sir Uvedale Price, who, in a series of elegant and masterly essays, pointed out the faults and follies of this Brown and his imitators, analysed the beautiful and picturesque in nature and art, and founded a new school, more spirited and free in its aim, deriving its principles directly from nature and painting. These, with Knight's elegant Poem, the *Landscape*, the *English Garden* by Mason, and Whately's *Observations on Modern Gardening*, all published between 1750 and the beginning of the year 1800, established the new style firmly in the public mind. On the Continent, especially in France, though the old fashioned gardens were not demolished, as in England, new ones were laid out in accordance with the dawning taste, and none of the antique establishments were thought perfect without a spot set apart as a *jardin Anglais*.

It is not a little remarkable that the Chinese taste in gardening, which was at first made known to the English public about this time, is by far the nearest previous approach to the modern style. Some critics, indeed, have asserted that the English are indebted to it for their ideas of the modern style. However this may be, and we confess it has very little weight with us, the harmonious system which the taste of the English has evolved in the modern style, is at the present day too far beyond the Chinese manner to admit of any comparison. The first is imbued with beauty of the most graceful and agreeable character, based upon nature, and refined by art; while the latter abounds in puerilities and whimsical conceits—rocky hills, five feet high—miniature bridges—dwarf oaks, a hundred years old and twenty inches in altitude—which, whatever may be our admiration for the curious ingenuity and skill tasked in their production, leave on our mind no very favorable impression of the taste which designed them.

The most distinguished English Landscape Gardeners of more recent date, are the late Humphrey Repton, who died in 1818; and since him John Claudius Loudon, better known in this country, as the celebrated gardening author. Repton's taste in Landscape gardening was cultivated and elegant, and many of the finest parks and pleasure grounds of England, at the present day, bear witness to the skill and harmony of his designs. His published works are full of instructive hints, and at Cobham Hall, one of the finest seats in Britain, is an inscription to his memory, by Lord Darnley.

Mr. Loudon's* writings and labors in tasteful gardening, are too well known, to render it necessary that we should do more than allude to them here. Much of what is known of the art in this country undoubtedly is, more or less directly, to be referred to the influence of his published works. Although he is, as it seems to us, somewhat deficient as an artist in imagination, no previous author ever deduced, so clearly, sound artistical principles in Landscape Gardening and Rural Architecture; and fitness, good sense, and beauty, are combined with much unity of feeling in all his works.

As the modern style owes its origin mainly to the English, so it has also been developed and carried to its greatest perfection in the British Islands. The law of primogeniture, which has there so long existed, in itself, contributes greatly to the continual improvement and embellishment of those vast landed estates, that remain perpetually in the hands of the same family. Magnificent

---

* While we are revising the second edition, we regret deeply to learn the death of Mr. Loudon. His herculean labors as an author have at last destroyed him; and in his death we lose one who has done more than any other person that ever lived, to popularize, and render universal, a taste for Gardening and Domestic Architecture.

buildings, added to by each succeeding generation, who often preserve also the older portions with the most scrupulous care; wide spread parks, clothed with a thick velvet turf, which, amid their moist atmosphere, preserves during great part of the year an emerald greenness— studded with noble oaks and other forest trees which number centuries of growth and maturity; these advantages, in the hands of the most intelligent and the wealthiest aristocracy in the world, have indeed made almost an entire landscape garden of "merry England." Among a multitude of splendid examples of these noble residences, we will only refer the reader to the celebrated Blenheim, the seat of the Duke of Marlborough, where the lake alone (probably the largest piece of artificial water in the world) covers a surface of two hundred acres: Chatsworth, the varied and magnificent seat of the Duke of Devonshire, where there are scenes illustrative of almost every style of the art: and Woburn Abbey, the grounds of which are full of the choicest specimens of trees and plants, and where the park, like that of Ashbridge, Arundel Castle, and several other private residences in England, is only embraced within a circumference of from ten to twenty miles.

On the continent of Europe, though there are a multitude of examples of the modern style of landscape gardening, which is there called the *English* or *natural* style, yet in the neighborhood of many of the capitals, especially those of the south of Europe, the taste for the geometric or ancient style of gardening still prevails to a considerable extent; partially, no doubt, because that style admits, with more facility, of those classical and architectural accompaniments of vases, statues, busts, etc., the passion for which pervades a people rich in ancient and modern sculptural works of art. Indeed many of the gardens on the continent are more striking from˙ their numerous sculpturesque ornaments, interspersed with fountains and jets-d'eau, than from the beauty or rarity of their vegetation, or from their arrangement.

In the United States, it is highly improbable that we shall ever witness such splendid examples of landscape gardens as those abroad, to which we have alluded. Here the rights of man are held to be equal; and if there are no enormous parks, and no class of men whose wealth is hereditary, there is, at least, what is more gratifying to the feelings of the philanthropist, the almost entire absence of a very poor class in the country; while we have, on the other hand, a large class of independent landholders, who are able to assemble around them, not only the useful and convenient, but the agreeable and beautiful, in country life.

The number of individuals among us who possess wealth and refinement sufficient to enable them to enjoy the pleasures of a country life, and who desire in their private residences so much of the beauties of landscape gardening and rural embellishment as may be had without any enormous expenditure of means, is every day increasing. And although, until lately, a very meagre plan of laying out the grounds of a residence, was all that we could lay claim to, yet the taste for elegant rural improvements is advancing now so rapidly, that we have no hesitation in predicting that in half a century more, there will exist a greater number of beautiful villas and country seats of moderate extent, in the Atlantic States, than in any country in Europe, England alone excepted. With us, a feeling, a taste, or an improvement, is contagious; and once fairly appreciated and established in one portion of the country, it is disseminated with a celerity that is indeed wonderful, to every other portion. And though it is necessarily the case where amateurs of any art are more numerous than its professors, that there will be, in devising and carrying plans into execution, many specimens of bad taste, and perhaps a sufficient number of efforts to improve without any real taste whatever, still we are convinced the effect of our rural embellishments will in the end be highly agreeable, as a false taste is not likely to be a permanent one in a community where everything is so much the subject of criticism.

With regard to the literature and practice of Landscape Gardening as an art, in North America, almost everything is yet before us, comparatively little having yet been done. Almost all the improvements of the grounds of our finest country residences, have been carried on under the direction of the proprietors themselves, suggested by their own good taste, in many instances improved by the study of European authors, or by a personal inspection of the finest places abroad. The only American work previously published which treats directly of Landscape Gardening, is the *American Gardener's Calendar*, by Bernard McMahon of Philadelphia. The only practitioner of the art, of any note, was the late M. Parmentier of Brooklyn, Long Island.

M. André Parmentier was the brother of that celebrated horticulturist, the Chevalier Parmentier, Mayor of Enghien, Holland. He emigrated to this country about the year 1824, and in the Horticultural Nurseries which he established at Brooklyn, he gave a specimen of the natural style of laying out grounds, combined with a scientific arrangement of plants, which excited public curiosity, and contributed not a little to the dissemination of a taste for the natural mode of landscape gardening.

During M. Parmentier's residence on Long Island, he was almost constantly applied to for plans for laying out the grounds of country seats, by persons in various parts of the Union, as well as in the immediate proximity of New York. In many cases he not only surveyed the demesne to be improved, but furnished the plants and trees necessary to carry out his designs. Several plans were prepared by him for residences of note in the Southern States; and two or three places in Upper Canada, especially near Montreal, were, we believe, laid out by his own hands and stocked from his nursery grounds. In his periodical catalogue, he arranged the hardy trees and shrubs that flourish in this latitude in classes, according to their height, etc., and published a short treatise on the superior claims of the natural, over the formal or geometric style of laying out grounds. In short, we consider M. Parmentier's labors and examples as having effected, directly, far more for landscape gardening in America, than those of any other individual whatever.

The introduction of tasteful gardening in this country is, of course, of a very recent date. But so long ago as from 25 to 50 years, there were several country residences highly remarkable for extent, elegance of arrangement, and the highest order and keeping. Among these, we desire especially to record here the celebrated seats of Chancellor Livingston, Wm. Hamilton, Esq., Theodore Lyman, Esq., and Judge Peters.

Woodlands, the seat of the Hamilton family, near

Philadelphia, was, so long ago as 1805, highly celebrated for its gardening beauties. The refined taste and the wealth of its accomplished owner, were freely lavished in its improvement and embellishment; and at a time when the introduction of rare exotics was attended with a vast deal of risk and trouble, the extensive green-houses and orangeries of this seat contained all the richest treasures of the exotic flora, and among other excellent gardeners employed, was the distinguished botanist Pursh, whose enthusiastic taste in his favorite science was promoted and aided by Mr. Hamilton. The extensive pleasure grounds were judiciously planted, singly and in groups, with a great variety of the finest species of trees. The attention of the visitor to this place is now arrested by two very large specimens of that curious tree, the Japanese Ginko (*Salisburia*), 60 or 70 feet high, perhaps the finest in Europe or America, by the noble magnolias, and the rich park-like appearance of some of the plantations of the finest native and foreign oaks. From the recent unhealthiness of this portion of the Schuylkill, Woodlands has fallen into decay, but there can be no question that it was, for a long time, the most tasteful and beautiful residence in America.

The seat of the late Judge Peters, about five miles from Philadelphia, was, 30 years ago, a noted specimen of the ancient school of landscape gardening. Its proprietor had a most extended reputation as a scientific agriculturist, and his place was also no less remarkable for the design and culture of its pleasure-grounds, than for the excellence of its farm. Long and stately avenues, with vistas terminated by obelisks, a garden adorned with marble vases, busts, and statues, and pleasure grounds filled with the rarest trees and shrubs, were conspicuous features here. Some of the latter are now so remarkable as to attract strongly the attention of the visitor. Among them, is the chestnut planted by Washington, which produces the largest and finest fruit; very large hollies; and a curious old box-tree much higher than the mansion near which it stands. But the most striking feature now, is the still remaining grand old avenue of hemlocks (*Abies canadensis*). Many of these trees, which were planted 100 years ago, are now venerable specimens, ninety feet high, whose huge trunks and wide spread branches are in many cases densely wreathed and draped with masses of English Ivy, forming the most picturesque sylvan objects we ever beheld.

Lemon Hill, half a mile above the Fairmount water-works of Philadelphia, was, 20 years ago, the most perfect specimen of the geometric mode in America, and since its destruction by the extension of the city, a few years since, there is nothing comparable with it, in that style, among us. All the symmetry, uniformity, and high art of the old school, were displayed here in artificial plantations, formal gardens with trellises, grottoes, spring-houses, temples, statues, and vases, with numerous ponds of water, jets-d'eau, and other water-works, parterres and an extensive range of hothouses. The effect of this garden was brilliant and striking; its position, on the lovely banks of the Schuylkill, admirable; and its liberal proprietor, Mr. Pratt, by opening it freely to the public, greatly increased the popular taste in the neighborhood of that city.

On the Hudson, the show place of the last age was the still interesting *Clermont*, then the residence of Chancellor Livingston. Its level or gently undulating lawn, four or five miles in length, the rich native woods, and the long vistas of planted avenues, added to its fine water view, rendered this a noble place. The mansion, the greenhouses, and the gardens, show something of the French taste in design, which Mr. Livingston's residence abroad, at the time when that mode was popular, no doubt, led him to adopt. The finest yellow locusts in America are now standing in the pleasure-grounds here, and the gardens contain many specimens of fruit trees, the first of their sorts introduced into the Union.

Waltham House, about nine miles from Boston, was, 25 years ago, one of the oldest and finest places, as regards Landscape Gardening. Its owner, the late Hon. T. Lyman, was a highly-accomplished man, and the grounds at Waltham House bear witness to a refined and elegant taste in rural improvement. A fine level park, a mile in length, enriched with groups of English limes, elms, and oaks, and rich masses of native wood, watered by a fine stream and stocked with deer, were the leading features of the place at that time; and this, and Woodlands, were the two best specimens of the modern style, as Judge Peters' seat, Lemon Hill, and Clermont, were of the ancient style, in the earliest period of the history of Landscape Gardening among us.

There is no part of the Union where the taste in Landscape Gardening is so far advanced, as on the middle portion of the Hudson. The natural scenery is of the finest character, and places but a mile or two apart often possess, from the constantly varying forms of the water, shores, and distant hills, widely different kinds of home landscape and distant view. Standing in the grounds of some of the finest of these seats, the eye beholds only the

Fig 1. View in the Grounds at Hyde Park.

Fig 2. The Manor of Livingston

soft foreground of smooth lawn, the rich groups of trees shutting out all neighboring tracts, the lake-like expanse of water, and, closing the distance, a fine range of wooded mountain. A residence here of but a hundred acres, so fortunately are these disposed by nature, seems to appropriate the whole scenery round, and to be a thousand in extent.

At the present time, our handsome villa residences are becoming every day more numerous, and it would require much more space than our present limits, to enumerate all the tasteful rural country places within our knowledge, many of which have been newly laid out, or greatly improved within a few years. But we consider it so important and instructive to the novice in the art of Landscape Gardening to examine, personally, country seats of a highly tasteful character, that we shall venture to refer the reader to a few of those which have now a reputation among us as elegant country residences.

Hyde Park, on the Hudson, formerly the seat of the late Dr. Hosack, now of W. Langdon, Esq., has been justly celebrated as one of the finest specimens of the modern style of Landscape Gardening in America. Nature has, indeed, done much for this place, as the grounds are finely varied, beautifully watered by a lively stream, and the views are inexpressibly striking from the neighborhood of the house itself, including, as they do, the noble Hudson for sixty miles in its course, through rich valleys and bold mountains. (See Fig. 1.) But the efforts of art are not unworthy so rare a locality; and while the native woods, and beautifully undulating surface, are preserved in their original state, the pleasure-grounds, roads, walks, drives, and new plantations, have been laid out in such a judicious manner as to heighten the charms of nature. Large and costly hot-houses were erected by Dr. Hosack, with also entrance lodges at two points· on the estate, a fine bridge over the stream, and numerous pavilions and seats commanding extensive prospects; in short, nothing was spared to render this a complete residence. The park, which at one time contained some fine deer, afforded a delightful drive within itself, as the whole estate numbered about seven hundred acres. The plans for laying out the grounds were furnished by Parmentier, and architects from New York were employed in designing and erecting the buildings. For a long time, this was the finest seat in America, but there are now many rivals to this claim.

The Manor of Livingston, the seat of Mrs. Mary Livingston, is seven miles east of the city of Hudson. The mansion stands in the midst of a fine park, rising gradually from the level of a rich inland country, and commanding prospects for sixty miles around. The park is, perhaps, the most remarkable in America, for the noble simplicity of its character, and the perfect order in which it is kept. The turf is, everywhere, short and velvet-like, the gravel-roads scrupulously firm and smooth, and near the house are the largest and most superb evergreens. The mansion is one of the chastest specimens of the Grecian style, and there is an air of great dignity about the whole demesne. (Fig. 2.)

Blithewood, the seat of R. Donaldson, Esq., near Barrytown on the Hudson, is one of the most charming villa residences in the Union. The natural scenery here, is nowhere surpassed in its enchanting union of softness and dignity—the river being four miles wide, its placid bosom broken only by islands and gleaming sails, and the horizon

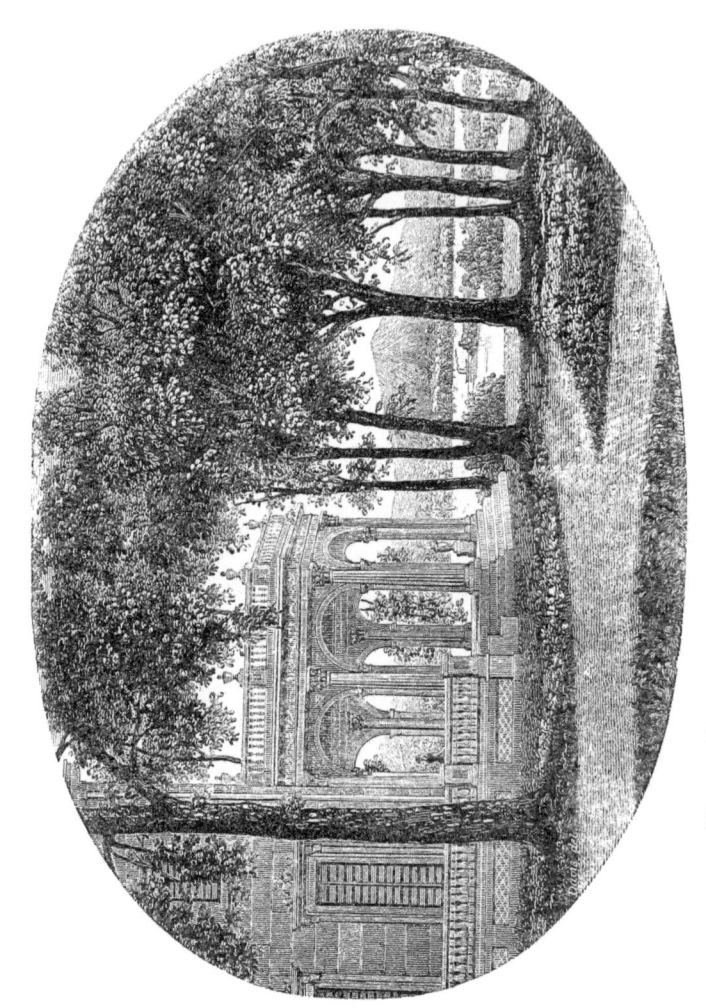

Fig. 3. Montgomery Place, Seat of Mrs. Edward Livingston.

grandly closing in with the tall blue summits of the distant Kaatskills. The smiling, gently varied lawn is studded with groups and masses of fine forest and ornamental trees, beneath which are walks leading in easy curves to rustic seats, and summer houses placed in secluded spots, or to openings affording most lovely prospects. (See Frontispiece.) In various situations near the house and upon the lawn, sculptured vases of Maltese stone are also disposed in such a manner as to give a refined and classic air to the grounds.

As a *pendant* to this graceful landscape, there is within the grounds scenery of an opposite character, equally wild and picturesque—a fine, bold stream, fringed with woody banks, and dashing over several rocky cascades, thirty or forty feet in height, and falling altogether a hundred feet in half a mile. (See view, Sect. VIII.) There are also, within the grounds, a pretty gardener's lodge, in the rural cottage style, and a new entrance lodge by the gate, in the bracketed mode; in short, we can recall no place of moderate extent, where nature and tasteful art are both so harmoniously combined to express grace and elegance.

Montgomery Place (see Fig. 3), the residence of Mrs. Edward Livingston, which is also situated on the Hudson near Barrytown, deserves a more extended notice than our present limits allow, for it is, as a whole, nowhere surpassed in America in point of location, natural beauty, or the landscape gardening charms which it exhibits.

It is one of our oldest improved country seats, having been originally the residence of Gen. Montgomery, the hero of Quebec. On the death of his widow it passed into the hands of her brother, Edward Livingston, Esq., the late minister to France, and up to the present moment has always received the most tasteful and judicious treatment.

The lover of the expressive in nature, or the beautiful in art, will find here innumerable subjects for his study. The natural scenery in many portions approaches the character of grandeur, and the foreground of rich woods and lawns, stretching out on all sides of the mountain, completes a home landscape of dignified and elegant seclusion, rarely surpassed in any country.

Among the fine features of this estate are the *wilderness*, a richly wooded and highly picturesque valley, filled with the richest growth of trees, and threaded with dark, intricate, and mazy walks, along which are placed a variety of rustic seats (Fig. 4). This valley is musical with the sound of waterfalls, of which there are several fine ones in the bold impetuous stream which finds its course through the lower part of the wilderness.

[Fig. 4. One of the Rustic Seats at Montgomery Place.]

Near the further end of the valley is a beautiful lake (Fig. 5), half of which lies cool and dark under the shadow of tall trees, while the other half gleams in the open sunlight.

In a part of the lawn, near the house, yet so surrounded by a dark setting of trees and shrubs as to form a rich picture by itself, is one of the most perfect flower gardens in the country, laid out in the arabesque manner, and glowing with masses of the gayest colors—each bed being composed wholly of a single hue. A large conservatory, an exotic garden, an arboretum, etc., are among the features of interest in this admirable residence. Including a *drive* through a fine bit of natural wood, south of the mansion, there are five miles of highly varied and picturesque private roads and walks, through the pleasure-grounds of Montgomery Place.

[Fig. 5. The Lake at Montgomery Place.]

Ellerslie is the seat of William Kelly, Esq. It is three miles below Rhinebeck. It comprises over six hundred acres, and is one of our finest examples of high keeping and good management, both in an ornamental and an agricultural point of view. The house is conspicuously placed on a commanding natural terrace, with a fair foreground of park surface below it, studded with beautiful groups of elms and oaks, and a very fine reach of river and distant hills. This is one of the most célebrated places on the Hudson, and there are few that so well pay the lover of improved landscape for a visit.

Just below Ellerslie are the fine mansion and pleasing grounds of Wm. Emmet, Esq.,—the former a stone edifice, in the castellated style, and the latter forming a most agreeable point on the margin of the river.

The seat of Gardiner Howland, Esq., near New Hamburgh, is not only beautiful in situation, but is laid out with great care, and is especially remarkable for the many rare trees and shrubs collected in its grounds.

Wodenethe, near Fishkill landing, is the seat of H. W. Sargent, Esq., and is a bijou full of interest for the lover of rural beauty; abounding in rare trees, shrubs, and plants, as well as vases, and objects of rural embellishment of all kinds.

Kenwood (Fig. 6), the residence of J. Rathbone, Esq., is one mile south of Albany. Ten years ago this spot was a wild and densely wooded hill, almost inaccessible. With great taste and industry Mr. Rathbone has converted it into a country residence of much picturesque beauty, erected in the Tudor style, one of the best villas in the country, with a gate-lodge in the same mode, and laid out the grounds with remarkable skill and good taste. There are about 1200 acres in this estate, and pleasure grounds, forcing houses, and gardens, are now flourishing where all was so lately in the rudest state of nature; while, by the judicious preservation of natural wood, the effect of a long cultivated demesne has been given to the whole.

The Manor House of the "*Patroon*" (as the eldest son of the Van Rensselaer family is called) is in the northern suburbs of the city of Albany. The mansion, greatly

Fig. 9. Kenwood, Residence of J Rathbone, Esq near Albany, N. Y.

Fig. 7. Beaverwyck, the Seat of Wm. P. Van Rensselaer, Esq.

Fig 8. Cottage Residence of Wm. H. Aspinwall, Esq.

enlarged and improved a few years since, from the designs of Upjohn, is one of the largest and most admirable in all respects, to be found in the country, and the pleasure-grounds in the rear of the house are tasteful and beautiful.

Beaverwyck, a little north of Albany, on the opposite bank of the river, is the seat of Wm. P. Van Rensselaer, Esq. (Fig. 7.) The whole estate is ten or twelve miles square, including the village of Bath on the river shore, and a large farming district. The home residence embraces several hundred acres, with a large level lawn, bordered by highly varied surface of hill and dale. The mansion, one of the first class, is newly erected from the plans of Mr. Diaper, and in its interior—its hall with mosaic floor of polished woods, its marble staircase, frescoed apartments, and spacious adjoining conservatory—is perhaps the most splendid in the Union. The grounds are yet newly laid out, but with much judgment; and *six or seven miles* of winding gravelled roads and walks have been formed—their boundaries now leading over level meadows, and now winding through woody dells. The drives thus afforded, are almost unrivalled in extent and variety, and give the stranger or guest, an opportunity of seeing the near and distant views to the best advantage.

At Tarrytown, is the cottage residence of Washington Irving, which is, in location and accessories, almost the beau ideal of a cottage ornée. The charming manner in which the wild foot-paths, in the neighborhood of this cottage, are conducted among the picturesque dells and banks, is precisely what one would look for here. A little below, Mr. Sheldon's cottage, with its pretty lawn and its charming brook, is one of the best specimens of this kind of residence on the river. At Hastings, four or five miles south, is the agreeable seat of Judge Constant.

About twelve miles from New York, on the Sound, is *Hunter's Island,* the seat of John Hunter, Esq., a place of much simplicity and dignity of character. The whole island may be considered an extensive park carpeted with soft lawn, and studded with noble trees. The mansion is simple in its exterior, but internally, is filled with rich treasures of art. The seat of James Munroe, Esq., on the East river in this neighborhood, abounds with beautiful trees, and many other features of interest.

The Cottage residence of William H. Aspinwall, Esq., on Staten Island, is a highly picturesque specimen of Landscape Gardening. The house is in the English cottage style, and from its open lawn in front, the eye takes in a wide view of the ocean, the Narrows, and the blue hills of Neversink. In the rear of the cottage, the surface is much broken and varied, and finely wooded and planted. In improving this picturesque site, a nice sense of the charm of natural expression has been evinced; and the sudden variations from smooth open surface, to wild wooden banks, with rocky, moss-covered flights of steps, strike the stranger equally with surprise and delight. A charming greenhouse, a knotted flower-garden, and a pretty, rustic moss-house, are among the interesting points of this spirited place. (See Fig. 8.)

The seat of the Wadsworth family, at *Geneseo,* is the finest in the interior of the state of New York. Nothing, indeed, can well be more magnificent than the *meadow park* at Geneseo. It is more than a thousand acres in extent, lying on each side of the Genesee river, and is filled with thousands of the noblest oaks and elms, many of which, but

Fig. 9. Seat of Wadsworth Family at Geneseo.

more especially the oaks, are such trees as we see in the pictures of Claude, or our own Durand; richly developed, their trunks and branches grand and majestic, their heads full of breadth and grandeur of outline. (See Fig. 9.) These oaks, distributed over a nearly level surface, with the trees disposed either singly or in the finest groups, as if most tastefully planted centuries ago, are solely the work of nature; and yet so entirely is the whole like the grandest planted park, that it is difficult to believe that all is not the work of some master of art, and intended for the accompaniment of a magnificent residence. Some of the trees are five or six hundred years old.

In Connecticut, *Monte Video*, the seat of Daniel Wadsworth, Esq., near Hartford, is worthy of commendation, as it evinces a good deal of beauty in its grounds, and is one of the most tasteful in the state. The residence of James Hillhouse, Esq., near New-Haven, is a pleasing specimen of the simplest kind of Landscape Gardening, where graceful forms of trees, and a gently sloping surface of grass, are the principal features. The villa of Mr. Whitney, near New-Haven, is one of the most tastefully managed in the state. In Maine, the most remarkable seat, as respects landscape gardening and architecture, is that of Mr. Gardiner, of Gardiner.

The environs of Boston are more highly cultivated than those of any other city in North America. There are here whole rural neighborhoods of pretty cottages and villas, admirably cultivated, and, in many cases, tastefully laid out and planted. The character of even the finest of these places is, perhaps, somewhat suburban, as compared with those of the Hudson river, but we regard them as furnishing admirable hints for a class of residence likely to become more numerous than any other in this country—the tasteful suburban cottage. The owner of a small cottage residence may have almost every kind of beauty and enjoyment in his grounds that the largest estate will afford, so far as regards the interest of trees and plants, tasteful arrangement, recreation, and occupation. Indeed, we have little doubt that he, who directs personally the curve of every walk, selects and plants every shrub and tree, and watches with solicitude every evidence of beauty and progress, succeeds in extracting from his tasteful grounds of half a dozen acres, a more intense degree of pleasure, than one who is only able to direct and enjoy, in a general sense, the arrangement of a vast estate.

Belmont, the seat of J. P. Cushing, Esq., is a residence of more note than any other near Boston; but this is, chiefly, on account of the extensive ranges of glass, the forced fruits, and the high culture of the gardens. A new and spacious mansion has recently been erected here, and the pleasure-grounds are agreeably varied with fine groups and masses of trees and shrubs on a pleasing lawn. (Fig. 10.)

The seat of Col. Perkins, at Brookline, is one of the most interesting in this neighborhood. The very beautiful lawn here, abounds with exquisite trees, finely disposed; among them, some larches and Norway firs, with many other rare trees of uncommon beauty of form. At a short distance is the villa residence of Theodore Lyman, Esq., remarkable for the unusually fine avenue of Elms leading to the house, and for the beautiful architectural taste displayed in the dwelling itself. The seat of the Hon. John

Fig. 10. Belmont Place, near Boston, the Seat of J. P. Cushing, Esq.

Fig. 11. Mr. Dunn's Cottage, Mount Holly, N. J.

Lowell, at Roxbury, possesses also many interesting gardening features.*

Pine Bank, the Perkins estate, on the border of Jamaica lake, is one of the most beautiful residences near Boston. The natural surface of the ground is exceedingly flowing and graceful, and it is varied by two or three singular little *dimples*, or hollows, which add to its effect. The perfect order of the grounds; the beauty of the walks, sometimes skirting the smooth open lawn, enriched with rare plants and shrubs, and then winding by the shadowy banks of the water; the soft and quiet character of the lake itself,—its margin richly fringed with trees, which conceal here and there a pretty cottage, its firm clean beach of gravel, and its water of crystal purity; all these features make this place a little gem of natural

---

* We Americans are proverbially impatient of delay, and a few years in prospect appear an endless futurity. So much is this the feeling with many, that we verily believe there are hundreds of our country places, which owe their bareness and destitution of foliage to the idea, so common, that it requires " an age" for forest trees to " *grow up*."

The middle-aged man hesitates about the good of planting what he imagines he shall never see arriving at maturity, and even many who are younger, conceive that it requires more than an ordinary lifetime to rear a fine wood of planted trees. About two years since, we had the pleasure of visiting the seat of the late Mr. Lowell, whom we found in a green old age, still enjoying, with the enthusiasm of youth, the pleasures of Horticulture and a country life. For the encouragement of those who are ever complaining of the tardy pace with which the growth of trees advances, we will here record that we accompanied Mr. L. through a belt of fine woods (skirting part of his residence), nearly half a mile in length, consisting of almost all our finer hardy trees, many of them apparently full grown, the whole of which had been planted by him when he was thirty-two years old. At that time, a solitary elm or two were almost the only trees upon his estate. We can hardly conceive a more rational source of pride or enjoyment, than to be able thus to walk, in the decline of years, beneath the shadow of umbrageous woods and groves, planted by our own hands, and whose growth has become almost identified with our own progress and existence.

and artistical harmony, and beauty. Mr. Perkins has just rebuilt the house, in the style of a French *maison de campagne;* and Pine Bank is now adorned with a most complete residence in the latest continental taste, from the designs of M. Lémoulnier.

On the other side of the lake is the *cottage of Thomas Lee, Esq.* Enthusiastically fond of botany, and gardening, in all its departments, Mr. Lee has here formed a residence of as much variety and interest as we ever saw in so moderate a compass—about 20 acres. It is, indeed, not only a most instructive place to the amateur of landscape gardening, but to the naturalist and lover of plants. Every shrub seems placed precisely in the soil and aspect it likes best, and native and foreign Rhododendrons, Kalmias, and other rare shrubs, are seen here in the finest condition. There is a great deal of variety in the surface here, and while the lawn-front of the house has a polished and graceful air, one or two other portions are quite picturesque. Near the entrance gate is an English oak, only fourteen years planted, now forty feet high.

The whole of this neighborhood of Brookline is a kind of landscape garden, and there is nothing in America, of the sort, so inexpressibly charming as the lanes which lead from one cottage, or villa, to another. No animals are allowed to run at large, and the open gates, with tempting vistas and glimpses under the pendent boughs, give it quite an Arcadian air of rural freedom and enjoyment. These lanes are clothed with a profusion of trees and wild shrubbery, often almost to the carriage tracks, and curve and wind about, in a manner quite bewildering to the stranger who attempts to thread them alone; and there are more hints here for the lover of the picturesque in lanes, than

View in the Grounds at Pine Bank.

View in the Grounds of James Arnold, Esq.

we ever saw assembled together in so small a compass.

In the environs of New Bedford are many beautiful residences. Among these, we desire particularly to notice the residence of James Arnold, Esq. There is scarcely a small place in New England, where the *pleasure-grounds* are so full of variety, and in such perfect order and keeping, as at this charming spot; and its winding walks, open bits of lawn, shrubs and plants grouped on turf, shady bowers, and rustic seats, all most agreeably combined, render this a very interesting and instructive suburban seat.

In New Jersey, the grounds of the Count de Survilliers, at Bordentown, are very extensive; and although the surface is mostly flat, it has been well varied by extensive plantations. At Mount Holly, about twenty miles from Camden, is Mr. Dunn's unique, semi-oriental cottage, with a considerable extent of pleasure ground, newly planted, after the designs of Mr. Notman. (Fig. 11.)

About Philadelphia there are several very interesting seats on the banks of the Delaware and Schuylkill, and the district between these two rivers.

The country seat of George Sheaff, Esq., one of the most remarkable in Pennsylvania, in many respects, is twelve miles north of Philadelphia. The house is a large and respectable mansion of stone, surrounded by pleasure-grounds and plantations of fine evergreen and deciduous trees. The conspicuous ornament of the grounds, however, is a magnificent white oak, of enormous size, whose wide stretching branches, and grand head, give an air of dignity to the whole place. (Fig. 12.) Among the sylvan features here, most interesting, are also the handsome evergreens, chiefly Balsam or Balm of Gilead firs, some of which are now much higher than the mansion. These trees were planted by Mr. Sheaff twenty-two years ago, and were then so small, that they were brought by him from Philadelphia, at various times, in his carriage—a circumstance highly encouraging to despairing planters, when we reflect how comparatively slow growing is this tree. This whole estate is a striking example of science, skill, and taste, applied to a country seat, and there are few in the Union, taken as a whole, superior to it.*

Cottage residence of Mrs. Camac. This is one of the most agreeable places within a few miles of Philadelphia. The house is a picturesque cottage, in the rural gothic style, with very charming and appropriate pleasure grounds, comprising many groups and masses of large and finely grown trees, interspersed with a handsome collection of shrubs and plants; the whole very tastefully arranged. (Fig. 13.) The lawn is prettily varied in surface, and there is a conservatory attached to the house, in which the plants in pots are hidden in beds of soft green moss, and which, in its whole effect and management, is more tasteful and elegant than any plant house, connected with a dwelling, that we remember to have seen.

* The farm is 300 acres in extent, and, in the time of De Witt Clinton, was pronounced by him the model farm of the United States. At the present time we know nothing superior to it; and Capt. Barclay, in his agricultural tour, says it was the only instance of regular, scientific system of husbandry in the English manner, he saw in America. Indeed, the large and regular fields, filled with luxuriant crops, everywhere of an exact evenness of growth, and everywhere free from weeds of any sort; the perfect system of manuring and culture; the simple and complete fences; the fine stock; the very spacious barns, every season newly whitewashed internally and externally, paved with wood, and as clean as a gentleman's stable (with stalls to fatten 90 head of cattle); these, and the masterly way in which the whole is managed, both as regards culture and profit, render this estate one of no common interest in an agricultural, as well as ornamental point of view.

Fig. 12. The Seat of George Sheaff, Esq.

Fig. 13. Mrs. Camac's Residence.

Stanton, near Germantown, four miles from Philadelphia, is a fine old place, with many picturesque features. The farm consists of 700 acres, almost without division fences— admirably managed—and remarkable for its grand old avenue of the hemlock spruce, 110 years old, leading to a family cemetery of much sylvan beauty. There is a large and excellent old mansion, with paved halls, built in 1731, which is preserved in its original condition. This place was the seat of the celebrated Logan, the friend of William Penn, and is now owned by his descendant, Albanus Logan.

The villa residence of Alexander Brown, Esq., is situated on the Delaware, a few miles from Philadelphia. There is here a good deal of beauty, in the natural style, made up chiefly by lawn and forest trees. A pleasing drive through plantations of 25 years' growth, is one of the most interesting features—and there is much elegance and high keeping in the grounds.

Below Philadelphia, the lover of beautiful places will find a good deal to admire in the country seat of John R. Latimer, Esq., near Wilmington, which enjoys the reputation of being the finest in Delaware. The place has all the advantages of high keeping, richly stocked gardens and conservatories, and much natural beauty, heightened by judicious planting, arrangement, and culture.

At the south are many extensive country residences remarkable for trees of unusual grandeur and beauty, among which the live oak is very conspicuous; but they are, in general, wanting in that high keeping and care, which is so essential to the charm of a landscape garden.

Of smaller villa residences, suburban chiefly, there are great numbers, springing up almost by magic, in the borders of our towns and cities. Though the possessors of these can scarcely hope to introduce anything approaching to a landscape garden style, in laying out their limited grounds, still they may be greatly benefited by an acquaintance with the beauties and the pleasures of this species of rural embellishment. When we are once master of the principles, and aware of the capabilities of an art, we are able to infuse an expression of tasteful design, or an air of more correct elegance, even into the most humble works, and with very limited means.

While we shall endeavor, in the following pages, to give such a view of modern Landscape Gardening, as will enable the improver to proceed with his fascinating operations, in embellishing the country residence, in a practical mode, based upon what are now generally received as the correct principles of the art, we would desire the novice, after making himself acquainted with all that can be acquired from written works within his reach, to strengthen his taste and add to his knowledge, by a practical inspection of the best country seats among us. In an infant state of society, in regard to the fine arts, much will be done in violation of good taste; but here, where nature has done so much for us, there is scarcely a large country residence in the Union, from which useful hints in Landscape Gardening may not be taken. And in nature, a group of trees, an accidental pond of water, or some equally simple object, may form a study more convincing to the mind of a true admirer of natural beauty, than the most carefully drawn plan, or the most elaborately written description.

SECTION II.

BEAUTIES AND PRINCIPLES OF THE ART.

Capacities of the art. The beauties of the ancient style. The modern style. The Beauti ful and the Picturesque: their distinctive characteristics. Illustrations drawn from Nature and Painting. Nature and principles of Landscape Gardening as an Imitative art. Distinction between the Beautiful and Picturesque. The principles of Unity, Harmony, and Variety.

" Here Nature in her unaffected dresse,
Plaited with vallies and imbost with hills,
Enchast with silver streams, and fringed with woods
Sits lovely."—
CHAMBERLAYNE.

" Il est des soins plus doux, un art plus enchanteur.
C'est peu de charmer l'œil, il faut parler au cœur.
Avez-vous donc connu ces rapports invisibles,
Des corps inanimés et des êtres sensibles?
Avez-vous entendu des eaux, des prés, des bois,
La muette éloquence et la secrète voix?
Rendez-nous ces effets." *Les Jardins, Book I.*

BEFORE we proceed to a detailed and more practical consideration of the subject, let us occupy ourselves for a moment with the consideration of the different results which are to be sought after, or, in other words, what kinds of beauty we may hope to produce by Landscape Gardening. To attempt the smallest work in any art, without knowing either the capacities of that art, or the schools, or modes, by which it has previously been characterized, is but to be groping about in a dim twilight, without the power of knowing, even should we be successful in our efforts, the real excellence of our production; or of judging its merit, comparatively, as a work of taste and imagination.

[Fig. 14. The Geometric style, from an old print.]

The beauties elicited by the ancient style of gardening were those of regularity, symmetry, and the display of labored art. These were attained in a merely mechanical manner, and usually involved little or no theory. The geometrical form and lines of the buildings were only extended and carried out in the garden. In the best classical models, the art of the sculptor conferred dignity and elegance on the garden, by the fine forms of marble vases and statues; in the more intricate and labored specimens of the

Dutch school, prevalent in England in the time of William IV. (Fig. 14), the results evince a fertility of odd conceits, rather than the exercise of taste or imagination. Indeed, as, to level ground naturally uneven, or to make an avenue, by planting rows of trees on each side of a broad walk, requires only the simplest perception of the beauty of mathematical forms, so, to lay out a garden in the geometric style, became little more than a formal routine, and it was only after the superior interest of a more natural manner was enforced by men of genius, that natural beauty of expression was recognised, and Landscape Gardening was raised to the rank of a fine art.

The ancient style of gardening may, however, be introduced with good effect in certain cases. In public squares and gardens, where display, grandeur of effect, and a highly artificial character are desirable, it appears to us the most suitable; and no less so in very small gardens, in which variety and irregularity are out of the question. Where a taste for imitating an old and quaint style of residence exists, the symmetrical and knotted garden would be a proper accompaniment; and pleached alleys, and sheared trees, would be admired, like old armor or furniture, as curious specimens of antique taste and custom.

The earliest professors of modern Landscape Gardening have generally agreed upon two variations, of which the art is capable—variations no less certainly distinct, on the one hand, than they are capable of intermingling and combining, on the other. These are the *beautiful* and the *picturesque:* or, to speak more definitely, the beauty characterized by simple and flowing forms, and that expressed by striking, irregular, spirited forms.

The admirer of nature, as well as the lover of pictures

and engravings, will at once call to mind examples of scenery distinctly expressive of each of these kinds of beauty. In nature, perhaps some gently undulating plain, covered with emerald turf, partially or entirely encompassed by rich, rolling outlines of forest canopy,—its wildest expanse here broken occasionally, by noble groups of round-headed trees, or there interspersed with single specimens whose trunks support heads of foliage flowing in outline, or drooping in masses to the very turf beneath them. In such a scene we often behold the azure of heaven, and its silvery clouds, as well as the deep verdure of the luxuriant and shadowy branches, reflected in the placid bosom of a silvan lake; the shores of the latter swelling out, and receding, in gentle curved lines; the banks, sometimes covered with soft turf sprinkled with flowers, and in other portions clothed with luxuriant masses of verdant shrubs. Here are all the elements of what is termed natural beauty,—or a landscape characterized by simple, easy, and flowing lines.

For an example of the opposite character, let us take a stroll to the nearest woody glen in your neighborhood—perhaps a romantic valley, half shut in on two or more sides by steep rocky banks, partially concealed and overhung by clustering vines, and tangled thickets of deep foliage. Against the sky outline breaks the wild and irregular form of some old, half decayed tree near by, or the horizontal and unique branches of the larch or the pine, with their strongly marked forms. Rough and irregular stems and trunks, rocks half covered with mosses and flowering plants, open glades of bright verdure opposed to dark masses of bold shadowy foliage, form prominent objects in the foreground. If water enlivens the scene, we shall hear the murmur of the noisy brook, or the cool dashing of the cascade, as it leaps over the rocky barrier. Let the stream turn the ancient and well-worn wheel of the old mill in the middle ground, and we shall have an illustration of the picturesque, not the less striking from its familiarity to every one.

To the lover of the fine arts, the name of Claude Lorraine cannot fail to suggest examples of beauty in some of its purest and most simple forms. In the best pictures of this master, we see portrayed those graceful and flowing forms in trees, foreground, and buildings, which delight so much the lover of noble and chaste beauty,—compositions emanating from a harmonious soul, and inspired by a climate and a richness of nature and art seldom surpassed.

On the other hand, where shall we find all the elements of the picturesque more graphically combined than in the vigorous landscapes of Salvator Rosa! In those rugged scenes, even the lawless aspects of his favorite robbers and banditti are not more spirited, than the bold rocks and wild passes by which they are surrounded. And in the productions of his pencil we see the influence of a romantic and vigorous imagination, nursed amid scenes teeming with the grand as well as the picturesque—both of which he embodied in the most striking manner.

In giving these illustrations of beautiful and of picturesque scenes, we have not intended them to be understood in the light of exact models for imitation in Landscape Gardening—only as striking examples of expression in natural scenery. Although in nature many landscapes partake in a certain degree of both these kinds of expression, yet it is no doubt true that the effect is more satisfactory, where either the one or the other character predominates. The accomplished amateur should be able to seize at once upon the characteristics of these two species of beauty in all scenery. To assist the reader in this kind of discrimination, we shall keep these expressions constantly in view, and we hope we shall be able fully to illustrate the difference in the expression of even single trees, in this respect. A few strongly marked objects, either picturesque or simply beautiful, will often confer their character upon a whole landscape; as the destruction of a single group of bold rocks, covered with wood, may render a scene, once picturesque, completely insipid.

The early writers on the modern style were content with trees allowed to grow in their natural forms, and with an easy assemblage of sylvan scenery in the pleasure-grounds, which resembled the usual woodland features of nature. The effect of this method will always be interesting, and an agreeable effect will always be the result of following the simplest hints derived from the free and luxuriant forms of nature. No residence in the country can fail to be pleasing, whose features are natural groups of forest trees, smooth lawn, and hard gravel walks.

But this is scarcely Landscape Gardening in the true sense of the word, although apparently so understood by many writers. By Landscape Gardening, we understand not only an imitation, in the grounds of a country residence, of the agreeable forms of nature, but *an expressive, harmonious, and refined imitation.** In Landscape Gardening,

---

* " Thus, there is a beauty of nature and a beauty of art. To copy the beauty of nature cannot be called being an artist in the highest sense of the word, as a mechanical talent only is requisite for this. The beautiful in art depends on ideas; and the true artist, therefore, must possess, together with the talent for technical execution, that genial power which revels freely in rich forms, and is capable of producing and animating them. It is by this, that the merit of the artist and his production is to be judged; and these cannot be

we should aim to separate the accidental and extraneous in nature, and to preserve only the spirit, or essence. This subtle essence lies, we believe, in the expression more or less pervading every attractive portion of nature. And it is by eliciting, preserving, or heightening this expression, that we may give our landscape gardens a higher charm, than even the polish of art can bestow.

Now, the two most forcible and complete expressions to be found in that kind of natural scenery which may be reproduced in Landscape Gardening, are the BEAUTIFUL and the PICTURESQUE. As we look upon these as quite distinct, and as success in practical embellishment must depend on our feeling and understanding these expressions beforehand, it is necessary that we should attach some definite meaning to terms which we shall be continually obliged to employ. This is, indeed, the more requisite, from the vague and conflicting opinions of most preceding writers on this branch of the subject; some, like Repton, insisting that they are identical; and others, like Price, that they are widely different.

Gilpin defines Picturesque objects to be "those which please from some quality capable of being illustrated in painting."

Nothing can well be more vague than such a definition. We have already described the difference between the beautiful landscapes of Claude and the picturesque scenes painted by Salvator. No one can deny their being essen-

properly estimated among those barren copyists which we find so many of our flower, landscape, and portrait painters to be. But the artist stands much higher in the scale, who, though a copyist of visible nature, is capable of seizing it with poetic feeling, and representing it in its more dignified sense; such, for example, as Raphael, Poussin, Claude, &c."—WEINBREUNER.

tially distinct in character; and no one, we imagine, will deny that they both please from "some quality capable of being illustrated in painting." The beautiful female heads of Carlo Dolce are widely different from those of the picturesque peasant girls of Gerard Douw, yet both are favorite subjects with artists. A symmetrical American elm, with its wide head drooping with garlands of graceful foliage, is very different in expression from the wild and twisted larch or pine tree, which we find on the steep sides of a mountain; yet both are favorite subjects with the painter. It is clear, indeed, that there is a widely different idea hidden under these two distinct types, in material forms.

Beauty, in all natural objects, as we conceive, arises from their expression of those attributes of the Creator— infinity, unity, symmetry, proportion, etc.—which he has stamped more or less visibly on all his works; and a beautiful living form is one in which the individual is a harmonious and well balanced development of a fine type. Thus, taking the most perfect specimens of beauty in the human figure, we see in them symmetry, proportion, unity, and grace—the presence of everything that could add to the idea of perfected existence. In a beautiful tree, such as a fine American elm, we see also the most complete and perfect balance of all its parts, resulting from its growth under the most favorable influences. It realizes, then, perfectly, the finest form of a fine type or species of tree.

But all nature is not equally Beautiful. Both in living things and in inorganized matter, we see on all sides evidences of nature struggling with opposing forces. Mountains are upheaved by convulsions, valleys are broken into fearful chasms. Certain forms of animal and vegetable life, instead of manifesting themselves in those more complete and perfect forms of existence where the matter and spirit are almost in perfect harmony, appear to struggle for the full expression of their character with the material form, and to express it only with difficulty at last. What is achieved with harmony, grace, dignity, almost with apparent repose, by existences whose type is the Beautiful, is done only with violence and disturbed action by the former. This kind of manifestation in nature we call the Picturesque.

More concisely, the Beautiful is nature or art obeying the universal laws of perfect existence (i. e. Beauty), easily, freely, harmoniously, and without the *display* of power. The Picturesque is nature or art obeying the same laws rudely, violently, irregularly, and often displaying power only.

Hence we find all Beautiful forms characterized by curved and flowing lines—lines expressive of infinity,* of grace, and willing obedience: and all Picturesque forms characterized by irregular and broken lines—lines expressive of violence, abrupt action, and partial disobedience, a struggling of the idea with the substance or the condition of its being. The Beautiful is an idea of beauty calmly and harmoniously expressed; the Picturesque an idea of beauty or power strongly and irregularly expressed. As an example of the Beautiful in other arts we refer to the Apollo of the Vatican; as an example of the Picturesque, to the Laocoon or the Dying Gladiator. In nature we would place before

* Hogarth called the curve the line of beauty, and all artists have felt instinctively its power, but Mr. Ruskin (in Modern Painters) was, we believe, the first to suggest the cause of that power—that it expresses in its varying tendencies, the infinite.

the reader a finely formed elm or chestnut, whose well balanced head is supported on a trunk full of symmetry and dignity, and whose branches almost sweep the turf in their rich luxuriance ; as a picturesque contrast, some pine or larch, whose gnarled roots grasp the rocky crag on which it grows, and whose wild and irregular branches tell of the storm and tempest that it has so often struggled against.*

In pictures, too, one often hears the Beautiful confounded with the Picturesque. Yet they are quite distinct; though in many subjects they may be found harmoniously combined. Some of Raphael's angels may be taken as perfect illustrations of the Beautiful. In their serene and heavenly countenances we see only that calm and pure existence of which perfect beauty is the outward type ; on the other hand, Murillo's beggar boys are only picturesque. What we admire in them (beyond admirable execution) is not their rags or their mean apparel, but a certain irregular struggling of a better feeling within, against this outward poverty of nature and condition.

Architecture borrows, partly perhaps by association, the same expression. We find the Beautiful in the most symmetrical edifices, built in the finest proportions, and of the purest materials. It is, on the other hand, in some irregular castle formed for defence, some rude mill nearly as wild as the glen where it is placed, some thatched cottage, weather stained and moss covered, that we find the Picturesque. The Temple of Jupiter Olympus in all its perfect proportions

---

* This also explains why trees, though they retain for the most part their characteristic forms, vary somewhat in expression according to their situation. Thus the larch, though always picturesque, is far more so in mountain ridges where it is exposed to every blast, than in sheltered lawns where it only finds soft airs and sunshine.

was prized by the Greeks as a model of beauty; we, who see only a few columns and broken architraves standing, with all their exquisite mouldings obliterated by the violence of time and the elements, find them Picturesque.

To return to a more practical view of the subject, we may remark, that though we consider the Beautiful and the Picturesque quite distinct, yet it by no means follows that they may not be combined in the same landscape. This is often seen in nature; and indeed there are few landscapes of large extent where they are not thus harmoniously combined.

But it must be remembered, that while Landscape Gardening is an imitation of nature, yet it is rarely attempted on so large a scale as to be capable of the same extended harmony and variety of expression; and also, that in Landscape Gardening as in the other fine arts, we shall be more successful by directing our efforts towards the production of a *leading* character or expression, than by endeavoring to join and harmonize several.

Our own views on this subject are simply these. When a place is small, and only permits a single phase of natural expression, always endeavor to heighten or to make that single expression predominate; it should clearly either aim only at the Beautiful or the Picturesque.

When, on the contrary, an estate of large size comes within the scope of the Landscape Gardener, he is at liberty to give to each separate scene its most fitting character; he will thus, if he is a skilful artist, be able to create great variety both of beautiful and picturesque expression, and he will also be able to give a higher proof of his power, viz. by uniting all those scenes into one whole, by bringing them all into harmony. An artist who can do this has reached the ultimatum of his art.

Again and again has it been said, that Landscape Gardening and Painting are allied. In no one point does it appear to us that they are so, more than in this—that in proportion to the limited nature of the subject should simplicity and unity of expression be remembered. In some of the finest smaller compositions of Raphael, or some of the Landscapes of Claude, so fully is this borne in mind, that every object, however small, seems to be instinct with the same expression ; while in many of the great historical pictures, unity and harmony are wrought out of the most complex variety of expression.

We must not be supposed to find in nature only the Beautiful and the Picturesque. Grandeur and Sublimity are also expressions strongly marked in many of the noblest portions of natural landscape. But, except in very rare instances, they are wholly beyond the powers of the landscape gardener, at least in the comparatively limited scale of his operations in this country. All that he has to do, is to respect them where they exist in natural landscape which forms part of his work of art, and so treat the latter, as to make it accord with, or at least not violate, the higher and predominant expression of the whole.

There are, however, certain subordinate expressions which may be considered as qualities of the Beautiful, and which may originally so prevail in natural landscape, or be so elicited or created by art, as to give a distinct character to a small country residence, or portions of a large one. These are simplicity, dignity, grace, elegance, gaiety, chasteness, &c. It is not necessary that we should go into a labored explanation of these expressions. They are more or less familiar to all. A few fine trees, scattered and grouped over any surface of smooth lawn, will give a

Fig. 15 Example of the beautiful in Landscape Gardening.

Fig. 16. Example of the Picturesque in Landscape Gardening

BEAUTIES AND PRINCIPLES OF THE ART. 73

character of simple beauty; lofty trees of great age, hills covered with rich wood, an elevation commanding a wide country, stamp a site with dignity; trees of full and graceful habit or gently curving forms in the lawn, walks, and all other objects, will convey the idea of grace; as finely formed and somewhat tall trees of rare species, or a great abundance of bright climbers and gay flowering shrubs and plants, will confer characters of elegance and gaiety.

He who would create in his pleasure grounds these more delicate shades of expression, must become a profound student both of nature and art; he must be able, by his own original powers, to seize the subtle essence, the half disclosed idea involved in the finest parts of nature, and to reproduce and develope it in his Landscape Garden.

Leaving such, however, to a broader range of study than a volume like this would afford, we may offer what, perhaps, will not be unacceptable to the novice—a more detailed sketch of the distinctive features of the Beautiful and the Picturesque, as these expressions should be embodied in Landscape Gardening.

THE BEAUTIFUL in Landscape Gardening (Fig. 15) is produced by outlines whose curves are flowing and gradual, surfaces of softness, and growth of richness and luxuriance. In the shape of the ground, it is evinced by easy undulations melting gradually into each other. In the form of trees, by smooth stems, full, round, or symmetrical heads of foliage, and luxuriant branches often drooping to the ground,—which is chiefly attained by planting and grouping, to allow free development of form; and by selecting trees of suitable character, as the elm, the ash, and the like. In walks and roads, by easy flowing curves, following natural shapes of the surface, with no sharp angles or abrupt turns. In water, by the smooth lake with curved margin, embellished with flowing outlines of trees, and full masses of flowering shrubs—or in the easy winding curves of a brook. The keeping of such a scene should be of the most polished kind,—grass mown into a softness like velvet, gravel walks scrupulously firm, dry, and clean; and the most perfect order and neatness should reign throughout. Among the trees and shrubs should be conspicuous the finest foreign sorts, distinguished by beauty of form, foliage, and blossom; and rich groups of shrubs and flowering plants should be arranged in the more dressed portions near the house. And finally, considering the house itself as a feature in the scene, it should properly belong to one of the classical modes; and the Italian, Tuscan, or Venetian forms are preferable, because these have both a polished and a domestic air, and readily admit of the graceful accompaniments of vases, urns, and other harmonious accessories. Or, if we are to have a plainer dwelling, it should be simple and symmetrical in its character, and its veranda festooned with masses of the finest climbers.

The Picturesque in Landscape Gardening (Fig. 16) aims at the production of outlines of a certain spirited irregularity, surfaces comparatively abrupt and broken, and growth of a somewhat wild and bold character. The shape of the ground sought after, has its occasional smoothness varied by sudden variations, and in parts runs into dingles, rocky groups, and broken banks. The trees should in many places be old and irregular, with rough stems and bark; and pines, larches, and other trees of striking, irregular growth, must appear in numbers sufficient to give character to the woody outlines. As, to produce the Beautiful, the trees are planted singly in open groups to allow full expansion, so for the Picturesque, the grouping takes every variety of form; almost every object should group with another; trees and shrubs are often planted closely together; and intricacy and variety—thickets—glades—and underwood—as in wild nature, are indispensable. Walks and roads are more abrupt in their windings, turning off frequently at sudden angles where the form of the ground or some inviting object directs. In water, all the wildness of romantic spots in nature is to be imitated or preserved; and the lake or stream with bold shore and rocky, wood-fringed margin, or the cascade in the secluded dell, are the characteristic forms. The keeping of such a landscape will of course be less careful than in the graceful school. Firm gravel walks near the house, and a general air of neatness in that quarter, are indispensable to the fitness of the scene in all modes, and indeed properly evince the recognition of art in all Landscape Gardening. But the lawn may be less frequently mown, the edges of the walks less carefully trimmed, where the Picturesque prevails; while in portions more removed from the house, the walks may sometimes sink into a mere footpath without gravel, and the lawn change into the forest glade or meadow. The architecture which belongs to the picturesque landscape, is the Gothic mansion, the old English or the Swiss cottage, or some other striking forms, with bold projections, deep shadows, and irregular outlines. Rustic baskets, and similar ornaments, may abound near the house, and in the more frequented parts of the place.

The *recognition of art*, as Loudon justly observes, is a first principle in Landscape Gardening, as in all other arts; and those of its professors have erred, who supposed that the object of this art is merely to produce a fac-simile of nature, that could not be distinguished from a wild scene. But we contend that this principle may be fully attained with either expression—the picturesque cottage being as well a work of art as the classic villa; its baskets, and seats of rustic work, indicating the hand of man as well as the marble vase and balustrade; and a walk, sometimes narrow and crooked, is as certainly recognised as man's work, as one always regular and flowing. Foreign trees of picturesque growth are as readily obtained as those of beautiful forms. The recognition of art is, therefore, always apparent in both modes. The evidences are indeed stronger and more multiplied in the careful polish of the Beautiful landscape,* and hence many prefer this species of landscape, not, as it deserves to be preferred, because it displays the most beautiful and perfect ideas in its outlines, the forms of its trees, and all that enters into its composition, but chiefly because it also is marked by that careful polish, and that completeness, which imply the expenditure of money, which they so well know how to value.

If we declare that the Beautiful is the more perfect expression in landscape, we shall be called upon to explain why the Picturesque is so much more attractive to many minds. This, we conceive, is owing partly to the imper-

* The *beau ideal* in Landscape Gardening, as a fine art, appears to us to be embraced in the creation of scenery full of expression, as the beautiful or picturesque, the materials of which are, to a certain extent, different from those in wild nature, being composed of the floral and arboricultural riches of *all climates*, as far as possible; uniting in the same scene, a richness and a variety never to be found in any one portion of nature;—a scene characterized as a work of art, by the variety of the materials, as foreign trees, plants, &c., and by the polish and keeping of the grounds in the natural style, as distinctly as by the uniform and symmetrical arrangement in the ancient style.

fection of our natures by which most of us sympathize more with that in which the struggle between spirit and matter is most apparent, than with that in which the union is harmonious and complete; and partly because from the comparative rarity of highly picturesque landscape, it affects us more forcibly when brought into contrast with our daily life. Artists, we imagine, find somewhat of the same pleasure in studying wild landscape, where the very rocks and trees seem to struggle with the elements for foothold, that they do in contemplating the phases of the passions and instincts of human and animal life. The manifestation of power is to many minds far more captivating than that of beauty.

All who enjoy the charms of Landscape Gardening, may perhaps be divided into three classes : those who have arrived only at certain primitive ideas of beauty which are found in regular forms. and straight lines; those who in the Beautiful seek for the highest and most perfect development of the idea in the material form ; and those who in the Picturesque enjoy most a certain wild and incomplete harmony between the idea and the forms in which it is expressed.

As the two latter classes embrace the whole range of modern Landscape Gardening, we shall keep distinctly in view their two governing principles—the Beautiful and the Picturesque, in treating of the practice of the art.

There are always circumstances which must exert a controlling influence over amateurs, in this country, in choosing between the two. These are, fixed locality, expense, individual preference in the style of building, and many others which readily occur to all. The great variety of attractive sites in the older parts of the country, afford an abundance of opportunity for either taste. Within the last five years, we think the Picturesque is beginning to be preferred. It has, when a suitable locality offers, great advantages for us. The raw materials of wood, water, and surface, by the margin of many of our rivers and brooks, are at once appropriated with so much effect, and so little art, in the picturesque mode; the annual tax on the purse too is so comparatively little, and the charm so great!

While, on one hand, the residences of a country of level plains usually allow only the beauty of simple and graceful forms; the larger demesne, with its swelling hills and noble masses of wood (may we not, prospectively, say the rolling prairie too?), should always, in the hands of the man of wealth, be made to display all the breadth, variety, and harmony of both the Beautiful and the Picturesque.

There is no surface of ground, however bare, which has not, naturally, more or less tendency to one or the other of these expressions. And the improver who detects the true character, and plants, builds, and embellishes, as he should, constantly aiming to elicit and strengthen it—will soon arrive at a far higher and more satisfactory result, than one who, in the common manner, works at random. The latter may succeed in producing pleasing grounds—he will undoubtedly add to the general beauty and tasteful appearance of the country, and we gladly accord him our thanks. But the improver who unites with pleasing forms an expression of sentiment, will affect not only the common eye, but much more powerfully, the imagination, and the refined and delicate taste.

But there are many persons with small cottage places, of little decided character, who have neither room, time, nor income, to attempt the improvement of their grounds fully, after either of those two schools. How shall they render their places tasteful and agreeable, in the easiest manner ? We answer, *by attempting only the simple and the natural;* and the unfailing way to secure this, is by employing as leading features only trees and grass. A soft verdant lawn, a few forest or ornamental trees well grouped, walks, and a few flowers, give universal pleasure ; they contain in themselves, in fact, the basis of all our agreeable sensations in a landscape garden (natural beauty, and the recognition of art) ; and they are the most enduring sources of enjoyment in any place. There are no country seats in the United States so unsatisfactory and tasteless, as those in which, without any definite aim, everything is attempted ; and a mixed jumble of discordant forms, materials, ornaments, and decorations, is assembled—a part in one style and a bit in another, without the least feeling of unity or congruity. These rural bedlams, full of all kinds of absurdities, without a leading character or expression of any sort, cost their owners a vast deal of trouble and money, without giving a tasteful mind a shadow of the beauty which it feels at the first glimpse of a neat cottage residence, with its simple, sylvan character of well kept lawn and trees. If the latter does not rank high in the scale of Landscape Gardening as an art, it embodies much of its essence as a source of enjoyment—the production of the Beautiful in country residences.

Besides the beauties of form and expression in the different modes of laying out grounds, there are certain universal and inherent beauties common to all styles, and, indeed, to every composition in the fine arts. Of these, we shall especially point out those growing out of the principles of UNITY, HARMONY, and VARIETY.

UNITY, or the *production of a whole*, is a leading principle of the highest importance, in every art of taste or design, without which no satisfactory result can be realized. This arises from the fact, that the mind can only attend, with pleasure and satisfaction, to one object, or one composite sensation, at the same time. If two distinct objects, or classes of objects, present themselves at once to us, we can only attend satisfactorily to one, by withdrawing our attention for the time from the other. Hence the necessity of a reference to this leading principle of unity.

To illustrate the subject, let us suppose a building, partially built of wood, with square windows, and the remainder of brick or stone, with long and narrow windows. However well such a building may be constructed, or however nicely the different proportions of the edifice may be adjusted, it is evident it can never form a satisfactory whole. The mind can only account for such an absurdity, by supposing it to have been built by two individuals, or at two different times, as there is nothing indicating unity of mind in its composition.

In Landscape Gardening, violations of the principle of unity are often to be met with, and they are always indicative of the absence of correct taste in art. Looking upon a landscape from the windows of a villa residence, we sometimes see a considerable portion of the view embraced by the eye, laid out in natural groups of trees and shrubs, and upon one side, or perhaps in the middle of the same scene, a formal avenue leading directly up to the house. Such a view can never appear a satisfactory whole, because we experience a confusion of sensations in contemplating it. There is an evident incongruity in bringing two modes of arranging plantations, so totally different, under the eye at one moment, which distracts, rather than pleases the mind. In this example, the avenue, taken by itself, may be a beautiful object, and the groups and connected masses may, in themselves, be elegant; yet if the two portions are seen together, they will not form a whole, because they cannot make a composite idea. For the same reason, there is something unpleasing in the introduction of fruit trees among elegant ornamental trees on a lawn, or even in assembling together, in the same beds, flowering plants and culinary vegetables—one class of vegetation suggesting the useful and homely alone to the mind, and the other, avowedly, only the ornamental.

In the arrangement of a large extent of surface, where a great many objects are necessarily presented to the eye at once, the principle of unity will suggest that there should be some grand or leading features to which the others should be merely subordinate. Thus, in grouping trees, there should be some large and striking masses to which the others appear to belong, however distant, instead of scattered groups, all of the same size. Even in arranging walks, a whole will more readily be recognised, if there are one or two of large size, with which the others appear connected as branches, than if all are equal in breadth, and present the same appearance to the eye in passing.

In all works of art which command universal admiration we discover an unity of conception and composition, an unity of taste and execution. To assemble in a single composition forms which are discordant, and portions dissimilar in plan, can only afford pleasure for a short time to tasteless minds, or those fond of trifling and puerile conceits. The production of an accordant whole is, on the contrary, capable of affording the most permanent enjoyment to educated minds, everywhere, and at all periods of time.

After unity, the principle of VARIETY is worthy of consideration, as a fertile source of beauty in Landscape Gardening. Variety must be considered as belonging more to the details than to the production of a whole, and it may be attained by disposing trees and shrubs in numerous different ways; and by the introduction of a great number of different species of vegetation, or kinds of walks, ornamental objects, buildings, and seats. By producing intricacy, it creates in scenery a thousand points of interest, and elicits new beauties, through different arrangements and combinations of forms and colors, light and shades. In pleasure-grounds, while the whole should exhibit a general plan, the different scenes presented to the eye, one after the other, should possess sufficient variety in the detail to keep alive the interest of the spectator, and awaken further curiosity.

HARMONY may be considered the principle presiding over variety, and preventing it from becoming discordant. It, indeed, always supposes *contrasts*, but neither so strong nor so frequent as to produce discord; and *variety*, but not so great as to destroy a leading expression. In plantations, we seek it in a combination of qualities, opposite in some respects, as in the color of the foliage, and similar in others more important, as the form. In embellishments, by a great variety of objects of interest, as sculptured vases, sun dials, or rustic seats, baskets, and arbors, of different forms, but all in accordance, or keeping with the spirit of the scene.

To illustrate the three principles, with reference to Landscape Gardening, we may remark, that, if unity only were consulted, a scene might be planted with but one kind of tree, the effect of which would be sameness; on the other hand, variety might be carried so far as to have every tree of a different kind, which would produce a confused effect. Harmony, however, introduces contrast and variety, but keeps them subordinate to unity, and to the leading expression; and is, thus, the highest principle of the three.

In this brief abstract of the nature of imitation in Landscape Gardening and the kinds of beauty which it is possible to produce by means of the art, we have endeavored to elucidate its leading principles, clearly, to the reader. These grand principles we shall here succinctly recapitulate, premising that a familiarity with them is of the very first importance in the successful practice of this elegant art, viz. :

THE IMITATION OF THE BEAUTY OF EXPRESSION, derived from a refined perception of the sentiment of nature : THE RECOGNITION OF ART, founded on the immutability of the true, as well as the beautiful : AND THE PRODUCTION OF UNITY, HARMONY, AND VARIETY, in order to render complete and continuous, our enjoyment of any artistical work.

Neither the professional Landscape Gardener, nor the amateur, can hope for much success in realizing the nobler effects of the art, unless he first make himself master of the natural character or prevailing expression of the place to be improved. In this nice perception, at a glance, of the natural expression, as well as the capabilities of a residence, lies the secret of the superior results produced even by the improver, who, to use the words of Horace Walpole, "is proud of no other art than that of softening nature's harshness, and copying her graceful touch." When we discover the *picturesque* indicated in the grounds of the residence to be treated, let us take advantage of it; and while all harshness incompatible with scenery near the house is removed, the original expression may in most cases be heightened, in all rendered more elegant and appropriate, without lowering it in force or spirit. In like manner good taste will direct us to embellish scenery expressive of *the Beautiful*, by the addition of forms, whether in trees, buildings, or other objects, harmonious in character, as well as in color and outline.

SECTION III.

ON WOOD.

The beauty of Trees in Rural Embellishments. Pleasure resulting from their cultivation. Plantations in the Ancient Style; their formality. In the Modern Style; grouping trees. Arrangement and grouping in the Graceful school; in the Picturesque school. Illustrations in planting villa, ferme ornée, and cottage grounds. General classification of trees as to forms, with leading characteristics of each class.

> " He gains all points, who pleasingly confounds,
> Surprises, varies, and conceals the bounds.
> Calls in the country, catches opening glades,
> Joins willing woods, and varies shades from shades;
> Now breaks, or now directs the intending lines;
> Paints as you plant, and, as you work, designs."
>
> POPE.

AMONG all the materials at our disposal for the embellishment of country residences, none are at once so highly ornamental, so indispensable, and so easily managed, as *trees*, or *wood*. We introduce them in every part of the landscape, —in the foreground as well as in the distance, on the tops of the hills and in the depths of the valleys. They are, indeed, like the drapery which covers a somewhat ungainly figure, and while it conceals its defects, communicates to it new interest and expression.

A tree, undoubtedly, is one of the most beautiful objects in nature. Airy and delicate in its youth, luxuriant and majestic in its prime, venerable and picturesque in its old age, it constitutes in its various forms, sizes, and developments, the greatest charm and beauty of the earth in all countries. The most varied outline of surface, the finest combination of picturesque materials, the stateliest country house would be comparatively tame and spiritless, without the inimitable accompaniment of foliage. Let those who have passed their whole lives in a richly wooded country, —whose daily visions are deep leafy glens, forest clad hills, and plains luxuriantly shaded,—transport themselves for a moment to the desert, where but a few stunted bushes raise their heads above the earth, or those wild steppes where the eye wanders in vain for some "leafy garniture,"—where the sun strikes down with parching heat, or the wind sweeps over with unbroken fury, and they may, perhaps, estimate, by contrast, their beauty and value.

We are not now to enumerate the great usefulness of trees,—their value in the construction of our habitations, our navies, the various implements of labor,—in short, the thousand associations which they suggest as ministering to our daily wants; but let us imagine the loveliest scene, the wildest landscape, or the most enchanting valley, despoiled of *trees*, and we shall find nature shorn of her fair proportions, and the character and expression of these favorite spots almost entirely destroyed.

Wood, in its many shapes, is then one of the greatest sources of interest and character in Landscapes. Variety, which we need scarcely allude to as a fertile source of beauty, is created in a wonderful degree by a natural arrangement of trees. To a pile of buildings, or even of ruins, to a group of rocks or animals, they communicate new life and spirit by their irregular outlines, which, by partially concealing some portions, and throwing others

ON WOOD AND PLANTATIONS. 87

into stronger light, contribute greatly to produce intricacy and variety, and confer an expression, which, without these latter qualities, might in a great measure be wanting. By shutting out some parts, and inclosing others, they divide the extent embraced by the eye into a hundred different landscapes, instead of one tame scene bounded by the horizon.

The different seasons of the year, too, are inseparably connected in our minds with the effects produced by them on woodland scenery. Spring is joyous and enlivening to us, as nature then puts on her fresh livery of green, and the trees bud and blossom with a renewed beauty, that speaks with a mute and gentle eloquence to the heart. In summer they offer us a grateful shelter under their umbrageous arms and leafy branches, and whisper unwritten music to the passing breeze. In autumn we feel a melancholy thoughtfulness as

"We stand among the fallen leaves,"

and gaze upon their dying glories. And in winter we see in them the silent rest of nature, and behold in their leafless spray, and seemingly dead limbs, an annual type of that deeper mystery—the deathless sleep of all being.

By the judicious employment of trees in the embellishment of a country residence, we may effect the greatest alterations and improvements within the scope of Landscape Gardening. Buildings which are tame, insipid, or even mean in appearance, may be made interesting, and often picturesque, by a proper disposition of trees. Edifices, or parts of them that are unsightly, or which it is desirable partly or wholly to conceal, can readily be hidden or improved by wood; and walks and roads, which otherwise would be but simple ways of approach from one point to another, are, by an elegant arrangement of trees on their margins, or adjacent to them, made the most interesting and pleasing portions of the residence.

In Geometric gardening, trees disposed in formal lines, exhibit as strongly art or design in the contriver, as regular architectural edifices; while, in a more elevated and enlightened taste, we are able to dispose them in our pleasure-grounds and parks, around our houses, in all the variety of groups, masses, thicket, and single trees, in such a manner as to rival the most beautiful scenery of general nature; producing a portion of landscape which unites with all the comforts and conveniences of rural habitation, the superior charm of refined arrangement, and natural beauty of expression.

If it were necessary to present any other inducement to the country gentleman to form plantations of trees, than the great beauty and value which they add to his estate, we might find it in the pleasure which all derive from their cultivation. Unlike the pleasure arising from the gratification of our taste in architecture, or any other of the arts whose productions are offered to us perfect and complete, the satisfaction arising from planting and rearing trees is never weakened. "We look," says a writer, "upon our trees as our offspring; and nothing of inanimate nature can be more gratifying than to see them grow and prosper under our care and attention,— nothing more interesting than to examine their progress, and mark their several peculiarities. In their progress from plants to trees, they every year unfold new and characteristic marks of their ultimate beauty, which not only compensate for past cares and troubles, but like the returns of gratitude, raise a most delightful train of sensations in the mind; so innocent and rational, that they may justly rank with the most exquisite of human enjoyments."

> " Happy is he, who in a country life
> Shuns more perplexing toil and jarring strife;
> Who lives upon the natal soil he loves,
> And sits beneath his old ancestral groves."

To this, let us add the complacent feelings with which a man in old age may look around him and behold these leafy monarchs, planted by his boyish hands and nurtured by him in his youthful years, which have grown aged and venerable along with him;

> " A wood coeval with himself he sees,
> And loves his own contemporary trees."

PLANTATIONS IN THE ANCIENT STYLE. In the arrangement and culture of trees and plants in the ancient style of Landscape Gardening, we discover the evidences of the formal taste,—abounding with every possible variety of quaint conceits, and rife with whimsical expedients, so much in fashion during the days of Henry and Elizabeth, and until the eighteenth century in England, and which is still the reigning mode in Holland, and parts of France. In these gardens, nature was tamed and subdued, or as some critics will have it, tortured into every shape which the ingenuity of the gardener could suggest; and such kinds of vegetation as bore the shears most patiently, and when carefully trimmed, assumed gradually the appearance of verdant statues, pyramids, crowing cocks, and rampant lions, were the especial favorites of the gardeners of the old school.* The stately etiquette and courtly precision of the manners of our English ancestors, extended into their gardens, and were reflected back by the very trees which lined their avenues, and the shrubs which surrounded their houses. "Nonsuch, Theobalds, Greenwich, Hampton Court, Hatfield, Moor-Park, Chatsworth, Beaconfield, Cashiobury, Ham, and many another," says William Howitt, "stood in all that stately formality which Henry and Elizabeth admired; and in which our Surreys, Leicesters, Essexes, the splendid nobles of the Tudor dynasty, the gay ladies and gallants of Charles II.'s court, had walked and talked,—fluttering in glittering processions, or flirting in green alleys and bowers of topiary work, and amid figures, in lead or stone, fountains, cascades,— copper-trees dropping sudden showers on the astonished passers under, stately terraces with gilded balustrades, and curious quincunxes, obelisks, and pyramids;—fitting objects of admiration of those who walked in high heeled shoes, ruffs, and fardingales, with fan in hand, or in trunk hose and laced doublet."

Symmetrical uniformity governed with despotic power even the trees and foliage, in the ancient style. In the more simple country residences, the plantations were always arranged in some regular lines or geometrical figures. Long parallel rows of trees were planted for groves and avenues along the principal roads and walks. The greatest care was taken to avoid any appearance of irregularity. A tree upon one side of the house was opposed by another *vis à vis*, and a row of trees at the

* The unique *ideal* of the "Garden of Eden," by one of the old Dutch painters, with sheared hedges, formal alleys, and geometric plots of flowers, for the entertainment of our first parents, is doubtless familiar to our readers.

ON WOOD AND PLANTATIONS. 91

right of the mansion had its always accompanying row on the left: or, as Pope in his Satire has more rhythmically expressed it—

> " Grove nods at grove, each alley has its brother,
> And half the platform just reflects the other."

In the interior of the park, the plantations were generally disposed either in straight avenues crossing each other, or clumped in the form of circles, stars, squares, etc.; and long vistas were obtained through the avenues divaricating from the house in various directions, over level surfaces. One of the favorite fancies of the geometric gardener was the Labyrinth (fig. 17), of which a few celebrated examples are still in existence in England, and which consisted of a multitude of trees thickly planted in impervious hedges, covering sometimes several acres of ground. These labyrinths were the source of much amusement to the family and guests, the trial of skill being to find the centre, and from that point to return again without assistance; and we are told by a historian of the garden of that period, that "the stranger having once entered, was sorely puzzled to get out."

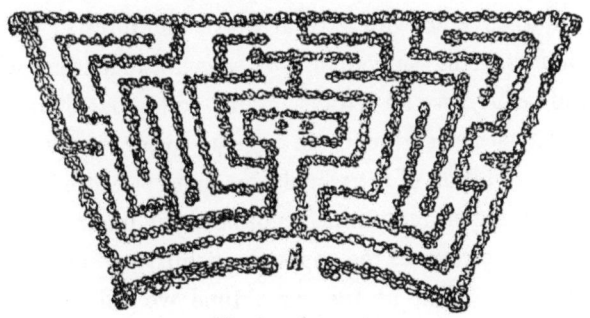

[Fig. 17. A Labyrinth.]

Since the days when these gardens were in their glory the taste in Landscape Gardening has undergone a great change. The beautiful and the picturesque are the new elements of interest, which, entering into the composition of our gardens and home landscapes, have to refined minds increased a hundred fold the enjoyment derived from this species of rural scenery. Still, there is much to admire in the ancient style. Its long and majestic avenues, the wide-spreading branches interlacing over our heads, and forming long, shadowy aisles, are, themselves alone, among the noblest and most imposing sylvan objects. Even the formal and curiously knotted gardens are interesting, from the pleasing associations which they suggest to the mind, as having been the favorite haunts of Shakspeare, Bacon, Spenser, and Milton. They are so inseparably connected, too, in our imaginations, with the quaint architecture of that era, that wherever that style of building is adopted (and we observe several examples already among us) this style of gardening may be considered as highly appropriate, and in excellent keeping with such a country house.

It has been remarked, that the geometric style would always be preferred in a new country, or in any country where the amount of land under cultivation is much less than that covered with natural woods and forests; as the inhabitants being surrounded by scenery abounding with natural beauty, would always incline to lay out their gardens and pleasure-grounds in regular forms, because the distinct exhibition of art would give more pleasure by contrast, than the elegant imitation of beautiful nature. That this is true as regards the mass of uncultivated minds, we do not deny. But at the same time we affirm that it evinces a meagre taste, and a lower state of the art, or a lower perception of beauty in the individual who employs the geometrical style in such cases. A person, whose place is surrounded by inimitably grand or sublime scenery, would undoubtedly fail to excite our admiration, by attempting a fac-simile imitation of such scenery on the small scale of a park or garden; but he is not, therefore, obliged to resort to right-lined plantations and regular grass plots, to produce something which shall be at once sufficiently different to attract notice, and so beautiful as to command admiration. All that it would be requisite for him to do in such a case, would be to employ rare and foreign ornamental trees; as for example, the horse-chestnut and the linden, in situations where the maple and the sycamore are the principal trees,—elegant flowering shrubs and beautiful creepers, instead of sumacs and hazels,—and to have his place kept in high and polished order, instead of the tangled wildness of general nature.

On the contrary, were a person to desire a residence newly laid out and planted, in a district where all around is in a high state of polished cultivation, as in the suburbs of a city, a species of pleasure would result from the imitation of scenery of a more spirited, natural character, as the picturesque, in his grounds. His plantations are made in irregular groups, composed chiefly of picturesque trees, as the larch, &c.—his walks would lead through varied scenes, sometimes bordered with groups of rocks overrun with flowering creepers and vines; sometimes with thickets or little copses of shrubs and flowering plants; sometimes through wild and comparatively neglected portions; the whole interspersed with open glades of turf.

In the majority of instances in the United States, the

modern style of Landscape Gardening, wherever it is appreciated, will, in practice, consist in arranging a demesne of from five to some hundred acres,—or rather that portion of it, say one half, one third, etc., devoted to lawn and pleasure-ground, pasture, etc.—so as to exhibit groups of forest and ornamental trees and shrubs, surrounding the dwelling of the proprietor, and extending for a greater or less distance, especially towards the place of entrance from the public highway. Near the house, good taste will dictate the assemblage of groups and masses of the rarer or more beautiful trees and shrubs; commoner native forest trees occupying the more distant portions of the grounds.*

PLANTATIONS IN THE MODERN STYLE. In the Modern Style of Landscape Gardening, it is our aim, in plantations, to produce not only what is called natural beauty, but even higher and more striking beauty of expression, and of individual forms, than we see in nature; to create variety

---

* Although we love planting, and avow that there are few greater pleasures than to see a darling tree, of one's own placing, every year stretching wider its feathery head of foliage, and covering with a darker shadow the soft turf beneath it, still, we will not let the ardent and inexperienced hunter after a location for a country residence, pass without a word of advice. This is, *always to make considerable sacrifice to get a place with some existing wood, or a few ready grown trees upon it;* especially near the site for the house. It is better to yield a little in the extent of prospect, or in the direct proximity to a certain locality, than to pitch your tent in a plain,—desert-like in its bareness—on which your leafy sensibilities must suffer for half a dozen years at least, before you can hope for any solace. It is doubtful whether there is not almost as much interest in studying from one's window the curious ramifications, the variety of form, and the entire harmony, to be found in a fine old tree, as in gazing from a site where we have no interruption to a panorama of the whole horizon; and we have generally found that no planters have so little courage and faith, as those who have commenced without the smallest group of large trees, as a nucleus for their plantations.

ON WOOD AND PLANTATIONS. 95

and intricacy in the grounds of a residence by various modes of arrangement; to give a highly elegant or polished air to places by introducing rare and foreign species; and to conceal all defects of surface, disagreeable views, unsightly buildings, or other offensive objects.

As uniformity, and grandeur of single effects, were the aim of the old style of arrangement, so variety and harmony of the whole are the results for which we labor in the modern landscape. And as the *Avenue*, or the straight line, is the leading form in the geometric arrangement of plantations, so let us enforce it upon our readers, the GROUP is equally the key-note of the Modern style. The smallest place, having only three trees, may have these pleasingly connected in a group; and the largest and finest park—the Blenheim or Chatsworth, of seven miles square, is only composed of a succession of groups, becoming masses, thickets, woods. If a demesne with the most beautiful surface and views has been for some time stiffly and awkwardly planted, it is exceedingly difficult to give it a natural and agreeable air; while many a tame level, with scarcely a glimpse of distance, has been rendered lovely by its charming groups of trees. How necessary, therefore, is it, in the very outset, that the novice, before he begins to plant, should know how to arrange a tasteful group!

Nothing, at first thought, would appear easier than to arrange a few trees in the form of a natural and beautiful group,—and nothing really is easier to the practised hand. Yet experience has taught us that the generality of persons, in commencing their first essays in ornamental planting, almost invariably crowd their trees into a close, regular *clump*, which has a most formal and unsightly appearance, as different as possible from the easy, flowing outline of the group.*

"Were it made the object of study," said Price, "how to invent something, which, under the name of ornament, should disfigure a whole park, nothing could be contrived to answer that purpose like a *clump*. Natural groups, being formed by trees of different ages and sizes, and at different distances from each other, often too by a mixture of those of the largest size with others of inferior growth, are full of variety in their outlines; and from the same causes, no two groups are exactly alike. But clumps, from the trees being generally of the same age and growth, from their being planted nearly at the same distance, in a circular form, and from each tree being equally pressed by his neighbor, are as like each other as so many puddings turned out of one common mould. Natural groups are full of openings and hollows, of trees advancing before, or retiring behind each other; all productive of intricacy, of variety, of deep shadows and brilliant lights: in walking about them the form changes at every step; new combinations, new lights and shades, new inlets present themselves in succession.

---

* A friend of ours, at Northampton, who is a most zealous planter, related to us a diverting expedient to which he was obliged to resort, in order to ensure *irregular groups*. Busily engaged in arranging plantations of young trees on his lawn, he was hastily obliged to leave home, and intrust the planting of the groups to some common garden laborers, whose ideas he could not raise to a point sufficiently high to appreciate any beauty in plantations, unless made in regular forms and straight lines. "Being well aware," says our friend, "that if left to themselves I should find all my trees, on my return, in hollow squares or circular clumps, I hastily *threw up a peck of potatoes* into the air, one by one, and directed my workmen to plant a tree where every potatoe fell! Thus, if I did not attain the maximum of beauty in grouping, I at least had something not so offensive as geometrical figures."

But *clumps*, like compact bodies of soldiers, resist attacks from all quarters; examine them in every point of view; walk round and round them; no opening, no vacancy, no stragglers; but in the true military character, *ils sont face partout !*"*

The chief care, then, which is necessary in the formation of groups, is, *not* to place them in any regular or artificial manner,—as one at each corner of a triangle, square, octagon, or other many-sided figure; but so to dispose them, as that the whole may exhibit the variety, connexion, and intricacy seen in nature. "The greatest beauty of a group of trees," says Loudon, "as far as respects their stems, is in the varied direction these take as they grow into trees; but as that is, for all practical purposes, beyond the influence of art, all we can do, is to vary as much as possible the ground plan of groups, or the relative positions which the stems have to each other where they spring from the earth. This is considerable, even where a very few trees are used, of which any person may convince himself by placing a few dots on paper. Thus two trees (fig. 18), or a tree and shrub, which is the smallest group (*a*), may be placed in three different positions with reference to a spectator in a fixed point; if he moves round them, they will first vary in form separately, and next unite in one or two groups, according to the position of the spectator. In like manner, three

---

* Those who peruse Price's " Essay on the Picturesque," cannot fail to be entertained with the vigor with which he advocates the picturesque, and attacks the *clumping* method of laying out grounds, so much practised in England on the first introduction of the modern style. Brown was the great practitioner at that time, and his favorite mode seems to have been to cover the whole surface of the grounds with an unmeaning assemblage of round bunchy clumps.

trees may be placed in four different positions; four trees may be placed in eight different positions (*b*); five trees may be grouped in ten different ways, as to ground plan; six may be placed in twelve different ways (*c*), and so on." (*Encyclopædia of Gard.*)

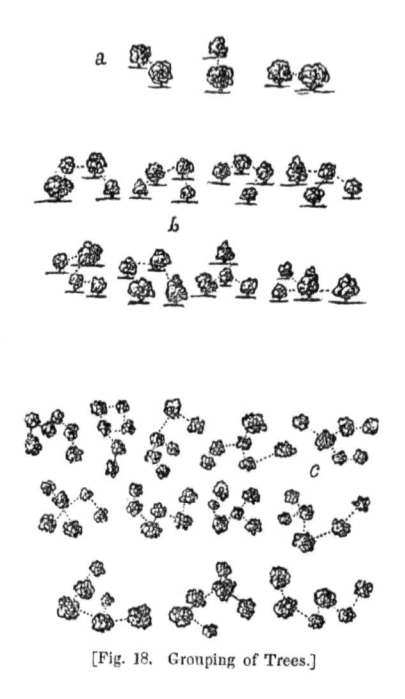

[Fig. 18. Grouping of Trees.]

In the composition of larger masses, similar rules must be observed as in the smaller groups, in order to prevent them from growing up in heavy, clumpish forms. The outline must be flowing, here projecting out into the grass, there receding back into the plantation, in order to take off all appearance of stiffness and regularity. Trees of medium and smaller size should be so interspersed with those of larger growth, as to break up all formal sweeps in the line produced by the tops of their summits, and occasionally, low trees should be planted on the outer edge of the mass, to connect it with the humble verdure of the surrounding sward.

In many parts of the union, where new residences are being formed, or where old ones are to be improved, the grounds will often be found, partially, or to a considerable extent, clothed with belts or masses of wood, either previously planted, or preserved from the woodman's axe. How easily we may turn these to advantage in the natural style of Landscape Gardening ; and by judicious trimming when too thick, or additions when too much scattered, elicit often the happiest effects, in a magical manner! In the accompanying sketch (fig. 19), the reader will recognise a portrait of a hundred familiar examples, existing with us, of the places of persons of considerable means and intelligence, where the house is not less meagre than the

[Fig. 19. View of a Country Residence, as frequently seen.]

stiff approach leading to it, bordered with a formal belt of trees. The succeeding sketch (fig. 20) exhibits this place as improved agreeably to the principles of modern Landscape Gardening, not only in the plantations, but in the house,—which appears tastefully altered from a plain unmeaning parallelogram, to a simple, old English cottage,— and in the more graceful approach. Effects like these

100   LANDSCAPE GARDENING.

are within the reach of very moderate means, and are peculiarly worth attention in this country, where so much has already been partially, and often badly executed.

[Fig. 20. View of the same Residence, improved.]

Where there are large masses of wood to regulate and arrange, much skill, taste, and judgment, are requisite, to enable the proprietors to preserve only what is really beautiful and picturesque, and to remove all that is superfluous. Most of our native woods, too, have grown so closely, and the trees are consequently so much drawn up, that should the improver thin out any portion, at once, to *single* trees, he will be greatly disappointed if he expects them to stand long; for the first severe autumnal gale will almost certainly prostrate them. The only method, therefore, is to allow them to remain in groups of considerable size at first, and to thin them out as is finally desired, when they have made stronger roots and become more inured to the influence of the sun and air.*

But to return to grouping; what we have already endeavored to render familiar to the reader, may be called

* When, in thinning woods in this manner, those left standing have a meagre appearance, a luxuriant growth may be promoted by the application of manure plentifully dug in about the roots. This will also, by causing an abundant growth of new roots, strengthen the trees in their position.

grouping in its simple meaning—for general effect, and with an eye only to the natural beauty of pleasing forms. Let us now explain, as concisely as we may, the mode of grouping in the two schools of Landscape Gardening heretofore defined, that is to say, grouping and planting for Beautiful effect, and for Picturesque effect; as we wish it understood that these two different expressions, in artificial landscape, are always to a certain extent under our control.

PLANTING AND GROUPING TO PRODUCE THE BEAUTIFUL. The elementary features of this expression our readers will remember to be fulness and softness of outline, and perfectly luxuriant development. To insure these in plantations, we must commence by choosing mainly trees of graceful habit and flowing outlines; and of this class of trees, hereafter more fully illustrated, the American elm and the maple may be taken as the type. Next, in disposing them, they must usually be planted rather distant in the groups, and often singly. We do not mean by this, that close groups may not occasionally be formed, but there should be a predominance of trees grouped at such a distance from each other, as to allow a full development of the branches on every side. Or, when a close group is planted, the trees composing it should be usually of the same or a similar kind, in order that they may grow up together and form one finely rounded head. Rich creepers and blossoming vines, that grow in fine luxuriant wreaths and masses, are fit accompaniments to occasional groups in this manner. Fig. 21 represents a plan of trees grouped along a road or walk, so as to develope the Beautiful.

It is proper that we should here remark, that a distinct species of after treatment is required for the two modes. Trees, or groups, where the Beautiful is aimed at, should be

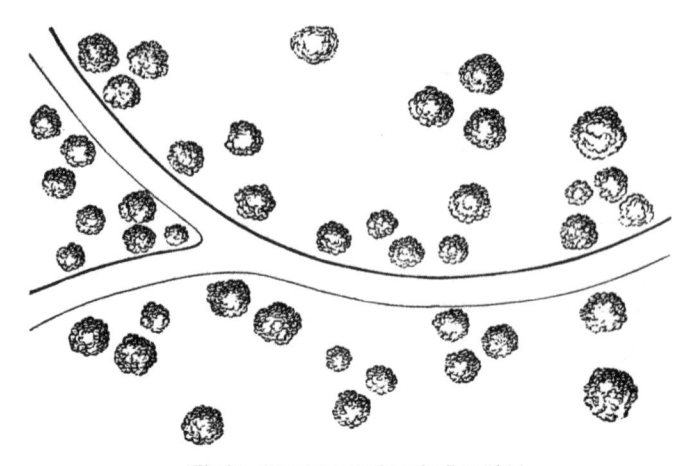

[Fig. 21. Grouping to produce the Beautiful.]

pruned with great care, and indeed scarcely at all, except to remedy disease, or to correct a bad form. Above all, the full luxuriance and development of the tree should be encouraged by good soil, and repeated manurings when necessary; and that most expressively elegant fall and droop of the branches, which so completely denotes the Beautiful in trees, should never be warred against by any trimming of the lower branches, which must also be carefully preserved against cattle, whose *browsing line* would soon efface this most beautiful disposition in some of our fine lawn trees. Clean, smooth stems, fresh and tender bark, and a softly rounded pyramidal or drooping head, are the characteristics of a Beautiful tree. We need not add that gently sloping ground, or surfaces rolling in easy undulations, should accompany such plantations.

PLANTING AND GROUPING TO PRODUCE THE PICTURESQUE. All trees are admissible in a picturesque place, but a predominance must be used by the planter of what are truly called picturesque trees, of which the larch and fir tribe, and some species of oak, may be taken as examples. In Picturesque plantations everything depends on *intricacy*

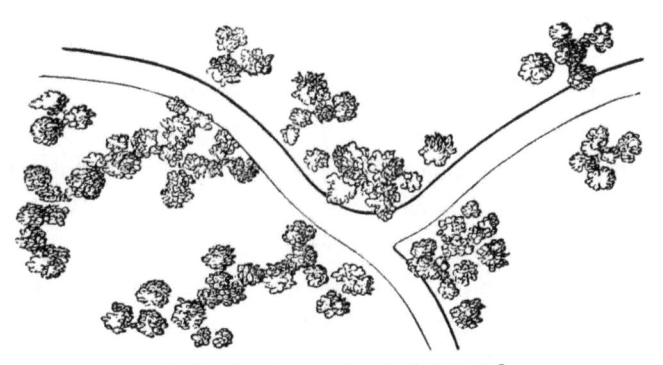

[Fig. 22. Grouping to produce the Picturesque.]

and *irregularity*, and grouping, therefore, must often be done in the most irregular manner—rarely, if ever, with single specimens, as every object should seem to connect itself with something else; but most frequently there should be irregular groups, occasionally running into thickets, and always more or less touching each other; trusting to after time for any thinning, should it be necessary. Fig. 22 may, as compared with Fig. 21, give an idea of picturesque grouping.

There should be more of the wildness of the finest and most forcible portions of natural woods or forests, in the disposition of the trees; sometimes planting them closely, even two or three in the same hole, at others more loose and scattered. These will grow up into wilder and more striking forms, the barks will be deeply furrowed and rough, the limbs twisted and irregular, and the forms and outlines distinctly varied. They should often be intermixed with smaller undergrowth of a similar character, as the hazel, hawthorn, etc., and formed into such picturesque and striking groups, as painters love to study and introduce into their pictures. Sturdy and bright vines, or such as are themselves picturesque in their festoons and hangings, should be allowed to clamber over occasional trees in a negligent manner; and the surface and grass, in parts of the scene not immediately in the neighborhood of the mansion, may be kept short by the cropping of animals, or allowed to grow in a more careless and loose state, like that of tangled dells and natural woods.

There will be the same open glades in picturesque as in beautiful plantations; but these openings, in the former, will be bounded by groups and thickets of every form, and of different degrees of intricacy, while in the latter the eye will repose on softly rounded masses of foliage, or single open groups of trees, with finely balanced and graceful heads and branches.

In order to know how a plantation in the Picturesque mode should be treated, after it is established, we should reflect a moment on what constitutes picturesqueness in any tree. This will be found to consist either in a certain natural roughness of bark, or wildness of form and outline, or in some accidental curve of a branch of striking manner of growth, or perhaps of both these conjoined. A broken or crooked limb, a leaning trunk, or several stems springing from the same base, are frequently peculiarities that at once stamp a tree as picturesque. Hence, it is easy to see that the excessive care of the cultivator of trees in the graceful school to obtain the smoothest trunks, and the most sweeping, perfect, and luxuriant heads of foliage, is quite the opposite of what is the picturesque arboriculturist's ambition. He desires to encourage a certain wildness of growth, and allows his trees to spring up occasionally in thickets to assist this effect; he delights in occasional irregularity of stem and outline, and he therefore suffers his trees here and there to crowd each other; he admires a twisted limb or a moss covered branch, and in pruning he therefore is careful to leave precisely what it would be the aim of the other to remove; and his pruning, where it is at all necessary, is directed rather towards increasing the naturally striking and peculiar habit of the picturesque tree, than assisting it in developing a form of unusual refinement and symmetry. From these remarks we think the amateur will easily divine, that planting, grouping, and culture to produce the Beautiful, require a much less artistic eye (though much more care and attention) than performing the same operations to elicit the Picturesque. The charm of a refined and polished landscape garden, as we usually see it in the Beautiful grounds with all the richness and beauty developed by high culture, arises from our admiration of the highest perfection, the greatest beauty of form, to which every object can be brought; and, in trees, a judicious selection, with high cultivation, will always produce this effect.

But in the Picturesque landscape garden there is visible a piquancy of effect, certain bold and striking growths and combinations, which we feel at once, if we know them to be the result of art, to be the production of a peculiar species of attention—not merely good, or even refined ornamental gardening. In short, no one can be a picturesque improver (if he has to begin with young plantations) who is not himself something of an artist—who has not studied nature with an artistical eye—and who is not capable of imitating, eliciting, or heightening, in his plantations or other portions of his residence, the picturesque in its many variations. And we may add here, that efficient and charming as is the assistance which all ornamental planters will derive from the study of the best landscape engravings and pictures of distinguished artists, they are indispensably necessary to the *picturesque* improver. In these he will often find embodied the choicest and most captivating studies from picturesque nature; and will see at a glance the effect of certain combinations of trees, which he might otherwise puzzle himself a dozen years to know how to produce

After all, as the picturesque improver here will most generally be found to be one who chooses a comparatively wild and wooded place, we may safely say that, if he has the true feeling for his work, he will always find it vastly easier than those who strive after the Beautiful; as the majority of the latter may be said to begin nearly anew—choosing places not for wildness and intricacy of wood, but for openness and the smiling, sunny, undulating plain, where they must of course to a good extent plant anew.

After becoming well acquainted with grouping, we should bring ourselves to regard those principles which govern our improvements as a whole. We therefore must call the attention of the improver to the two following principles, which are to be constantly in view: *the production of a whole, and the proper connexion of the parts.*

Any person who will take the trouble to reflect for a moment on the great diversity of surface, change of position, aspects, views, etc., in different country residences, will at once perceive how difficult, or, indeed, how impossible it is, to lay down any fixed or exact rules for arranging plantations in the modern style. What would be precisely adapted to a hilly rolling park, would often be found entirely unfit for adoption in a smooth, level surface, and the contrary. Indeed, the chief beauty of the modern style is the variety produced by following a few leading principles, and applying them to different and varied localities; unlike the geometric style, which proceeded to level, and arrange, and erect its avenues and squares, alike in every situation, with all the precision and certainty of mathematical demonstration.

In all grounds to be laid out, however, which are of a lawn or park-like extent, and call for the exercise of judgment and taste, the mansion or dwelling-house, being itself the chief or leading object in the scene, should form, as it were, the central point, to which it should be the object of the planter to give importance. In order to do this effectually, the large masses or groups of wood should cluster round, or form the back-ground to the main edifice; and where the offices or out-buildings approach the same neighborhood, they also should be embraced. We do not mean by this to convey the idea, that a thick wood should be planted around and in the close neighborhood of the mansion or villa, so as to impede the free circulation of air; but its appearance and advantages may be easily produced by a comparatively loose plantation of groups well connected by intermediate trees, so as to give all the effect of a large mass. The front, and at least that side nearest the approach road, will be left open, or nearly so; while the plantations on the *back-ground* will give dignity and importance to the house, and at the same time effectually screen the approach to the farm buildings, and other objects which require to be kept out of view; and here, both for the purposes of shelter and richness of effect, a good proportion of evergreens should be introduced.

From this principal mass, the plantations must break off in groups of greater or less size, corresponding to the extent covered by it; if large, they will diverge into masses of considerable magnitude, if of moderate size, in groups made up of a number of trees. In the lawn front of the house, appropriate places will be found for a number of the most elegant single trees, or small groups of trees, remarkable for the beauty of their forms, foliage, or blossoms. Care must be taken, however, in disposing these, as well as many of the groups, that they are not placed so as, at some future time, to interrupt or disturb the finest points of prospect.

In more distant parts of the plantations will also appear masses of considerable extent, perhaps upon the boundary line, perhaps in particular situations on the sides, or in the interior of the whole; and the various groups which are distributed between should be so managed as, though in most cases distinct, yet to appear to be the connecting links which unite these distant shadows in the composition, with the larger masses near the house. Sometimes several small groups will be almost joined together; at others the effect may be kept up by a small group, aided by a few neighboring single trees. This, for a park-like place. Where the place is small, a pleasure-ground character is all that can be obtained. But by employing chiefly shrubs, and only a few trees, very similar and highly beautiful effects may be attained.

The grand object in all this should be to open to the eye, from the windows or front of the house, a wide surface, partially broken up and divided by groups and masses of trees into a number of pleasing lawns or openings, differing in size and appearance, and producing a charming *variety* in the scene, either when seen from a given point or when examined in detail. It must not be forgotten that, as a general rule, the grass or surface of the lawn answers as the principal light, and the woods or plantations as the shadows, in the same manner in nature as in painting; and that these should be so managed as to lead the eye to the mansion as the most important object when seen from without, or correspond to it in grandeur and magnitude, when looked upon from within the house. If the surface is too much crowded with groups of foliage, *breadth* of light will be found wanting; if left too bare, there will be felt, on the other hand, an absence of the noble effect of deep and broad shadows.

One of the loveliest charms of a fine park is, undoubtedly, variation or undulation of surface. Everything, accordingly, which tends to preserve and strengthen this pleasing character, should be kept constantly in view. Where, therefore, there are no obvious objections to such a course, the eminences, gentle swells, or hills, should be planted, in preference to the hollows or depressions. By planting the elevated portions of the grounds, their apparent height is increased; but by planting the hollows, all distinction is lessened and broken up. Indeed, where there is but a trifling and scarcely perceptible undulation, the importance of the swells of surface already existing is surprisingly increased, when this course of planting is adopted; and the whole, to the eye, appears finely varied.

Where the grounds of the residence to be planted are level, or nearly so, and it is desirable to confine the view, on any or all sides, to the lawn or park itself, the boundary groups and masses must be so connected together as, from the most striking part or parts of the prospect (near the house for example) to answer this end. This should be done, not by planting a continuous, uniformly thick belt of trees round the outside of the whole; but by so arranging the various outer groups and thickets, that when *seen from the given points* they shall appear connected in one whole. In this way, there will be an agreeable variation in the margin, made by the various bays, recesses, and detached projections, which could not be so well effected if the whole were one uniformly unbroken strip of wood.

But where the house is so elevated as to command a more extensive view than is comprised in the demesne itself, another course should be adopted. The grounds planted must be made to connect themselves with the surrounding scenery, so as not to produce any violent contrast to the eye, when compared with the adjoining country. If then, as is most frequently the case, the lawn or pleasure-ground join, on either side or sides, cultivated farm lands, the proper connexion may be kept up by advancing a few groups or even scattered trees into the neighboring fields. In the middle states there are but few cultivated fields, even in ordinary farms, where there is not to be seen, here and there, a handsome cluster of saplings or a few full grown trees; or if not these, at least some tall growing bushes along the fences, all of which, by a little exercise of this leading principle of *connexion*, can, by the planter of taste, be made to appear with few or trifling additions, to divaricate from, and ramble out of the park itself. Where the park joins natural woods, connexion is still easier, and where it bounds upon one of our noble rivers, lakes, or other large sheets of water, of course connexion is not expected; for sudden contrast and transition is there both natural and beautiful.

In all cases good taste will suggest that the more polished parts of the lawns and grounds should, whatever character is attempted, be those nearest the house. There the most rare and beautiful sorts of trees are displayed, and the entire plantations agree in elegance with the style of art evinced in the mansion itself. When there is much extent, however, as the eye wanders from the neighborhood of the residence, the whole evinces less polish; and gradually, towards the furthest extremities, grows ruder, until it assimilates itself to the wildness of general nature around. This, of course, applies to grounds of large extent, and must not be so much enforced where the lawn embraced is but moderate, and therefore comes more directly under the eye.

It will be remembered that, in the foregoing section, we stated it as one of the leading principles of the art of Landscape Gardening, that in every instance where the grounds of a country residence have a marked natural character, whether of beautiful or picturesque expression, the efforts of the improver will be most successful if he contributes by his art to aid and strengthen that expression. This should ever be borne in mind when we are commencing any improvements in planting that will affect the general expression of the scene, as there are but few country residences in the United States of any importance which have not naturally some distinct landscape character; and the labors of the improver will be productive of much greater satisfaction and more lasting pleasure, when they aim at effects in keeping with the whole scene, than if no regard be paid to this important point. This will be felt almost intuitively by persons who, perhaps, would themselves be incapable of describing the cause of their gratification, but would perceive the contrary at once; as many are unable to analyse the pleasure derived from harmony in music, while they at once perceive the introduction of discordant notes.

We do not intend that this principle should apply so closely, that extensive grounds naturally picturesque shall have nothing of the softening touches of more perfect beauty; or that a demesne characterized by the latter expression should not be occasionally enlivened with a few "*smart touches*" of the former. This is often necessary, indeed, to prevent tame scenery from degenerating into insipidity, or picturesque into wildness, too great to be appropriate in a country residence. Picturesque trees give new spirit to groups of highly beautiful ones, and the latter sometimes heighten by contrast the value of the former. All of which, however, does not prevent the *predominance* of the leading features of either style, sufficiently strong to mark it as such; while, occasionally, something of zest or elegance may be borrowed from the opposite character, to suit the wishes or gratify the taste of the proprietor.

GROUND PLANS OF ORNAMENTAL PLANTATIONS. To illustrate partially our ideas on the arrangement of plantations we place before the reader two or three examples, premising, that the small scale to which they are reduced prevents our giving to them any character beyond that of the general one of the design. The first (Fig. 23) represents a portion, say one third or one half of a piece of property selected for a country seat, and which has hitherto been kept in tillage as ordinary farm land. The public

[Fig. 23. Plan of a common Farm, before any improvements.]

road, *a*, is the boundary on one side : *dd* are prettily wooded dells or hollows, which, together with a few groups near the proposed site of the house, *c*, and a few scattered single trees, make up the aggregate of the original woody embellishments of the locality.

In the next figure (Fig. 24) a ground plan of the place is given, as it would appear after having been judiciously laid out and planted, with several years' growth. At *a*, the approach road leaves the public highway and leads to the house at *c* : from whence paths of smaller size, *b*, make the circuit of the ornamental portion of the residence, taking advantage of the wooded dells, *d*, originally existing, which offer some scope for varied walks concealed from each other by the intervening masses of thicket. It will

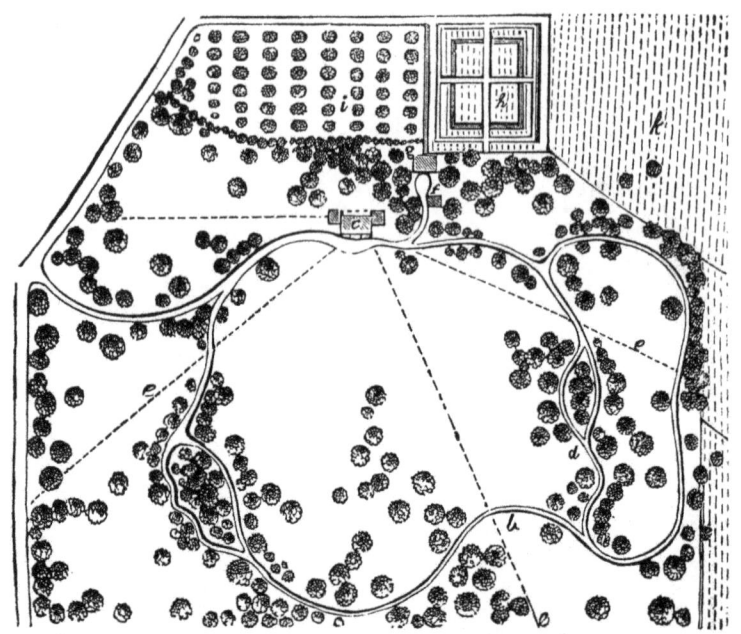

[Fig. 24. Plan of the foregoing grounds as a Country Seat, after ten years' improvement.]

be seen here, that one of the largest masses of wood forms a background to the house, concealing also the out-buildings; while, from the windows of the mansion itself, the trees are so arranged as to group in the most pleasing and effective manner; at the same time broad masses of turf meet the eye, and fine distant views are had through the vistas in the lines, *e e*. In this manner the lawn appears divided into four distinct lawns or areas bounded by groups of trees, instead of being dotted over with an unmeaning confusion of irregular masses of foliage. The form of these areas varies also with every change of position in the spectator, as seen from different portions of the grounds, or different points in the walks; and they can be still further varied at pleasure by adding more single trees or small groups, which should always, to produce variety of outline, be

ON WOOD AND PLANTATIONS.   115

placed opposite the *salient* parts of the wood, and not in the recesses, which latter they would appear to diminish or clog up. The stables are shown at f; the barn at g, and the kitchen garden adjacent at h; the orchard at i; and a small portion of the farm lands at k; a back entrance to the out-buildings is shown in the rear of the orchard. The plan has been given for a place of seventy acres, thirty of which include the pleasure grounds, and forty the adjoining farm lands.

Figure 25 is the plan of an American mansion

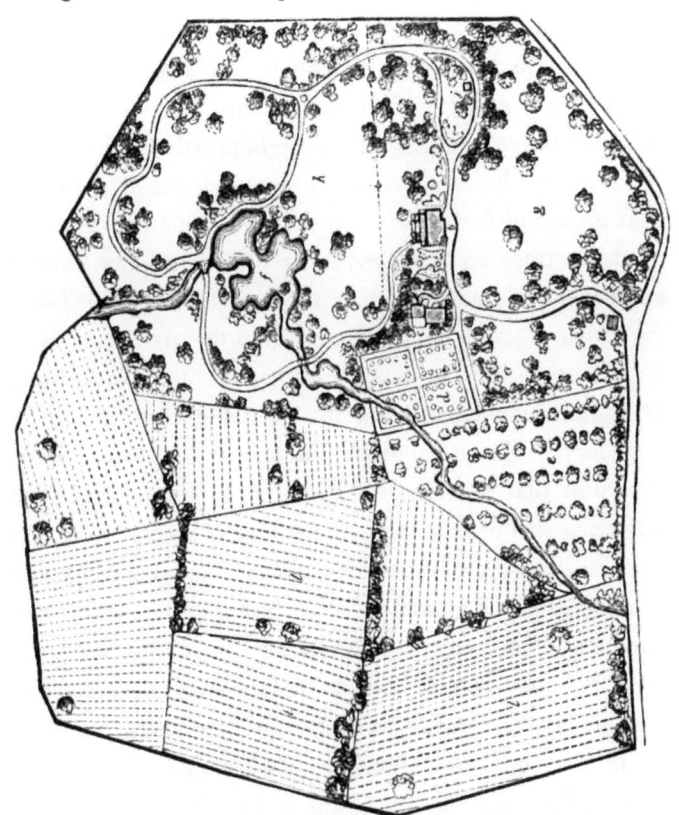

[Fig. 25. Plan of a Mansion Residence, laid out in the natural style.]

residence of considerable extent, only part of the farm lands, *l*, being here delineated. In this residence, as there is no extensive view worth preserving beyond the bounds of the estate, the pleasure grounds are surrounded by an irregular and picturesque belt of wood. A fine natural stream or rivulet, which ran through the estate, has been formed into a handsome pond, or small lake, *f*, which adds much to the interest of the grounds. The approach road breaks off from the highway at the entrance lodge, *a*, and proceeds in easy curves to the mansion, *b*; and the groups of trees on the side of this approach nearest the house, are so arranged that the visitor scarcely obtains more than a glimpse of the latter, until he arrives at the most favorable position for a first impression. From the windows of the mansion, at either end, the eye ranges over groups of flowers and shrubs; while, on the entrance front, the trees are arranged so as to heighten the natural expression originally existing there. On the other front, the broad mass of light reflected from the green turf at *h*, is balanced by the dark shadows of the picturesque plantations which surround the lake, and skirt the whole boundary. At *i*, a light, inconspicuous wire fence separates that portion of the ground, *g*, ornamented with flowering shrubs and kept mown by the scythe, from the remainder, of a park-like character, which is kept short by the cropping of animals. At *c*, are shown the stables, carriage house, etc., which, though near the approach road, are concealed by foliage, though easily accessible by a short curved road, *returning from* the house, so as not to present any road leading in the same direction, to detract from the dignity of the approach in going to it. A prospect tower, or rustic pavilion, on a little eminence overlooking the whole estate, is shown at *j*. The small arabesque beds near the house are filled with masses of choice flowering shrubs and plants; the kitchen garden is shown at *d*, and the orchard at *e*.

Suburban villa residences are, every day, becoming more numerous; and in laying out the grounds around them, and disposing the sylvan features, there is often more ingenuity, and as much taste required, as in treating a country residence of several hundred acres. In the small area of from one half an acre to ten or twelve acres, surrounding often a villa of the first class, it is desirable to assemble many of the same features, and as much as possible of the enjoyment, which are to be found in a large and elegant estate. To do this, the space allotted to various purposes, as the kitchen garden, lawn, etc., must be judiciously portioned out, and so characterized and divided by plantations, that the whole shall appear to be much larger than it really is, from the fact that the spectator is never allowed to see the whole at a single glance; but while each portion is complete in itself, the plan shall present nothing incongruous or ill assorted.

An excellent illustration of this species of residence, is afforded the reader in the accompanying plan (Fig. 26) of the grounds of *Riverside Villa*. This pretty villa at Burlington, New Jersey (to which we shall again refer), was lately built, and the grounds, about six or eight acres in extent, laid out, from the designs of John Notman, Esq., architect, of Philadelphia; and while the latter promise a large amount of beauty and enjoyment, scarcely anything which can be supposed necessary for the convenience or wants of the family, is lost sight of.

The house, *a*, stands quite near the bank of the river,

118   LANDSCAPE GARDENING.

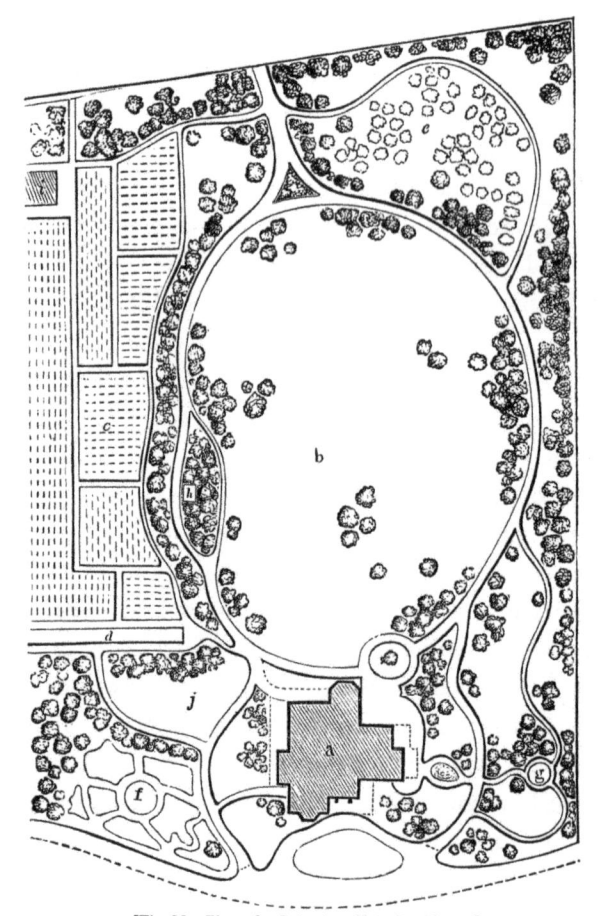

[Fig. 26. Plan of a Suburban Villa Residence.]

while one front commands fine water views, and the other looks into the lawn or pleasure grounds, *b*. On one side of the area is the kitchen garden, *c*, separated and concealed from the lawn by thick groups of evergreen and deciduous trees. At *e*, is a picturesque orchard, in which the fruit trees are planted in groups instead of straight lines, for the sake of effect. Directly under the windows of the drawing-room is the flower garden, *f*; and at *g*, is a seat. The walk around the lawn is also a carriage road, affording entrance and egress from the rear of the grounds, for garden purposes, as well as from the front of the house. At *h*, is situated the ice-house; *d*, hot-beds; *j*, bleaching green; *i*, gardener's house, etc. In the rear of the latter are the stables, which are not shown on the plan.

The embellished farm (*ferme ornée*) is a pretty mode of combining something of the beauty of the landscape garden with the utility of the farm, and we hope to see small country seats of this kind become more general. As regards profit in farming, of course, all modes of arranging or distributing land are inferior to simple square fields; on account of the greater facility of working the land in rectangular plots. But we suppose the owner of the small ornamental farm to be one with whom profit is not the first and only consideration, but who desires to unite with it something to gratify his taste, and to give a higher charm to his rural occupations. In Fig. 27, is shown part of an embellished farm, treated in the picturesque style throughout. The various trees, under grass or tillage, are divided and bounded by winding roads, *a*, bordered by hedges of buckthorn, cedar, and hawthorn, instead of wooden fences; the roads being wide enough to afford a pleasant drive or walk, so as to allow the owner or visitor to enjoy at the same time an agreeable circuit, and a glance at all the various crops and modes of culture. In the plan before us, the approach from the public road is at *b*; the dwelling at *c*; the barns and farm-buildings at *d*; the kitchen garden at *e*; and the orchard at *f*. About the house are distributed some groups of trees, and here the fields, *g*, are kept in grass, and are either mown

120   LANDSCAPE GARDENING.

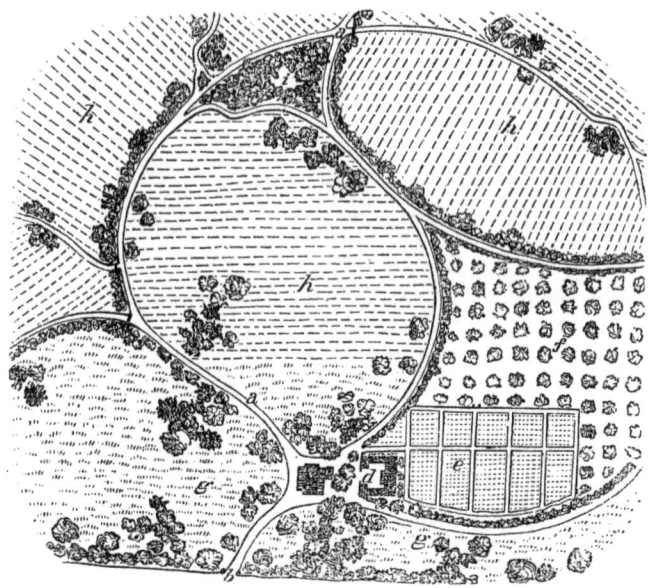

[Fig. 27. View of a Picturesque farm (*ferme ornée*).]

or pastured. The fields in crops are designated *h*, on the plan; and a few picturesque groups of trees are planted, or allowed to remain, in these, to keep up the general character of the place. A low dell, or rocky thicket, is situated at *i*,. Exceedingly interesting and agreeable effects may be produced, at little cost, in a picturesque farm of this kind. The hedges may be of a great variety of suitable shrubs, and, in addition to those that we have named, we would introduce others of the sweet brier, the Michigan or prairie rose (admirably adapted for the purpose), the flowering crab, and the like—beautiful and fragrant in their growth and blossoms. These hedges we would cause to grow thick, rather by interlacing the branches, than by constant shearing or trimming, which would give them a less formal, and a more free and natural air. The winding lanes traversing the farm need only be gravelled near the house, in other portions being left in grass, which will need little care, as it will generally be kept short enough by the passing of men and vehicles over it.

A picturesque or ornamental farm like this would be an agreeable residence for a gentleman retiring into the country on a small farm, desirous of experimenting for himself with all the new modes of culture. The small and irregular fields would, to him, be rather an advantage, and there would be an air of novelty and interest about the whole residence. Such an arrangement as this would also be suitable for a fruit farm near one of our large towns, the fields being occupied by orchards, vines, grass, and grain. The house and all the buildings should be of a simple, though picturesque and accordant character.

The *cottage ornée* may have more or less ground attached to it. It is the ambition of some to have a great house and little land, and of others (among whom we remember the poet Cowley) to have a little house and a large garden. The latter would seem to be the more natural taste. When the grounds of a cottage are large, they will be treated by the landscape gardener nearly like those of a villa residence, when they are smaller a more quiet and simple character must be aimed at. But even where they consist of only a rood or two, something tasteful and pretty may be arranged.* In Fig. 28, is shown a small piece of ground on one side of a cottage, in which a picturesque character is attempted to be maintained. The plantations here are made mostly with shrubs instead of trees, the latter being

---

* For a variety of modes of treating the grounds of small places, see our *Designs for Cottage Residences.*

only sparingly introduced for the want of room. In the disposition of these shrubs, however, the same attention to picturesque effect is paid as we have already pointed out in our remarks on grouping; and by connecting the thickets and groups here and there, so as to conceal one walk from the other, a surprising variety and effect will frequently be produced in an exceedingly limited spot.

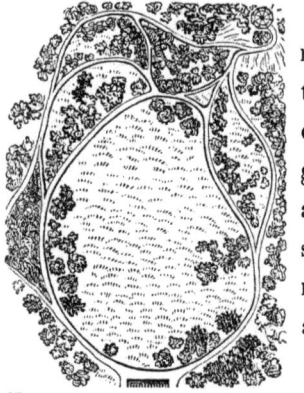

The same limited grounds might be planted so as to produce the Beautiful; choosing, in this case, shrubs of symmetrical growth and fine forms, planting and grouping them somewhat singly, and allowing every specimen to attain its fullest luxuriance of development.

[Fig. 28. Grounds of a Cottage *ornée*.]

In making these arrangements, even in the small area of a fourth of an acre, we should study the same principles and endeavor to produce the same harmony of effects, as if we were improving a mansion residence of the first class. The extent of the operations, and the sums lavished, are not by any means necessarily connected with successful and pleasing results. The man of correct taste will, by the aid of very limited means and upon a small surface, be able to afford the mind more true pleasure, than the improver who lavishes thousands without it, creating no other emotion than surprise or pity at the useless expenditure incurred; and the Abbé Delille says nothing more true than that,

"Ce noble emploi demand un artiste qui pense,
Prodigue de génie, et non pas de dépense."

From the inspection of plans like these, the tyro may learn something of the manner of arranging plantations, and of the general effect of the natural style in particular cases and situations. But the knowledge they afford is so far below that obtained by an inspection of the effects in reality, that the latter should in all cases be preferred where it is practicable. In this style, unlike the ancient, it is almost impossible that the same plan should exactly suit any other situation than that for which it was intended, for its great excellence lies in the endless variety produced by its application to different sites, situations, and surfaces; developing the latent capacities of one place and heightening the charms of another.

But the leading principles as regards the formation of plantations, which we have here endeavored briefly to elucidate, are the same in all cases. After becoming familiar with these, should the amateur landscape gardener be at a loss how to proceed, he can hardly do better, as we have before suggested, than to study and recur often to the beautiful compositions and combinations of nature, displayed in her majestic groups, masses, and single trees, as well as open glades and deep thickets; of which, fortunately, in most parts of our country, checkered here and there as it is with beautiful and picturesque scenery, there is no dearth or scarcity. Keeping these few principles in his mind, he will be able to detect new beauties and transfer them to his own estate; for nature is truly inexhaustible in her resources of the Beautiful.

CLASSIFICATION OF TREES AS TO EXPRESSION. The amateur who wishes to dispose his plantations in the natural style of Landscape Gardening so as to produce graceful or picturesque landscape, will be greatly aided by a study of the peculiar expression of trees individually and in composition. The effect of a certain tree singly is often exceedingly different from that of a group of the same trees. To be fully aware of the effect of groups and masses requires considerable study, and the progress in this study may be greatly facilitated by a recurrence from groups in nature to groups in pictures.

As a further aid to this most desirable species of information we shall offer a few remarks on the principal varieties of character afforded by trees in composition.

Almost all trees, with relation to forms, may be divided into three kinds, viz. *round-headed* trees, *oblong* or *pyramidal* trees, and *spiry-topped* trees ; and so far as the expressions of the different species comprised in these distinct classes are concerned, they are, especially when viewed at a distance (as much of the wood seen in a prospect of any extent necessarily must be), productive of nearly the same general effects.

Round-headed trees compose by far the largest of these divisions. The term includes all those trees which have an irregular surface in their boughs, more or less varied in outline, but exhibiting in the whole a top or head comparatively round ; as the oak, ash, beech, and walnut. They are generally beautiful when young, from their smoothness, and the elegance of their forms; but often grow picturesque when age and time have had an opportunity to produce their wonted effects upon them. In general, however, the different round-headed trees may be considered as the most appropriate for introduction in highly-cultivated scenery, or landscapes where the character is that of graceful or polished beauty ; as they harmonize with almost all scenes,

[Fig. 29. Round-headed Trees.

buildings, and natural or artificial objects, uniting well with other forms and doing violence to no expression of scenery. From the numerous breaks in the surface of their foliage, which reflect differently the lights and produce deep shadows, there is great intricacy and variety in the heads of many round-topped trees; and therefore, as an outer surface to meet the eye in a plantation, they are much softer and more pleasing than the unbroken line exhibited by the sides of oblong or spiry-topped trees. The sky outline also, or the upper part of the head, varies greatly in round-topped trees from the irregularity in the disposition of the upper branches in different species, as the oak and ash, or even between individual specimens of the same kind of tree, as the oak, of which we rarely see two trees alike in form and outline, although they have the same characteristic expression; while on the other hand no two verdant objects can bear a greater general resemblance to each other and show more sameness of figure than two Lombardy poplars.

"In a tree," says Uvedale Price, "of which the foliage is everywhere full and unbroken, there can be but little variety of *form;* then, as the sun strikes only on the surface, neither can there be much variety of *light* and *shade;* and as the apparent color of objects changes according to the different degrees of light or shade in which they are placed, there can be as little variety of *tint;* and lastly, as there are none of these openings that excite and nourish curiosity, but the eye is everywhere opposed by one uniform leafy screen, there can be as little intricacy as variety." From these remarks, it will be perceived that even among round-headed trees there may be great difference in the comparative beauty of different sorts ; and judging from the excellent standard here laid down, it will also be seen how much in the eye of a painter a tree with a beautifully diversified surface, as the oak, surpasses in the composition of a scene one with a very regular and compact surface and outline, as the horse-chestnut. In planting large masses of wood, therefore, or even in forming large groups in park scenery, round-headed trees of the ordinary loose and varied manner of growth common in the majority of forest trees, are greatly to be preferred to all others. When they cover large tracts, as several acres, they convey an emotion of *grandeur* to the mind ; when they form vast forests of thousands of acres, they produce a feeling of *sublimity;* in the landscape garden when they stand alone, or in fine groups, they are *graceful* or *beautiful.* While young they have an elegant appearance ; when old they generally become majestic or picturesque. Other trees may suit scenery or scenes of particular and decided characters, but *round-headed* trees are decidedly the chief adornment of general landscape.

Spiry-topped trees (Fig. 30) are distinguished by straight leading stems and horizontal branches, which are comparatively small, and taper gradually to a point. The foliage is generally evergreen, and in most trees of this class

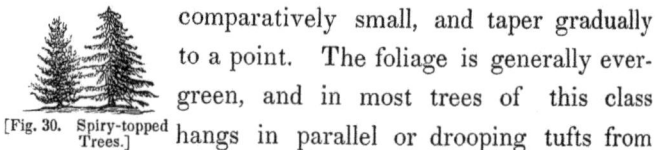

[Fig. 30. Spiry-topped Trees.]

hangs in parallel or drooping tufts from the branches. The various evergreen trees, composing the spruce and fir families, most of the pines, the cedar, and among deciduous trees, the larch, belong to this division. Their hue is generally much darker than that of deciduous trees, and there is a strong similarity, or almost sameness, in the different kinds of trees which may properly be called spiry-topped.

From their sameness of form and surface this class of trees, when planted in large tracts or masses, gives much less pleasure than round-headed trees; and the eye is soon wearied with the *monotony* of appearance presented by long rows, groups, or masses, of the same form, outline, and appearance; to say nothing of the effect of the uniform dark color, unrelieved by the warmer tints of deciduous trees. Any one can bear testimony to this, who has travelled through a pine, hemlock, or fir forest, where he could not fail to be struck with its gloom, tediousness, and monotony, especially when contrasted with the variety and beauty in a natural wood of deciduous, round-headed trees.

Although spiry-topped trees in large masses cannot be generally admired for ornamental plantations, yet they have a character of their own, which is very striking and peculiar, and we may add, in a high degree valuable to the Landscape Gardener. Their general expression when single or scattered is extremely spirited, wild, and picturesque; and when judiciously introduced into artificial scenery, they produce the most charming and unique effects. "The situations where they have most effect is among rocks and in very irregular surfaces, and especially on the steep sides of high mountains, where their forms and the direction of their growth seem to harmonize with the pointed rocky summits." Fir and pine forests are extremely dull and monotonous in sandy plains and smooth surfaces (as in the pine barrens of the southern states); but among the broken rocks, craggy precipices, and otherwise endlessly varied surfaces (as in the Alps, abroad, and the various rocky heights in the Highlands of the Hudson and the Alleghanies, at home) they are full of variety. It will readily be seen, therefore, that spiry-topped trees should always be planted in considerable quantities in wild, broken, and picturesque scenes, where they will appear perfectly in keeping, and add wonderfully to the peculiar beauty of the situation. In all grounds where there are abruptly varied surfaces, steep banks, or rocky precipices, this class of trees lends its efficient aid to strengthen the prevailing beauty, and to complete the finish of the picture. In smooth, level surfaces, though spiry-topped trees cannot be thus extensively employed, they are by no means to be neglected or thought valueless, but may be so combined and mingled with other round-headed and oblong-headed trees, as to produce very rich and pleasing effects. A tall larch or two, or a few spruces rising out of the centre of a group, give it life and spirit, and add greatly, both by contrast of form and color, to the force of round-headed trees. A stately and regular white pine or hemlock, or a few thin groups of the same trees peeping out from amidst, or bordering a large mass of deciduous trees, have great power in adding to the interest which the same awakens in the mind of the spectator. Care must be taken, however, that the very spirited effect which is here aimed at, is not itself defeated by the over anxiety of the planter, who, in scattering too profusely these very strongly marked trees, makes them at last so plentiful, as to give the whole a mingled and confused look, in which neither the graceful and sweeping outlines of the round-headed nor the picturesque summits of the spiry-topped trees predominate; as the former decidedly should, in all scenes where an expression of peculiarly irregular kind is not aimed at.

The larch, to which we shall hereafter recur at some length, may be considered one of the most picturesque trees of this division; and being more rapid in its growth than most evergreens, it may be used as a substitute for, or in conjunction with them, where effect is speedily desired.

Oblong-headed trees show heads of foliage more lengthened out, more formal, and generally more tapering, than round-headed ones. They differ from spiry-topped trees in having upright branches instead of horizontal ones, and in forming a conical or pyramidal mass of foliage, instead of a spiry, tufted one. They are mostly deciduous; and approaching more nearly to round-headed trees than spiry-topped ones do, they may perhaps be more frequently introduced. The Lombardy poplar may be considered the representative of this division, as the oak is of the first, and the larch and fir of the second. Abroad, the oriental cypress, an evergreen, is used to produce similar effects in scenery.

[Fig. 31. Oblong-headed trees.]

The great use of the Lombardy poplar, and other similar trees in composition, is to relieve or break into groups, large masses of wood. This it does very effectually, when its tall summit rises at intervals from among round-headed trees, forming pyramidal centres to groups where there was only a swelling and flowing outline. Formal rows, or groups of oblong-headed trees, however, are tiresome and monotonous to the last degree; a straight line of them being scarcely better in appearance than a tall, stiff, gigantic hedge. Examples of this can be easily found in many parts of the Union where the crude and formal taste of proprietors, by leading them to plant long lines of Lombardy poplars, has had the effect of destroying the beauty of many a fine prospect and building.

Conical or oblong-headed trees, when carefully employed, are very effective for purposes of *contrast*, in conjunction with horizontal lines of buildings such as we see in Grecian or Italian architecture. Near such edifices, *sparingly introduced, and mingled in small proportion with round-headed trees,* they contrast advantageously with the long cornices, flat roofs, and horizontal lines that predominate in their exteriors. Lombardy poplars are often thus introduced in pictures of Italian scenery, where they sometimes break the formality of a long line of wall in the happiest manner. Nevertheless, if they should be indiscriminately employed, or even used in any considerable portion in the decoration of the ground immediately adjoining a building of any pretensions, they would inevitably defeat this purpose, and by their tall and formal growth diminish the apparent magnitude, as well as the elegance of the house.

Drooping trees, though often classed with oblong-headed trees, differ from them in so many particulars, that they deserve to be ranked under a separate head. To this class belong the weeping willow, the weeping birch, the drooping elm, etc. Their prominent characteristics are gracefulness and elegance; and we consider them as unfit, therefore, to be employed *to any extent* in scenes where it is desirable to keep up the expression of a wild or highly picturesque character. As single objects, or tastefully grouped in beautiful landscape, they are in excellent keeping, and contribute much to give value to the leading expression.

When drooping trees are mixed indiscriminately with other round-headed trees in the composition of groups or masses, much of their individual character is lost, as it depends not so much on the top (as in oblong and spiry trees) as upon the side branches, which are of course concealed by those of the adjoining trees. Drooping trees, therefore, as elms, birches, etc., are shown to the best advantage on the *borders* of groups or the *boundaries* of plantations. It must not be forgotten, but constantly kept in mind, that all strongly marked trees, like bright colors in pictures, only admit of occasional employment; and that the very object aimed at in introducing them will be defeated if they are brought into the lawn and park in masses, and distributed heedlessly on every side. An English author very justly remarks, therefore, that the poplar, the willow, and the drooping birch, are "most dangerous trees in the hands of a planter who has not considerable knowledge and good taste in the composition of a landscape." Some of them, as the native elm, from their abounding in oui own woods, may appear oftener; while others which have a peculiar and exotic look, as the weeping willow, should only be seen in situations where they either do not disturb the prevailing expression, or (which is better) where they are evidently in good keeping. "The weeping willow," says Gilpin, with his usual good taste, "is not adapted to sublime objects. We wish it not to screen the broken buttress and Gothic windows of an abbey, or to overshadow the battlements of a ruined castle. These offices it resigns to the oak, whose dignity can

132   LANDSCAPE GARDENING.

support them. The weeping willow seeks an humble scene—some romantic footpath bridge, which it half conceals, or some grassy pool over which it hangs its streaming foliage,

> ——'And dips
> Its pendent boughs, as if to drink.'" *

The manner in which a picturesque bit of landscape can be supported by picturesque spiry-topped trees, and its expression degraded by the injudicious employment of graceful drooping trees, will be apparent to the reader in the two accompanying little sketches. In the first (Fig. 32), the abrupt hill, the rapid mountain torrent, and the distant Alpine summits, are in fine keeping with the tall spiry larches and firs, which, shooting up on either side of the old bridge, occupy the foreground. In the second (Fig. 33), there is evidently something discordant in the scene which strikes the spectator at first sight; this is the misplaced introduction of the large willows, which belong to a scene very different in character. Imagine a removal of the surrounding hills, and let the rapid stream spread out into a smooth peaceful lake with gradually retiring shores, and the blue summits in the distance, and then the willows will harmonize admirably.

[Fig. 32. Trees in keeping.]

[Fig. 33. Trees out of keeping.]

Having now described the peculiar characteristics of these different classes of round-headed, spiry-topped, oblong, and drooping trees, we should consider the proper

* Forest Scenery, p. 133.

method by which a harmonious combination of the different forms composing them may be made so as not to violate correct principles of taste. An indiscriminate mixture of their different forms would, it is evident, produce anything but an agreeable effect. For example, let a person plant together in a group, three trees of totally opposite forms and expressions, viz. a weeping willow, an oak, and a poplar; and the expression of the whole would be destroyed by the confusion resulting from their discordant forms. On the other hand, the mixture of trees that exactly correspond in their forms, if these forms, as in oblong or drooping trees, are similar will infallibly create sameness. In order then to produce beautiful variety which shall neither on the one side run into confusion, nor on the other verge into monotony, it is requisite to give some little attention to the harmony of form and color in the composition of trees in artificial plantations.

The only rules which we can suggest to govern the planter are these : First, if a certain leading expression is desired in a group of trees, together with as great a variety as possible, such species must be chosen as harmonize with each other in certain leading points. And, secondly, in occasionally intermingling trees of opposite characters, discordance may be prevented, and harmonious expression promoted, by interposing other trees of an intermediate character.

In the first case, suppose it is desired to form a group of trees, in which gracefulness must be the leading expression. The willow alone would have the effect; but in groups, willows alone produce sameness : in order, therefore, to give variety, we must choose other trees which, while they differ from the willow in some particulars, agree in others. The elm has much larger and darker foliage, while it has also a drooping spray; the weeping birch differs in its leaves, but agrees in the pensile flow of its branches; the common birch has few pendent boughs, but resembles in the airy lightness of its leaves; and the three-thorned acacia, though its branches are horizontal, has delicate foliage of nearly the same hue and floating lightness as the willow. Here we have a group of five trees, which is, in the whole, full of gracefulness and variety, while there is nothing in the composition inharmonious to the practised eye.

To illustrate the second case, let us suppose a long sweeping outline of maples, birches, and other light, mellow-colored trees, which the improver wishes to vary and break into groups, by spiry-topped, evergreen trees. It is evident, that if these trees were planted in such a manner as to peer abruptly out of the light-colored foliage of the former trees, in dark or almost black masses of tapering verdure, the effect would be by no means so satisfactory and pleasing, as if there were a partial transition from the mellow, pale green of the maples, etc., to the darker hues of the oak, ash, or beech, and finally the sombre tint of the evergreens. Thus much for the coloring; and if, in addition to this, oblong-headed trees or pyramidal trees were also placed near and partly intermingled with the spiry-topped ones, the unity of the whole composition would be still more complete.*

---

* We are persuaded that very few persons are aware of the beauty, varied and endless, that may be produced by arranging trees with regard to their *coloring.* It requires the eye and genius of a Claude or a Poussin, to develope all these hidden beauties of harmonious combination. Gilpin rightly

Contrasts, again, are often admissible in woody scenery; and we would not wish to lose many of our most superb trees, because they could not be introduced in particular portions of landscape. Contrasts in trees may be so violent as to be displeasing; as in the example of the groups of the three trees, the willow, poplar, and oak: or they may be such as to produce spirited and pleasing effects. This must be effected by planting the different divisions of trees, first, in small leading groups, and then by effecting a union between the groups of different character, by intermingling those of the nearest similarity into and near the groups: in this way, by easy transitions from the drooping to the round-headed, and from these to the tapering trees, the whole of the foliage and forms harmonize well.

[Fig. 34. Example in grouping.]

"Trees," observes Mr. Whately, in his elegant treatise on this subject, "which differ in but one of these circumstances, of shape, green, or growth, though they agree in every other, are sufficiently distinguished for the

says, in speaking of the dark Scotch fir, "with regard to color in general, I think I speak the language of painting, when I assert that the picturesque eye makes little distinction in this matter. It has no attachment to one color in preference to another, but considers the beauty of all coloring as resulting, not from the colors themselves, but almost entirely from their harmony with other colors in their neighborhood. So that as the Scotch fir tree is combined or stationed, it forms a beautiful umbrage or a murky spot."

purpose of *variety* ; if they differ in two or three, they become *contrasts :* if in all, they are opposite, and seldom group well together. Those, on the contrary, which are of one character, and are distinguished only as the characteristic mark is strongly or faintly impressed upon them, form a beautiful mass, and unity is preserved without sameness."*

There is another circumstance connected with the color of trees, that will doubtless suggest itself to the improver of taste, the knowledge of which may sometimes be turned to valuable account. We mean the effects produced in the apparent coloring of a landscape by distance, which painters term *aërial perspective.* Standing at a certain position in a scene, the coloring is deep, rich, and full in the foreground, more tender and mellow in the middle-ground, and softening to a pale tint in the distance.

> " Where to the eye three well marked distances
> Spread their peculiar coloring, vivid green,
> Warm brown, and black opake the foreground bears
> Conspicuous: sober olive coldly marks
> The second distance; thence the third declines
> In softer blue, or lessening still, is lost
> In fainted purple. When thy taste is call'd
> To deck a scene where nature's self presents
> All these distinct gradations, then rejoice
> As does the Painter, and like him apply
> Thy colors; plant thou on each separate part
> Its proper foliage."

Advantage may occasionally be taken of this peculiarity in the gradation of color, in Landscape Gardening, by the creation, as it were, of an *artificial distance.* In grounds

---

* Observations on Modern Gardening.

and scenes of limited extent, the apparent size and breadth may be increased, by planting a majority of the trees in the foreground, of dark tints, and the boundary with foliage of a much lighter hue. In the same way, the apparent breadth of a piece of water will be greatly added to, by placing the paler colored trees on the shore opposite to the spectator. These hints will suggest other ideas and examples of a similar nature, to the minds of those who are alive to the more minute and exquisite beauties of the landscape.

An acquaintance, individually, with the different species of trees of indigenous and foreign growth, which may be cultivated with success in this climate, is absolutely essential to the amateur or the professor of Landscape Gardening. The tardiness or rapidity of their growth, the periods at which their leaves and flowers expand, the soils they love best, and their various habits and characters, are all subjects of the highest interest to him. In short, as a love of the country almost commences with a knowledge of its peculiar characteristics, the pure air, the fresh enamelled turf, and the luxuriance and beauty of the whole landscape; so the taste for the embellishment of Rural Residences must grow out of an admiration for beautiful trees, and the delightful effects they are capable of producing in the hands of persons of taste and lovers of nature.

Admitting this, we think, in the comparatively meagre state of general information on this subject among us, we shall render an acceptable service to the novice, by giving a somewhat detailed description of the character and habits of most of the finest hardy forest and ornamental trees. Among those living in the country, there are many who care little for the beauties of Landscape Gardening, who are yet interested in those trees which are remarkable for the beauty of their forms, their foliage, their blossoms, or their useful purposes. This, we hope, will be a sufficient explanation for the apparently disproportionate number of pages which we shall devote to this part of our subject.

SECTION IV.

DECIDUOUS ORNAMENTAL TREES.

The History and Description of all the finest hardy Deciduous Trees. REMARKS ON THEIR EFFECTS IN LANDSCAPE GARDENING, INDIVIDUALLY AND IN COMPOSITION. Their Cultivation, etc. The Oak. The Elm. The Ash. The Linden. The Beech. The Poplar. The Horse-chestnut. The Birch. The Alder. The Maple. The Locust. The Three-thorned Acacia. The Judas-tree. The Chestnut. The Osage Orange. The Mulberry. The Paper Mulberry. The Sweet Gum. The Walnut. The Hickory The Mountain Ash. The Ailantus. The Kentucky Coffee. The Willow. The Sassafras. The Catalpa. The Persimon. The Pepperidge. The Thorn. The Magnolia. The Tulip. The Dogwood. The Salisburia. The Paulonia. The Virgilia. The Cypress. The Larch, etc.

> O gloriosi spiriti de gli boschi,
> O Eco, o antri foschi, o chiare linfe,
> O faretrate ninfe, o agresti Pani,
> O Satiri e Silvani, o Fauni e Driadi,
> Naiadi ed Amadriadi, o Semidee
> Oreadi e Napee.—
>   SANNAZZARO.

> "O spirits of the woods,
> Echoes and solitudes, and lakes of light ;
> O quivered virgins bright, Pan's rustical
> Satyrs and sylvans all, dryads and ye
> That up the mountains be ; and ye beneath
> In meadow or in flowery heath."

THE OAK. *Quercus.*

Nat. Ord. Corylaceæ.   *Lin. Syst.* Monœcia, Polyandria.

H E Arcadians believed the oak to have been the first created of all trees ; and when we consider its great and surpassing utility and beauty, we are fully disposed to concede it the first rank among the denizens of the forest. Springing up with a noble trunk, and stretching out its broad limbs over the soil,

> "These monarchs of the wood,
> Dark, gnarled, centennial oaks,"

seem proudly to bid defiance to time; and while generations of man appear and disappear, they withstand the storms of a thousand winters, and seem only to grow more venerable and majestic. They are mentioned in the oldest histories; we are told that Absalom was caught by his hair in "the thick boughs of a great oak;" and Herodotus informs us that the first oracle was that of Dodona, set up in the celebrated oak grove of that name. There, at first, the oracles were delivered by the priestesses, but, as was afterwards believed, by the inspired oaks themselves—

> "Which in Dodona did enshrine,
> So faith too fondly deemed, a voice divine."

Acorns, the fruit of the oak, appear to have been held in considerable estimation as an article of food among the ancients. Not only were the swine fattened upon them, as in our own forests, but they were ground into flour, with which bread was made by the poorer classes. Lucretius mentions, that before grain was known they were the common food of man; but we suppose the fruit of the chestnut may also have been included under that term.

> "That oake whose acornes were our foode before
> The Cerese seede of mortal man was knowne."
>
> SPENSER.

The civic crown, given in the palmy days of Rome to the most celebrated men, was also composed of oak leaves.

It should not be forgotten that the oak was worshipped by the ancient Britons. Baal or Yiaoul (whence Yule) was the god of fire, whose symbol was an oak. Hence at his festival, which was at Christmas, the ceremony of kindling the Yule log was performed among the ancient Druids. This fire was kept perpetual throughout the year, and the hearths of all the people were annually lighted from these sacred fires every Christmas. We believe the curious custom is still extant in some remote parts of England, where the "Yule log" is ushered in with much glee and rejoicing once a year.

As an ornamental object we consider the oak the most varied in expression, the most beautiful, grand, majestic, and picturesque of all deciduous trees. The enormous size and extreme old age to which it attains in a favorable situation, the great space of ground that it covers with its branches, and the strength and hardihood of the tree, all contribute to stamp it with the character of dignity and grandeur beyond any other compeer of the forest. When young its fine foliage (singularly varied in many of our native species) and its thrifty form render it a beautiful tree. But it is not until the oak has attained considerable size that it displays its true character, and only when at an age that would terminate the existence of most other trees that it exhibits all its magnificence. Then its deeply furrowed trunk is covered with mosses; its huge branches, each a tree, spreading out horizontally from the trunk with great boldness, its trunk of huge dimension, and its "high top, bald with dry antiquity;" all these, its true characteristics, stamp the oak, as Virgil has expressed it in his Georgics—

> "Jove's own tree,
> That holds the woods in awful sovereignty;
> For length of ages lasts his happy reign,
> And lives of mortal man contend in vain.
> Full in the midst of his own strength he stands,
> Stretching his brawny arms and leafy hands,
> His shade protects the plains, his head the hills commands."
> DRYDEN'S TRANS.

"The oak," says Gilpin, "is confessedly the most picturesque tree in itself, and the most accommodating in composition. It refuses no subject either in natural or in artificial landscape. It is suited to the grandest, and may with propriety be introduced into the most pastoral. It adds new dignity to the ruined tower and the Gothic arch; and by stretching its wild, moss-grown branches athwart their ivied walls, it gives them a kind of majesty coeval with itself; at the same time its propriety is still preserved if it throws its arms over the purling brook or the mantling pool, where it beholds

> "Its reverend image in the expanse below."

Milton introduces it happily even in the lowest scene—

> "Hard by a cottage chimney smokes,
> From between two aged oaks."

The oak is not only one of the grandest and most picturesque objects as a single tree upon a lawn, but it is equally unrivalled for groups and masses. There is a breadth about the lights and shadows reflected and embosomed in its foliage, a singular freedom and boldness in its outline, and a pleasing richness and intricacy in its huge ramification of branch and limb, that render it highly adapted to these purposes. Some trees, as the willow or the spiry poplar,

DECIDUOUS ORNAMENTAL TREES.   143

though pleasing singly, are monotonous to the last degree when planted in quantities. Not so, however, with the oak, as there is no tree, when forming a wood entirely by itself, which affords so great a variety of form and disposition, light and shade, symmetry and irregularity, as this king of the forests.

To arrive at its highest perfection, ample space on every side must be allowed the oak. A free exposure to the sun and air, and a deep mellow soil, are highly necessary to its fullest amplitude. For this reason, the oaks of our forests,

[Fig. 35. The Charter Oak, Hartford.]

being thickly crowded, are seldom of extraordinary size; and there are more truly majestic oaks in the parks of England than are to be found in the whole cultivated portion of the United States. Here and there, however, throughout our country may be seen a solitary oak of great age and immense size, which attests the fitness of the soil and climate, and displays the grandeur of our native species. The Wadsworth Oak near Geneseo, N. Y., of extraordinary dimensions, the product of one of our most fertile valleys, has attracted the admiration of hundreds of travellers on the route to Niagara. Its trunk measures thirty-six feet in circumference. The celebrated *Charter Oak* at Hartford, which has figured so conspicuously in the history of New England, is still existing in a green old age, one of the most interesting monuments of the past to be found in the country.*

Near the village of Flushing, Long Island, on the farm of Judge Lawrence, is growing one of the noblest oaks in the country. It is truly park-like in its dimensions, the circumference of the trunk being nearly thirty feet, and its majestic head of corresponding dignity. In the deep alluvial soil of the western valleys, the oak often assumes a grand aspect, and bears witness to the wonderful fertility of the soil in that region.†

---

* The house seen in the engraving represents the old "Wyllis House." This family, its former occupants, furnished the Secretary of State for Connecticut for more than a century. Near the Charter Oak are some of the apple trees *planted by the Pilgrims,* evidently Pearmains. Some of these, lately felled, have been examined, and are found to be more than 200 years old.

† The following well authenticated description of a famous English oak, is worth a record here. " Close by the gate of the water walk of Magdalen College, Oxford, grew an oak which perhaps stood there a sapling when Alfred the Great founded the University. This period only includes a space of 900 years, which is no great age for an oak. About 500 years after the time of Alfred, Dr. Stukely tells us, William of Waynefleet expressly ordered his college (Magdalen College) to be founded near the Great Oak ; and an oak could not, I think, be less than 500 years of age to merit that title, together with the honor of fixing the site of a college. When the magnificence of Cardinal Wolsey erected that handsome tower which is so ornamental to the whole

DECIDUOUS ORNAMENTAL TREES.

As beauty is often closely connected in our minds with utility, we must be allowed a word on the great value of this tree. For its useful properties the oak has scarcely any superior. "To enumerate," says old Evelyn in his quaint *Sylva*, "the incomparable uses of this wood were needless; but so precious was the esteem of it of old, there was an express law among the Twelve Tables concerning the very gathering of the acorns, though they should be found fallen on another man's ground. The land and the sea do sufficiently speak for the improvement of this excellent material, for houses and ships, cities and navies, are builded with it." In almost all the finest buildings of

building, this tree might probably be in the meridian of its glory. It was afterwards much injured in the reign of Charles II., when the present walks were laid out. Its roots were disturbed, and from that time it declined fast and became a mere trunk. The oldest members of the University can hardly recollect it in better plight; but the faithful records of history have handed down its ancient dimensions. Through a space of 16 yards on every side it once flung its branches; and under its magnificent pavilion could have sheltered with ease 3,000 men. In the summer of 1778 this magnificent ruin fell to the ground. From a part of its ruins a chair has been made for the President of the College, which will long continue its memory."—*Gilpin's Forest Scenery*.

The *King Oak*, Windsor Forest, once the favorite tree of William the Conqueror, is now more than 1,000 years old, and the interior of the trunk is quite hollow. Professor Burnet, who described it, lunched inside this tree with a party, and says it is capable of accommodating ten or twelve persons comfortably at dinner, sitting.

The *Beggar's Oak* in Bagot's Park is twenty feet in girth five feet from the ground. The roots rise above the surface in a very extraordinary manner, so as to furnish a natural seat for the beggars chancing to pass along the pathway near it; and the circumference taken there is 68 feet. The branches extend from the tree 48 feet in every direction.

The *Wallace Oak* at Edenslee, near where Wallace was born, is a noble tree 21 feet in circumference. It is 67 feet high, and its branches extend 45 feet east, 36 west, 30 south, and 25 north. Wallace and 300 of his men are said to have hid themselves from the English among the branches of this tree, which was then in full leaf.

Europe, particularly the vast Gothic edifices of the middle ages, oak was the chief material for the interior. The rich old wainscot, the innumerable carvings and decorations of those days were executed in this material. In America the vast pine forests produce a wood easily wrought, which has in a great measure superseded the use of this fine timber, and the exportation of immense quantities of the former to the eastern continent, has even in some degree lessened its consumption abroad. But for certain purposes where great strength and durability are required, the oak will always take the precedence claimed for it by Evelyn.* The English oak is probably rather superior in these qualities to most of our American species; but for ship-building the Live oak of the southern states is not exceeded by any timber in the world.

Different species of Oak. This country is peculiarly rich in various kinds of oak; Michaux enumerating no less than forty species indigenous to North America. Of these the most useful are the Live oak (*Quercus virens*), of such inestimable value for ship-building; the Spanish oak (*Q. falcata*); the Red oak (*Q. rubra*), etc., the bark of which is extensively used in tanning; the Quercitron or Black oak, which is highly valuable as affording a fine yellow or brown dye for wool, silks, paper-hangings, etc.; and the White oak, which is chiefly used for timber. We shall

---

* The doors of the inner chapels of Westminster, it is stated, are of the same age as the original building; and as the original ancient edifice was founded in 611 they must consequently be more than 1,200 years old. Professor Burnet, in his curious *Amenitates Quercineæ*, observes, that many of the stakes driven into the Thames by the Ancient Britons, to impede the progress of Julius Cæsar, are in a good state of preservation, "having withstood the destroyer time nearly 2,000 years."

here describe only a few of those which are most entitled to the consideration of the planter, either for their valuable properties or as ornamental trees, and calculated for planting in woods or single masses.

The White oak. (*Quercus alba.*) This is one of the most common of the American oaks, being very generally distributed over the country, from Canada to the southern states. In good strong soils it forms a tree 70 or 80 feet high, with wide extending branches; but its growth depends much upon this circumstance. It may readily be known even in winter by its whitish bark, and by the dry and withered leaves which often hang upon this species through the whole of that season. The leaves are about four inches wide and six in length, divided uniformly into rounded lobes without points; these lobes are deeper in damp soils. When the leaves first unfold in the spring they are downy beneath, but when fully grown they are quite smooth, and pale green on the upper surface and whitish or glaucous below. The acorn is oval and the cup somewhat flattened at the base. This is the most valuable of all our native oaks, immense quantities of the timber being used for various purposes in building; and staves of the white oak for barrels are in universal use throughout the Union. The great occasional size and fine form of this tree, in some natural situations, prove how noble an object it would become when allowed to expand in full vigor and majesty in the open air and light of the park. It more nearly approaches the English oak in appearance than any other American species.

Rock Chestnut oak. (*Q. Prinus Monticola.*) This is one of the most ornamental of our oaks, and is found in considerable abundance in the middle states. It has the peculiar advantage of growing well on the most barren and rocky soils, and can therefore be advantageously employed by the landscape gardener, when a steep, dry, rocky bank is to be covered with trees. In deep, mellow soil, its growth is wonderfully vigorous, and it rapidly attains a height of 50 or 60 feet, with a corresponding diameter. The head is rather more symmetrical in form and outline than most trees of this genus, and the stem, in free, open places, shoots up into a lofty trunk. The leaves are five or six inches long, three or four broad, oval and uniformly denticulated, with the teeth more regular but less acute than the Chestnut white oak. When beginning to open in the spring they are covered with a thick down; but when fully expanded they are perfectly smooth and of a delicate texture. *Michaux.*

Chestnut White oak. (*Quercus Prinus palustris.*) This species much resembles the last, but differs in having longer leaves, which are obovate, and deeply toothed. It is sparingly found in the northern states, and attains its greatest altitude in the south, where it is often seen 90 feet in height. Though generally found in the neighborhood of swamps and low grounds, it grows with wonderful rapidity in a good, moderately dry soil, and from the beauty of its fine spreading head, and the quickness of its growth, is highly deserving of introduction into our plantations.

The Yellow oak. (*Q. Prinus acuminata.*) The Yellow oak may be found scattered through our woods over nearly the whole of the Union. Its leaves are lanceolate, and regularly toothed, light green above, and whitish beneath; the acorns small. It forms a stately tree, 70 feet high; and the branches are more upright in their growth, and more clustering, as it were, round the central trunk, than other species. The beauty of its long pointed leaves, and their peculiar mode of growth, recommend it to mingle with other trees, to which it will add variety.

The Pin oak. (*Q. palustris.*) The Pin oak forms a tree in moist situations, varying in height from 60 to 80 feet. The great number of small branches intermingled with the large ones, have given rise to the name of this variety. It is a hardy, free growing species, particularly upon moist soils. Loudon considers it, from its " far extending, drooping branches, and light and elegant foliage," among the most graceful of oaks. It is well adapted to small groups, and is one of the most thrifty growing and easily obtained of all our northern oaks.

The Willow oak. (*Q. Phellos.*) This remarkable species of oak may be recognised at once by its narrow, entire leaves, shaped almost like those of the willow, and about the same size, though thicker in texture. It is not found wild north of the barrens of New Jersey, where it grows plentifully, but thrives well in cultivation much further north. The stem of this tree is remarkably smooth in every stage of its growth. It is so different in appearance and character from the other species of this genus, that in plantations it would never be recognised by a person not conversant with oaks, as one of the family. It deserves to be introduced into landscapes for its singularity as an oak, and its lightness and elegance of foliage individually.

The Mossy-cup oak. (*Q. olivæformis.*) This is so called because the scales of the cups terminate in a long, moss-like fringe, nearly covering the acorn. It is quite a rare species, being only found on the upper banks of the Hudson, and on the Genesee river. The foliage is fine, large, and deeply cut, and the lower branches of the tree droop in a beautiful manner when it has attained some considerable size. *Quercus macrocarpa*, the Over-cup White oak, is another beautiful kind found in the western states, which a good deal resembles the Mossy-cup oak in the acorn. The foliage, however, is uncommonly fine, being the largest in size of any American species; fifteen inches long, and eight broad. It is a noble tree, with fine deep green foliage ; and the growth of a specimen planted in our grounds has been remarkably vigorous.

Scarlet oak. (*Quercus coccinea*.) A native of the middle states ; a noble tree, often eighty feet high. The leaves, borne on long petioles, are a bright lively green on both surfaces, with four deep cuts on each side, widest at the bottom. The great and peculiar beauty of this tree, we conceive to be its property of assuming a deep scarlet tint in autumn. At that period it may, at a great distance, be distinguished from all other oaks, and indeed from every other forest tree. It is highly worthy of a place in every plantation.

The Live oak. (*Quercus virens*.) This fine species will not thrive north of Virginia. Its imperishable timber is the most valuable in our forests ; and, at the south, it is a fine park tree, when cultivated, growing about 40 feet high, with, however, a rather wide and low head. The thick oval leaves are evergreen, and it is much to be regretted that this noble tree will not bear our northern winters.

The English Royal oak. (*Q. robur*.) This is the great representative of the family in Europe, and is one of the most magnificent of the genus, growing often in the fine old woods and parks of England, to eighty and one hundred feet in height. The branches spread over a great surface. "The leaves are petiolated, smooth, and of a uniform color on both sides, enlarged towards the summit, and very coarsely toothed." As a single tree for park scenery, this equals any American species in *majesty* of form, though it is deficient in individual beauty of foliage to some of our oaks. It is to be found for sale in our nurseries, and we hope will become well known among us. The timber is closer grained and more durable, though less elastic than the best American oak; and Michaux, in his Sylva, recommends its introduction into this country largely, on these accounts.

The Turkey oak. (*Q. Cerris.*) There are two beautiful hybrid varieties of this species, which have been raised in England by Messrs. Lucombe and Fulham, which we hope will yet be found in our ornamental plantations. They are partially evergreen in winter, remarkably luxuriant in their growth, attaining a height of seventy or eighty feet, and elegant in foliage and outline. The Lucombe and Fulham oaks grow from one to five feet in a season; the trees assume a beautiful pyramidal shape, and as they retain their fine glossy leaves till May, they would form a fine contrast to other deciduous trees.

We might here enumerate a great number of other fine foreign oaks; among which the most interesting are the Holly or Holm oak (*Quercus Ilex*); and the Cork oak (*Q. Suber*), of the south of France, which produces the cork of commerce (both rather too tender for the north); the Kermes oak (*Q. coccifera*), from which a scarlet dye is obtained; and the Italian Esculent oak (*Q. Esculus*), with sweet nutritious acorns. Those, however, who wish to investigate them, will pursue this subject further in European works; while that splendid treatise on our forest trees, the North American Sylva of Michaux, will be found to give full and accurate descriptions of all our numerous indigenous varieties, of which many are peculiar to the southern states.

The oak flourishes best on a strong loamy soil, rather moist than dry. Here at least the growth is most rapid, although, for timber, the wood is generally not so sound on a moist soil as a dry one, and the tree goes to decay more rapidly. Among the American kinds, however, some may be found adapted to every soil and situation, though those species which grow on upland soils, in stony, clayey, or loamy bottoms, attain the greatest size and longevity. When immense trees are desired, the oak should either be transplanted very young, or, which is preferable, raised from the acorn sown where it is finally to remain. This is necessary on account of the very large *tap roots* of this genus of trees, which are either entirely destroyed or greatly injured by removal. Transplanting this genus of trees should be performed either early in autumn, as soon as the leaves fall or become brown, or in spring before the abundant rains commence.

The Elm. *Ulmus.*

Nat. Ord. Ulmaceæ.   *Lin. Syst.* Pentandria, Digynia.

We have ascribed to the oak the character of pre-

DECIDUOUS ORNAMENTAL TREES.   153

eminent dignity and majesty among the trees of the forest. Let us now claim for the elm the epithets graceful and elegant. This tree is one of the noblest in the size of its trunk, while the branches are comparatively tapering and slender, forming themselves, in most of the species, into long and graceful curves. The flowers are of a chocolate or purple color, and appear in the month of April, before the leaves. The latter are light and airy, of a pleasing light green in the spring, growing darker, however, as the season advances. The elm is one of the most common trees in both continents, and has been well known for its beauty and usefulness since a remote period. In the south of Europe, particularly in Lombardy, elm trees are planted in vineyards, and the vines are trained in festoons from tree to tree in the most picturesque manner. Tasso alludes to this in the following stanza:

> "Come olmo, a cui la pampinosa pianta
> Cupida s'avviticchi e si marite;
> Se ferro il tronca, o fulmine lo schianta
> Trae seco a terra la compagna vite."
> *Gerusalemme Liberata*, 2. 326.

It is one of the most common trees for public walks and avenues, along the highways in France and Germany, growing with great rapidity, and soon forming a widely extended shade. In Europe, the elm is much used for keels in ship-building, and is remarkably durable in water; more extensive use is made of it there than of the American kinds in this country, though the wood of the Red American elm is more valuable than any other in the United States for the blocks used in ship rigging.

For its graceful beauty the elm is entitled to high

regard. Standing alone as a single tree, or in a group of at most three or four in number, it developes itself in all its perfection. The White American elm we consider the most beautiful of the family, and to this we more particularly allude. In such situations as we have just mentioned, this tree developes its fine ample form in the most perfect manner. Its branches first spring up embracing the centre, then bend off in finely diverging lines, until in old trees they often sweep the ground with their loose pendent foliage. With all this lightness and peculiar gracefulness of form, it is by no means a meagre looking tree in the body of its foliage, as its thick tufted masses of leaves reflect the sun and embosom the shadows as finely as almost any other tree, the oak excepted. We consider it peculiarly adapted for planting, in scenes where the expression of elegant or classical beauty is desired. In autumn the foliage assumes a lively yellow tint, contrasting well with the richer and more glowing colors of our native woods. Even in winter it is a pleasing object, from the minute division of its spray and the graceful droop of its branches. It is one of the most generally esteemed of our native trees for ornamental purposes, and is as great a favorite here as in Europe for planting in public squares and along the highways. Beautiful specimens may be seen in Cambridge, Mass., and very fine avenues of this tree are growing with great luxuriance in and about New Haven.* The charming villages of New England, among which Northampton and Springfield are pre-eminent, borrow from the superb and wonderfully luxuriant elms which decorate their fine

---

* The great elm of Boston Common is 22 feet in circumference.

streets and avenues, the greater portion of their peculiar loveliness. The elm should not be chosen where large groups and masses are required, as the similarity of its form in different individuals might then create a monotony; but as we have before observed, it is peculiarly well calculated for small groups, or as a single object. The roughness of the bark, contrasting with the lightness of its foliage and the easy sweep of its branches, adds much also to its effect as a whole.

We shall briefly describe the principal species of the elm.

The American White elm. (*Ulmus Americana.*) This is the best known and most generally distributed of our native species, growing in greater or less profusion over the whole of the country included between Lower Canada and the Gulf of Mexico. It often reaches 80 feet in height in fine soils, with a diameter of 4 or 5 feet. The leaves are alternate, 3 or 4 inches long, unequal in size at the base, borne on petioles half an inch to an inch in length, oval, acuminate, and doubly denticulated. The seeds are contained in a flat, oval, winged seed-vessel, fringed with small hairs on the margin. The flowers, of a dull purple color, are borne in small bunches on short footstalks at the end of the branches, and appear very early in the spring. This tree prefers a deep rich soil, and grows with greater luxuriance if it be rather moist, often reaching in such situations an altitude of nearly 100 feet. It is found in the greatest perfection in the alluvial soils of the fertile valleys of the Connecticut, the Mississippi, and the Ohio rivers.

The Red or Slippery elm. (*U. fulva.*) A tree of

lower size than the White elm, attaining generally only 40 or 50 feet. According to Michaux, it may be distinguished from the latter even in winter, by its buds, which are larger and rounder, and which are covered a fortnight before their development with a russet down. The leaves are larger, rougher, and thicker than those of the White elm; the seed-vessels larger, destitute of fringe; the stamens short, and of a pale rose color. This tree bears a strong likeness to the Dutch elm, and the bark abounds in mucilage, whence the name of Slippery elm. The branches are less drooping than those of the White elm.

The Wahoo elm (*U. alata*) is not found north of Virginia. It may at once be known in every stage of its growth by the fungous cork-like substance which lines the branches on both sides. It is a very singular and curious tree, of moderate stature, and grows rapidly and well when cultivated in the northern states.

The common European elm. (*U. campestris.*) This is the most commonly cultivated forest tree in Europe, next to the oak. It is a more upright growing tree than the White elm, though resembling it in the easy disposition and delicacy of its branches. The flowers, of a purple color, are produced in round bunches close to the stem. The leaves are rough, doubly serrated, and much more finely cut than those of our elms. It is a fine tree, 60 or 70 feet high, growing with rapidity, and is easily cultivated. The timber is more valuable than the American sort, though the tree is inferior to the White elm in beauty. There are some dozen or more fine varieties of this species cultivated in the English nurseries, among which the most remarkable are the Twisted elm (*U. c. tortuosa*), the trunk of which is singularly marked with hollows and protuberances, and the grain of the wood curiously twisted together: the Kidbrook elm (*U. c. virens*), which is a sub-evergreen: the Gold and Silver striped elms, with variegated leaves, and the Narrow-leaved elm (*U. c. viminalis*), which resembles the birch: the Cork-barked elm (*U. c. suberosa*), the young branches of which are covered with cork, etc.

The latter is one of the hardiest and most vigorous of all ornamental trees in this climate. It thrives in almost every soil, and its rich, dark foliage, which hangs late in autumn, and its somewhat picturesque form, should recommend it to every planter.

The Scotch or Wych elm. (*U. montana.*) This is a tree of lower stature than the common European elm, its average height being about 40 feet. The leaves are broad, rough, pointed, and the branches extend more horizontally, drooping at the extremities. The bark on the branches is comparatively smooth. It is a grand tree, "the head is so finely massed and yet so well broken as to render it one of the noblest of park trees; and when it grows wild amid the rocky scenery of its native Scotland, there is no tree which assumes so great or so pleasing a variety of character."* In general appearance, the Scotch elm considerably resembles our White elm, and it is a very rapid grower. Its most ornamental varieties are the Spiry-topped elm (*U. m. fastigiata*), with singularly twisted leaves, and a very upright growth: the weeping Scotch elm (*U. m. pendula*), a very remarkable variety, the branches of which droop in a

---

* Sir Thos. Lauder, in Gilpin, 1. 91.

fan-like manner : and the Smooth-leaved Scotch elm (*U. m. glabra*).

There is scarcely any soil to which some of the different elms are not adapted. The European species prefer a deep, dry soil; the Scotch or Wych elm will thrive well even in very rocky places; and the White elm grows readily in all soils, but most luxuriantly in moist places. All the species attain their maximum size when planted in a deep loam, rather moist than dry. They bear transplanting remarkably well, suffering but little even from the mistaken practice of those persons who reduce them in transplanting to the condition of bare poles, as they shoot out a new crop of branches, and soon become beautiful young trees in spite of the mal-treatment. As the elm scarcely produces a tap root, even large trees may be removed, when the operation is skilfully performed. In such cases, the recently-removed tree should be carefully and plentifully supplied with water until it is well established in its new situation. The elm is also easily propagated by seed, layers, or, in some species, by suckers from the root.

The Plane or Buttonwood Tree. *Platanus.*

Nat. Ord. Platanaceæ.   Lin. Syst. Monœcia, Polyandria.

The plane, *Platanus*, derives its name from πλατυς, *broad*, on account of the broad, umbrageous nature of its branches. It is a well known tree of the very largest size, common to both hemispheres, and greatly prized for the fine shade afforded by its spreading head, in the warmer parts of Europe and Asia. No tree was in greater esteem with the ancients for this purpose; and we are told that the Academic groves, the neighborhood of the public schools, and all those favorite avenues where the Grecian philosophers were accustomed to resort, were planted with these trees; and beneath their shade Aristotle, Plato, and Socrates, delivered the choicest wisdom and eloquence of those classic days. The Eastern plane (*Platanus orientalis*) was first brought to the Roman provinces from Persia, and so highly was it esteemed that according to Pliny, the Morini paid a tribute to Rome for the privilege of enjoying its shade. To that author we are also indebted for the history of the great plane tree that grew in the province of Lycia, which was of so huge a size, that the governor of the province, Licinius Mutianus, together with eighteen of his retinue, feasted in the hollow of its trunk.

In the United States, the plane is not generally found growing in great quantities in any one place, but is more or less scattered over the whole country. In deep, moist, alluvial soils, it attains a size scarcely, if at all, inferior to that of the huge trees of the eastern continent; forming at least, in the body of its trunk, a larger circumference than any other of our native trees. The younger Michaux (*Sylva*, 1, 325) measured a tree near Marietta, Ohio, which at four feet from the ground was found to be forty-seven feet in circumference; and a specimen has lately been cut on the banks of the Genesee river, of such enormous size, that a section of the trunk was hollowed out and furnished as a small room, capable of containing fourteen persons.* On the margins of the great western rivers it sometimes rises up seventy feet, and then expands into a fine, lofty head, surpassing in grandeur all its neighbors of the forest. The large branches of the plane shoot out in a horizontal direction; the trunk generally ascending in a regular, stately, and uninterrupted manner. The blossoms are small greenish balls appearing in spring, and the fertile ones grow to an inch in diameter, assuming a deep brownish color, and hang upon the tree during the whole winter. A striking and peculiar characteristic of the plane, is its property of throwing off or shedding continually the other coating of bark here and there in patches. Professor Lindley (*Introduction to the Natural System*, 2d ed. 187) says this is owing to its deficiency in the expansive power of the fibre common to the bark of other trees, or, in other words, to the rigidity of its tissue: being therefore incapable of stretching with the growth of the tree, it bursts open on different parts of the trunk, and is cast off. This gives the trunk quite a lively and picturesque look, extending more or less even to the extremity of the branches; and makes this tree quite conspicuous in winter. Bryant, in his address to Green River, says:

> "Clear are the depths where its eddies play,
> And dimples deepen and whirl away,
> And the plane tree's *speckled arms* o'ershoot
> The swifter current that mines its root."

The great merit of the plane, or buttonwood, is its

* A buttonwood on the Montezuma estate, Jefferson, Cayuga Co., N. Y., is forty-seven and a half feet in circumference; and the diameter of the hollow two feet from the ground, is fifteen feet. (*N. Y. Med. Repository*, IV. 427.)

extreme vigor and luxuriance of growth. In a good soil it will readily reach a height of thirty-five or forty feet in ten years. It is easily transplanted; and in new residences, bare of trees, where an effect is desired speedily, we know of nothing better adapted quickly to produce abundance of foliage, shelter, and shade. When the requisite foliage is obtained, and other trees of slower growth have reached a proper size, the former may be thinned out. As the plane tree grows to the largest size, it is only proper for situations where there is considerable ground, and where it can without inconvenience to its fellows have ample room for its full development. Then soaring up, and extending its wide-spread branches on every side, it is certainly a very majestic tree. The color of the foliage is of a paler green than is usual in forest trees; and although of large size, is easily wafted to and fro by the wind, thereby producing an agreeable diversity of light pleasing to the eye in summer. In winter the branches are beautifully hung, even to their furthest ends, with the numerous round russet-balls, or seed-vessels, each suspended by a slender cord, and swinging about in the air. The outline of the head is pleasingly irregular, and its foliage against a sky outline is bold and picturesque. It is not a tree to be planted in thick groves by itself, but to stand alone and detached, or in a group with two or three. In avenues it is often happily employed, and produces a grand effect. It also grows with great vigor in close cities, as some superb specimens in the square of the State-house, Pennsylvania Hospital, and other places in Philadelphia fully attest.

There is but a trifling difference in general effect between our plane or buttonwood and the Oriental plane. For the purposes of shade and shelter, the American is the finest, as its foliage is the longest and broadest. The Oriental plane (*Platanus orientalis*) has the leaves lobed like our native kind (*P. occidentalis*), but the segments are much more deeply cut; the footstalks of its leaves are green, while those of the American are of a reddish hue, and the fruit or ball is much smaller and rougher on the outer surface when fully grown. Both species are common in the nurseries, and are worthy the attention of the planter; the Oriental, as well for the interesting associations connected with it, being the favorite shade-tree of the east, etc., as for its intrinsic merits as a lofty and majestic tree.

Two of the varieties of P. occidentalis are sometimes cultivated, the chief of which is the Maple-leaved plane (*P. O. acerifolia*).

THE ASH TREE. *Fraxinus*.

Nat. Ord. Oleaceæ. *Lin. Syst.* Polygamia, Diœcia.

The name of the ash, one of the finest and most useful of forest trees, is probably derived from the Celtic *asc*, a pike—as its wood was formerly in common use for spears and other weapons. Homer informs us that Achilles was slain with an ashen spear. In modern times the wood is in universal use for the various implements of husbandry, for the different purposes of the wheelwright and carriage-maker, and in short for all purposes where great strength and elasticity are required; for in these qualities the ash is second to no tree in the forest, the hickory alone excepted. The ash is a large and lofty tree, growing, when surrounded by other trees, sixty or seventy feet high, and three or more in diameter. When exposed on all sides it forms a beautiful, round, compact head of loose, pinnated, light green foliage, and is one of the most vigorous growers among the hard-wooded trees. The American species of ash are found in the greatest luxuriance and beauty on the banks and margins of rivers where the soil is partially dry, yet where the roots can easily penetrate down to the moisture. The European ash is remarkable for its hardy nature, being often found in great vigor on steep rocky hills, and amid crevices where most other trees flourish badly. Southey alludes to this in the following lines :—

> " Grey as the stone to which it clung, half root,
> Half trunk, the young ash rises from the rock."

As the ash grows strongly, and the roots, which extend to a great distance, ramify near the surface, it exhausts the soil underneath and around it to an astonishing degree. For this reason the grass is generally seen in a very meagre and starved condition in a lawn where the ash tree abounds. Here and there a single tree of the ash will have an excellent effect, seen from the windows of the house; but we would chiefly employ it for the grand masses, and to intermingle with other large groups of trees in an extensive plantation. When the ash is young it forms a well rounded head; but when older the lower branches bend towards the ground, and then slightly turn up in a very graceful manner. We take pleasure in quoting what that great lover and accurate delineator of forest beauties, Mr. Gilpin, says of the ash. " The ash generally carries its principal stem higher than the oak, and rises in an easy flowing line. But its chief beauty consists in the lightness of its whole appearance. Its branches at first keep close to the trunk and form acute angles with it; but as they begin to lengthen they generally take an easy sweep, and the looseness of the leaves corresponding with the lightness of the spray, the whole forms an elegant depending foliage. Nothing can have a better effect than an old ash hanging from the corner of a wood, and bringing off the heaviness of the other foliage with its loose pendent branches."—(*Forest Scenery, p. 82.*)

The highest and most characteristic beauty of the American White ash (and we consider it the finest of all the species) is the coloring which its leaves put on in autumn. Gilpin complains that the leaf of the European ash "decays in a dark, muddy, unpleasing tint." Not so the White ash. In an American wood, such as often lines and overhangs the banks of the Hudson, the Connecticut, and many of our noble northern streams, the ash assumes peculiar beauty in autumn, when it can often be distinguished from the surrounding trees for four or five miles, by the peculiar and beautiful deep brownish purple of its fine mass of foliage. This color, though not lively, is so full and rich as to produce the most pleasing harmony with the bright yellows and reds of the other deciduous trees, and the deep green of the pines and cedars.

The ash, unlike the elm, starts into vegetation late in the spring, which is an objection to planting it in the immediate vicinity of the house. In winter the long greyish white or ash-colored branches are pleasing in tint, compared with those of other deciduous trees.

The White ash. (*Fraxinus Americana.*) This species, according to Michaux, is common to the colder parts of the Union, and is most abundant north of the Hudson. It owes its name to the light color of the bark, which on large stocks is deeply furrowed, and divided into squares of one to three inches in diameter. The trunk is perfectly straight, and in close woods is often undivided to the height of more than 40 feet. The leaves are composed of three or four pairs of leaflets, terminated by an odd one; the whole twelve or fourteen inches long. Early in spring they are covered with a light down which disappears as summer advances, when they become quite smooth, of a light green color above and whitish beneath. The foliage, as well as the timber of our White ash, is finer than that of the common European ash, and the tree is much prized in France and Germany.

The Black ash (*F. sambucifolia*), sometimes called the Water ash, requires a moist soil to thrive well, and is seen in the greatest perfection on the borders of swamps. Its buds are of a deep blue; the young shoots of a bright green, sprinkled with dots of the same color, which disappear as the season advances. It may readily be distinguished from the White ash by its bark, which is of a duller hue and less deeply furrowed. The Black ash is altogether a tree of less stature than the preceding.

The other native sorts are the Red ash (*F. tomentosa*), with the bark of a deep brown tint, found in Pennsylvania: the Green ash (*F. viridis*), which also grows in Pennsylvania, and is remarkable for the brilliant green of both sides of the leaves: the Blue ash (*F. Quadrangulata*), a beautiful tree of Kentucky, 70 feet high, distinguished by the four opposite membranes of a greenish color, found on the young shoots: and the Carolina ash (*F. platycarpa*), a small tree, the leaves of which are covered with a thick down in spring.

The common European ash (*F. excelsior*) strongly resembles the White ash. It may, however, easily be known by its very black buds, and longer, more serrated leaflets, which are sessile, instead of being furnished with petioles like the White ash. This fine tree, as well as the White ash, grows to 80 or 90 feet in height, with a very handsome head.

The Weeping ash, Fig. 36, is a very remarkable variety

[Fig. 36. The Weeping Ash.]

of the European ash, with pendulous or weeping branches; and is worthy a place in every lawn for its curious ramification, as well as for its general beauty. It is generally propagated by grafting on any common stock, as the White ash, 7 or 8 feet high, when the branches immediately begin to turn down in a very striking and peculiar manner. The droop of the branches is hardly a graceful one, yet it is so unique, either when leafless, or in full foliage, that it has long been one of our greatest favorites.

The Flowering ash (*Fraxinus Ornus**) is a small tree of about 20 feet, growing plentifully in the south of Europe, and is also found sparingly in this country. Its chief beauty lies in the beautiful clusters of pale or greenish-white flowers, borne on the terminal branches in May and June. The foliage and general appearance of the tree are much like those of the common ash; but when in blossom it resembles a good deal the Carolina Fringe tree. In Italy a gummy substance called manna exudes from the bark, which is used in medicine.

The Lime or Linden Tree. *Tilia.*

Nat. Ord. Tilaceæ.   *Lin. Syst.* Polyandria, Monogynia.

This tree, or rather the American sort, is well known among us by the name of *basswood*. It is a rapidly growing, handsome, upright, and regularly shaped tree; and all the species are much esteemed, both in Europe and this country, for planting in avenues and straight lines, wherever the taste is in favor of geometric plantations. In Germany and Holland it is a great favorite for bordering their wide and handsome streets, and lining their long and straight canals. "In Berlin," Granville says in his travels, "there is a celebrated street called '*unter der Linden,*' (under the lime trees,) a gay and splendid avenue, planted with double

---

* *Ornus Europæus* of Persoon, and the European botanists. Beck remarks that the American kind is so little known, that it is difficult to determine whether it is a different species or only a mere variety of the European.

rows of this tree, which presented to my view a scene far more beautiful than I had hitherto witnessed in any town, either in France, Flanders, or Germany." In this country the European lime is also much planted in our cities; and some avenues of it may be seen in Philadelphia, particularly before the State-house in Chestnut-street. The basswood is a very abundant tree in some parts of the middle states, and is seen growing in great profusion, forming thick woods by itself in the interior of this state. With us the wood is considered too soft to be of much value, but in England it was formerly in high repute as an excellent material for the use of carvers. Some very beautiful specimens of old carving in lime wood may be seen in Windsor Castle and Trinity College.* The Russian bass mats, which find their way to every commercial country, are prepared from the inner bark of this tree. The sap affords a sugar like the maple, although in less quantities; and it is stated in the Encyclopædia of Plants (p. 467) "that the honey made from the flowers of the lime tree is reckoned the finest in the world. Near Knowno, in Lithuania, there are large forests chiefly of this tree, and probably a distinct variety. The honey produced in these forests sells at more than double the price of any other, and is used extensively in medicine and for liqueurs."

* The art of carving in wood, brought to such perfection by Gibbons, is now, we believe, much given up; therefore the lime has lost a most important branch of its usefulness. Perhaps the finest specimens of the works of Gibbons are to be seen at Chatsworth, the seat of the Duke of Devonshire, in Derbyshire. The execution of the flowers, fish, game, nets, etc., on the panelling of the walls is quite wonderful. It was of him that Walpole justly said, 'that he was the first artist who gave to wood the loose and airy lightness of flowers, and chained together the various productions of the elements, with a free disorder natural to each species.' The lime tree is still, however, used by the carver, and we hope that the art of wood carving may gradually be restored."—*Sir T. D. Lauder.*

The leaves of the lime are large and handsome, heart-shaped in form, and pleasing in color. The flowers, which open in June, hang in loose, pale yellow cymes or clusters, are quite ornamental and very fragrant.

————Sometimes
A scent of violets and blossoming limes
Loitered around us; then of honey cells,
Made delicate from all white flower bells.

KEATS.

It was a favorite tree in the ancient style of gardening, as it bore the shears well, and was readily clipt into all manner of curious and fantastic shapes. When planted singly on a lawn, and allowed to develope itself fully on every side, the linden is one of the most beautiful of trees. Its head then forms a fine pyramid of verdure, while its lower branches sweep the ground and curve upwards in the most pleasing form. For this reason, though the linden is not a picturesque tree, it is very happily adapted for the graceful landscape, as its whole contour is full, flowing, and agreeable. The pleasant odor of its flowers is an additional recommendation, as well as its free growth and handsome leaves. Were it not that of late it is so liable to insects, we could hardly say too much in its praise as a fine ornament for streets and public parks. There, its regular form corresponds well with the formality of the architecture; it's shade affords cool and pleasant walks, and the delightful odor of its blossoms is doubly grateful in the confined air of the city. Our basswood has rather less of uniformity in its outline than the European lindens but the general form is the same.

The American lime, or basswood (*Tilia Americana*), is the most robust tree of the genus, and produces much more vigorous shoots than the European species. It prefers a deep and fertile soil, where the trunk grows remarkably straight, and the branches form a handsome, well-rounded summit. The flowers are borne on long stalks, and are pendulous from the branches. The leaves are large, heart-shaped, finely cut on the margin, and terminated by a point at the extremity. The seeds, which ripen in autumn, are like small peas, round and greyish.

The white lime (*T. alba*) is rare in the eastern states, but common in Pennsylvania and the states south of it. It is not a tree of the largest size, but its flowers are the finest of our native sorts. The leaves are also very large, deep green on the upper surface, and white below; they are more obliquely heart-shaped than those of the common basswood. The young branches are covered with a smooth silvery bark. This species is very common on the Susquehannah river.

The Downy lime tree. (*T. pubescens.*) The under side of the leaves, and the fruits of this species, are, as its name denotes, covered with a short down. Its flowers are nearly white; the serratures of the leaves wider apart, and the base of the leaf obliquely truncated. It is a handsome large tree, a native of Florida, though hardy enough, as experience proves, to bear our northern winters.

The European lime (*T. Europæa*) is distinguished from the American sorts, by its smaller and more regularly cordate and rounded leaves. Unlike our native species, the flowers are not furnished with inner scale-like petals. The foliage is rather deeper in hue than the native sorts, and the branches of the head rather more regular in form and disposition. There are two pretty varieties of the English lime which are well known in this country, viz. the Red-barked, or corallina (*var. rubra*), with red branches; and the Golden-barked (*var. aurea*), with handsome yellow branches. These trees are peculiarly beautiful in winter, when a few of them mingled with other deciduous trees make a pleasing variety of coloring in the absence of foliage. The broad-leaved European lime is the finest for shade and ornament. The whitish foliage of *Tilia alba*, which probably is also a variety, has a beautiful appearance, somewhat like the Abele tree, in a gentle breeze.

These trees grow well on any good friable soil, and readily endure transplantation. They bear trimming remarkably well; and when but little root is obtained the head may be shortened in proportion, and the tree will soon make vigorous shoots again. All the species are easily increased by layers.

The Beech Tree. *Fagus.*

Nat. Ord. Corylaceæ.   *Lin. Syst.* Monœcia, Polyandria.

The Beech is a large, compact, and lofty tree, with a greyish bark and finely divided spray, and is a common inhabitant of the forest in all temperate climates. In the United States, this tree is generally found congregated in very great quantities, wherever the soil is most favorable; hundreds of acres being sometimes covered with this single kind of timber. Such tracts are familiarly known as "beech woods." The leaves of the beech are remarkably thin in texture, glazed and shining on the upper surface, and so thickly set upon the numerous branches, that it forms the darkest and densest shade of any of our deciduous forest trees. It appears to have been highly valued by the ancients as a shade tree; and Virgil says in its praise, in a well-known Eclogue:

> " Tityre, tu patulæ recubans sub tegmine *fagi*,
> Sylvestrem tenui musam meditaris avena."

It bears a small compressed nut or mast, oily and sweet, which once was much valued as an article of food. The most useful purpose to which we have heard of their being applied, is in the manufacture of an oil, scarcely inferior to olive oil. This is produced from the mast of the beech forests in the department of Oise, France, in immense quantities; more than a million of sacks of the nuts having been collected in that department in a single season. They are reduced, when perfectly ripe, to a fine paste, and the oil is extracted by gradual pressure. The product of oil, compared with the crushed nuts, is about sixteen per cent. (*Michaux, N. American Sylva.*)

In Europe, the wood of the beech is much used in the manufacture of various utensils; but here, where our forests abound in woods vastly superior in strength. durability, and firmness, that of the beech is comparatively little esteemed.

For ornamental purposes, the beech, from its comparatively slow growth, and its abundance in various parts of the country, does not command the admiration here which it does in Europe. Campbell, the poet, has produced so eloquent and beautiful an appeal in favor of an old denizen of the forest, entitled the "Beech Tree's Petition," that we gladly quote it, hoping it may perchance stay the hand of some *soi-disant* improver, who would despoil our native woods of their proudest glories:

> "Oh, leave this barren spot to me!
> Spare, woodman, spare the beechen tree!
> Though bush or floweret never grow
> My dark, unwarming shade below;
> Nor summer bud perfume the dew
> Of rosy blush or yellow hue!
> Nor fruits of autumn, blossom-born,
> My green and glossy leaves adorn;
> Nor murmuring tribes from me derive
> Th' ambrosial amber of the hive;
> Yet leave this barren spot to me—
> Spare, woodman, spare the beechen tree!
>
> Thrice twenty summers I have seen
> The sky grow bright, the forest green;
> And many a wintry wind have stood
> In bloomless, fruitless solitude,
> Since childhood in my pleasant bower
> First spent its sweet and sportive hour;
> Since youthful lovers in my shade
> Their vows of truth and rapture made;
> And on my trunk's surviving frame
> Carved many a long-forgotten name.
> Oh! by the sighs of gentle sound
> First breathed upon this sacred ground,
> By all that Love has whispered here,
> Or beauty heard with ravished ear;
> As Love's own altar, honor me—
> Spare, woodman, spare the beechen tree!"

The beech is quite handsome and graceful when young, and when large it forms one of the heaviest and grandest of *beautiful* park trees. From this massy quality, however, it is excellently adapted to mingle with other trees when a thick and impenetrable mass of foliage is desired:

and, on account of its density, it is also well suited to shut out unsightly buildings, or other objects.

The leaves of many beech trees hang on the tree, in a dry and withered state, during the whole winter. This is chiefly the case with young trees; but we consider it as greatly diminishing its beauty at that season, as the tree is otherwise very pleasing to the eye, with its smooth, round, grey stem, and small twisted spray. A deciduous tree, we think, should as certainly drop its leaves at the approach of cold weather, as an evergreen should retain them; more especially if its leaves have a dead and withered appearance, as is the case with those of the beech in this climate.

The White beech (*Fagus Sylvatica*) is the common beech tree of the middle and western states. It is found in the greatest perfection in a cool situation and a moist soil. The bark is smooth and grey, even upon the oldest stocks. The leaves oval, smooth, and shining, coarsely cut on the edges, and margined with a soft down in the spring.

The Red beech (*F. ferruginea*), so called on account of the color of its wood, loves a still colder climate than the other, and is found in the greatest perfection in British America. The leaves are divided into coarser teeth on the margin than the foregoing species. The nuts are much smaller, and the whole tree forms a lower and more spreading head.

The European beech (*F. sylvatica*) is thought by many botanists to be the same species as our white beech, or at most only a variety. Its average height in Europe is about fifty feet; the buds are shorter, and the leaves not so coarsely toothed as our native sorts. The Purple beech is a very ornamental variety of the European beech, common in the gardens. Both surfaces of the leaves, and even the young shoots, are deep purple; and although the growth is slow, yet it is in every stage of its progress, and more particularly when it reaches a good size, one of the strangest anomalies among trees, in the hue of its foliage. There is also a variety called the copper-colored beech, with paler purple leaves ;* and a more rare English variety (*F. s. pendula*), the Weeping beech, with graceful pendent branches.

THE HORNBEAM (*Carpinus Americana*), and the IRONWOOD (*Ostrya Virginica*), are both well known small trees, belonging to the same natural family as the beech. They are of little value in ornamental plantations; but from their thick foliage, they might perhaps be employed to advantage in making thick verdant screens for shelter or concealment.

THE POPLAR TREE. *Populus*.

Nat. Ord. Salicaceæ.   *Lin. Syst.* Diœcia, Octandria.

Arbor Populi, or the people's tree, was the name given in the ancient days of Rome to this tree, as being peculiarly appropriated to those public places most frequented by the people: some ingenious authors have still further justified the propriety of the name, by adding, that its trembling leaves are like the *populace*, always in motion.

The poplars are light-wooded, rapid-growing trees; many

---

* The finest Copper Beech in America is growing in the grounds of Thomas Ash, Esq., Throgs Neck, Westchester Co., N. Y. It is more than fifty feet high, with a broad and finely formed head.

of them of huge size, and all with pointed, heart-shaped leaves. The tassel-like catkins, or male blossoms, of a red or brownish hue, appear early in the spring. Some of the American kinds, as the Balsam and Balm of Gilead poplars, have their buds enveloped in a fragrant gum; others, as the Silver poplar, or Abele, are remarkable for the snowy whiteness of the under side of the foliage; and the Lombardy poplar, which

> "Shoots up its spire, and shakes its leaves in the sun,"
>
> PROCTOR.

for its remarkably conical or spire-like manner of growth. The leaves of all the species, being suspended upon long and slender footstalks, are easily put in motion by the wind. This, however, is peculiarly the case with the aspen, the leaves of which may often be seen trembling in the slightest breeze, when the foliage of the surrounding trees is motionless. There is a popular legend in Scotland respecting this tree, which runs thus:

> "Far off in the Highland wilds 'tis said
> (But truth now laughs at fancy's lore),
> That of this tree the cross was made,
> Which erst the Lord of Glory bore;
> And of that deed its leaves confess,
> E'er since, a troubled consciousness."

In Landscape Gardening the poplar is not highly esteemed; but it is a valuable tree when judiciously employed, and produces a given quantity of foliage and shade sooner perhaps than any other. Some of the American kinds are majestic and superb trees when old, particularly the Cottonwood and Balsam poplars.* One of the handsomest sorts

* There is a noble specimen of the Cottonwood, or, as it is here called, the

is the Silver poplar, which is much valued in our ornamental plantations; the more so, perhaps, because it is an

[Fig. 37. The Cottonwood.]

exotic. At some distance, the downy under surfaces of the leaves, turned up by the wind, give it very much the aspect of a tree covered with white blossoms. This effect is the more striking, when it is situated in front of a group or

Balm of Gilead poplar, about two miles north of Newburgh, on the Hudson, which gives its name to the small village (Balmville) near it. The branches cover a surface of one hundred feet in diameter, the trunk girths twenty feet, and the branches stretch over the public road in a most majestic manner. (*See Fig.* 37.)

mass of the darker foliage of other trees. It is valuable for retaining its leaves in full beauty to the latest possible period in the autumn, even when all the other deciduous trees are either brown, or have entirely lost their leafy honors. Its growth is extremely rapid, forming a fine rounded head of thirty feet in height, in six or eight years.

The Lombardy poplar is a beautiful tree, and in certain situations produces a very elegant effect; but it has been planted so indiscriminately, in some parts of this country, in close monotonous lines before the very doors of our houses, and in many places in straight rows along the highways for miles together, to the neglect of our fine native trees, that it has been tiresome and disgusting. This tree may, however, be employed with singular advantage in giving life, spirit, and variety to a scene composed entirely of round-headed trees, as the oak, ash, etc.,—when a tall poplar, emerging here and there from the back or centre of the group, often imparts an air of elegance and animation to the whole. It may, also, from its marked and striking contrast to other trees, be employed to fix or direct the attention to some particular point in the landscape. When large poplars of this kind are growing near a house of but moderate dimensions, they have a very bad effect by completely overpowering the building, without imparting any of that grandeur of character conferred by an old oak, or other spreading tree. It should be introduced but sparingly in landscape composition, as the moment it is made common in any scene, it gives an air of sameness and formality, and all the spirited effect is lost which its sparing introduction among other trees produces. The Lombardy poplar is so well adapted to confined situations, as its branches require less lateral room than those of almost any other large deciduous tree.

It is an objection to some of the poplars, that in any cultivated soil they produce an abundance of suckers. For this reason they should be planted only in grass ground, or in situations where the soil will not be disturbed, or where the suckers will not be injurious. Indeed, we conceive them to be chiefly worthy of introduction in grounds of large extent, to give variety to plantations of other and more valuable trees. They grow well in almost every soil, moist or dry, and some species prefer quite wet and springy places.

The chief American poplars are the Tachamahaca or Balsam poplar (*Populus balsamifera*), chiefly found in Northern America; a large tree, 80 feet high, with fragrant gummy buds and lanceolate-oval leaves; the Balm of Gilead poplar (*P. candicans*), resembling the foregoing in its buds, but with very large, broad, heart-shaped foliage. From these a gum is sometimes collected, and used medicinally for the cure of scurvy. The American aspen (*P. tremuloides*), about 30 feet high, a common tree with very tremulous leaves and greenish bark; the large American aspen (*P. grandidentata*), 40 feet high, with large leaves bordered with coarse teeth or denticulations; the Cotton tree (*P. argentea*), 60 or 70 feet, with leaves downy in a young state; the American Black poplar of smaller size, having the young shoots covered with short hair; the Cottonwood (*P. Canadensis*), found chiefly in the western part of this state, a fine tree, with smooth, unequally-toothed, wide cordate leaves; and the Carolina poplar (*P. angulata*), an enormous tree of the swamps of the south and west, considerably resembling the Cotton tree, but without the resinous buds of that species.

Among the European kinds, the most ornamental, as we have already remarked, is the Silver aspen, White poplar, or Abele tree (*P. alba*), which grows to a great size on a deep loamy soil in a very short time. The leaves are divided into lobes, and toothed on the margin, smooth and very deep green above, and densely covered with a soft, close, white down beneath. There are some varieties of this species known abroad, with leaves more or less downy, etc. Sir J. E. Smith remarks in his English Flora, that the wood, though but little used, is much firmer than that of any other British poplar; making as handsome floors as the best Norway fir, with the additional advantage that they will not readily take fire, like any resinous wood.

The English aspen (*P. tremula*) considerably resembles our native aspen; but the buds are somewhat gummy. The Athenian poplar (*P. Græca*) is a tree about 40 feet high, with smaller, more rounded, and equally serrated foliage. The common Black European poplar (*P. nigra*) is also a large, rapidly growing tree, with pale-green leaves slightly notched: the buds expand later than most other poplars, and the young leaves are at first somewhat reddish in color. The Necklace-bearing poplar (*P. monilifera*), so called from the circumstance of the catkins being arranged somewhat like beads in a necklace, is supposed to have been derived from Canada, but there are some doubts respecting its origin: in the south it is generally called the Virginia poplar.

The Lombardy poplar (*P. dilatata*), a native of the banks of the Po, where it is sometimes called the Cypress poplar, from its resembance to that tree, is too well known among us to need any description. Only one sex, the female, has hitherto been introduced into this country; and it has consequently produced no seeds here, but has been entirely propagated by suckers from the root.

THE HORSE-CHESTNUT TREE. *Æsculus.*

Nat. Ord. Æsculaceæ.   *Lin. Syst.* Heptandria, Monogynia.

A large, showy, much admired, ornamental tree, bearing large leaves composed of seven leaflets, and, in the month of May, beautiful clusters of white flowers, delicately mottled with red and yellow. It is a native of Middle Asia, but flourishes well in the temperate climates of both hemispheres. It was introduced into England, probably from Turkey, about the year 1575 : in that country the nuts are often ground into a coarse flour, which is mixed with other food and given to horses that are broken-winded; and from this use the English name of the tree was derived.

A starch has been extracted in considerable quantity from the nuts. The wood is considered valueless in the United States.

The Horse-chestnut is by no means a picturesque tree, being too regularly rounded in its outlines, and too compact and close in its surface, to produce a spirited effect in light and shade. But it is nevertheless one of the most *beautiful* exotic trees which will bear the open air in this climate. The leaves, each made of clusters of six or seven leaflets, are of a fine dark-green color; the whole head of foliage has much grandeur and richness in its depth of hue and massiness of outline; and the regular, rounded, pyramidal shape, is something so different from that of most of our indigenous trees, as to strike the spectator with an air of novelty and distinctness. The great beauty of the Horse-chestnut is the splendor of its inflorescence, surpassing that of almost all our native forest trees: the huge clusters of gay blossoms, which every spring are distributed with such luxuriance and profusion over the surface of the foliage, and at the extremity of the branches, give the whole tree the aspect rather of some monstrous flowering shrub, than of an ordinary tree of the largest size. At that season there can be no more beautiful object to stand singly upon the lawn, particularly if its branches are permitted to grow low down the trunk, and (as they naturally will as the tree advances) sweep the green sward with their drooping foliage. Like the lime tree, however, care must be taken, in the modern style, to introduce it rather sparingly in picturesque plantations, and then only as a single tree, or upon the margin of large groups, masses, or plantations; but it may be more freely used in grounds in the graceful style, for which it is highly suitable. When handsome avenues or straight lines are wanted, the Horse-chestnut is again admirably suited, from its symmetry and regularity. It is, therefore, much and justly valued for these purposes in our towns and cities, where its deep shade and beauty of blossom are peculiarly desirable, the only objection to it being the early fall of its leaves. The Horse-chestnut is very interesting in its mode of growth. The large buds are thickly covered in winter with a resinous gum, to protect them from the cold and moisture; in the spring these burst open, and the whole growth of the young shoots, leaves, flowers, and all, is completed in about three or four weeks. When the leaves first unfold, they are clothed with a copious cotton-like down, which falls off when they have attained their full size and development.

The growth of the Horse-chestnut is slow for a soft-wooded tree, when the trees are young; after five or six years, however, it advances with more rapidity, and in twenty years forms a beautiful and massy tree. It prefers a strong, rich, loamy soil, and is easily raised from the large nuts, which are produced in great abundance.

There are several species of Horse-chestnut, but the common one (*Æsculus Hippocastanum*) is incomparably the finest. The American sorts are the following: (*Æsculus Ohioensis*,) or Ohio Buckeye, as it is called in the western states; a small sized tree, with palmated leaves consisting of *five* leaflets, and pretty, bright yellow flowers, with red stamens. The fruit is about half the size of the exotic species. The Red-flowered Horse-chestnut (*Æsculus rubicunda*) is a small tree with scarlet flowers; and the Smooth-leaved (*Æ. glabra*) has pale yellow flowers. All the foregoing have prickly fruit. Besides these are two small Horse-chestnuts with smooth fruit, which thence properly belong to the genus *Pavia*, viz. the Yellow-flowered Pavia (*P. lutea*) of Virginia and the southern states; and the Red-flowered (*P. rubra*), with pretty clusters of reddish flowers; both these have leaves resembling those of the Horse-chestnut, except in being divided into five leaflets, instead of seven. There are some other species, which are, however, rather shrubs than trees.

The Birch Tree. *Betula.*

Nat. Ord. Betulaceæ. *Lin. Syst.* Monœcia, Polyandria.

The Birch trees are common inhabitants of the forests of all cold and elevated countries. They are remarkable for their smooth, silvery-white, or reddish colored stems, delicate and pliant spray, and small, light foliage. There is no deciduous tree which will endure a more rigorous climate, or grow at a greater elevation above the level of the sea. It is found growing in Greenland and Kamschatka, as far north as the 58th and 60th degree of latitude, and on the Alps in Switzerland, according to that learned botanist, M. DeCandolle, at the elevation of 4,400 feet. It is undoubtedly the most useful tree of northern climates. Not only are cattle and sheep sometimes fed upon the leaves, but the Laplander constructs his hut of the branches; the Russian forms the bark into shoes, baskets, and cordage for harnessing his reindeer; and the inhabitants of Northern Siberia, in times of scarcity, grind it to mix with their oatmeal for food. In this country the birch is no less useful. The North American Indian, and all who are obliged to travel the wild, unfrequented portions of British America,—who have to pass over rapids, and make their way through the wilderness from river to river,—find the canoe made of the birch bark, the lightest, the most durable, and convenient vessel, for these purposes, in the world.*

---

* The following interesting description of their manufacture, we quote from Michaux. "The most important purpose to which the Canoe birch is applied, and one in which its place is supplied by no other tree, is the construction of

The wood of our Black birch is by far the finest; and, as it assumes a beautiful rosy color when polished, and is next in texture to the wild Cherry tree, it is considerably esteemed among cabinet-makers in the eastern states, for chairs, tables, and bedsteads.

In Europe, the sap of the birch is collected in the spring, in the same manner as that of the maple in this country, boiled with sugar and hops, and fermented with the aid of yeast. The product of the fermentation is called *birch wine*, and is described as being a remarkably pleasant and healthy beverage.

Though perhaps too common in some districts of our country to be properly regarded as an ornamental tree, yet in others where it is less so, the birch will doubtless be esteemed as it deserves. With us it is a great favorite; and we regard it as a very elegant and graceful tree, not less on account of the silvery white bark of several species, than from the extreme delicacy of the spray, and the pleasing lightness and airiness of the foliage. In all the species, the branches have a tendency to form those graceful curves which contribute so much to the beauty

canoes. To procure proper pieces, the largest and smoothest trunks are selected; in the spring, two circular incisions are made several feet apart, and two longitudinal ones, on opposite sides of the tree: after which, by introducing a wedge, the bark is easily detached. These plates are usually ten or twelve feet long, and two feet nine inches broad. To form canoes, they are stitched together with fibrous roots of the white spruce, about the size of a quill, which are deprived of the bark, split, and suppled in water. The seams are coated with resin of the Balm of Gilead. Great use is made of these canoes by the savages, and the French Canadians, in their long journeys through the interior of the country: they are light, and very easily transported on the shoulders from one lake to another, which is called the portage. A canoe calculated for four persons, with their baggage, weighs from forty to fifty pounds; and some of them are made to carry fifteen passengers."

of trees; but the European weeping birch is peculiarly pleasing as it grows old, on that account. It is this variety which Coleridge pronounces,

> "————Most beautiful
> Of forest trees—the Lady of the woods."

And Bernard Barton, speaking of our native species, says,

> ————" See the beautiful Birch tree fling
> Its shade on the grass beneath—
> Its glossy leaf, and its silvery stem;
> Dost thou not love to look on them?"

The American sorts, and particularly the Black birch, start into leaf very early in the spring, and their tender green is agreeable to the eye at that season; while the swelling buds and young foliage in many kinds, give out a delicious, though faint perfume. Even the blossoms, which hang like little brown tassels from the drooping branches, are interesting to the lover of nature.

> " The fragrant birch above him hung
> Her tassels in the sky,
> And many a vernal blossom sprung,
> And nodded careless by."
>
> BRYANT.

Nothing can well be prettier, seen from the windows of the drawing-room, than a large group of trees, whose depth and distance is made up by the heavy and deep masses of the ash, oak, and maple; and the portions nearest the eye or the lawn terminated by a few birches, with their sparkling white stems, and delicate, airy, drooping foliage. Our White birch, being a small tree, is very handsome in such situations, and offers the most pleasing variety to the eye, when seen in connexion with other foliage. Several kinds, as the Yellow and the Black birches, are really stately trees, and form fine groups by themselves. Indeed, most beautiful and varied masses might be formed by collecting together all the different kinds, with their characteristic barks, branches, and foliage.

As an additional recommendation, many of these trees grow on the thinnest and most indifferent soils, whether moist or dry; and in cold, bleak, and exposed situations, as well as in warm and sheltered places.

We shall enumerate the different kinds as follows :—

The Canoe birch, *Boleau à Canot*, of the French Canadians (*B. papyracea*), sometimes also called the Paper birch, is, according to Michaux, most common in the forests of the eastern states, north of latitude 43°, and in the Canadas. There it attains its largest size, sometimes seventy feet in height, and three in diameter. Its branches are slender, flexible, covered with a shining brown bark, dotted with white; and on trees of moderate size, the bark of the trunk is of a brilliant white; it is often used for roofing houses, for the manufacture of baskets, boxes, etc., besides its most important use for canoes, as already mentioned. The leaves, borne on petioles four or five lines long, are of a middling size, oval, unequally denticulated, smooth, and of a dark green color.

The White birch (*B. populifolia*) is a tree of much smaller size, generally from twenty to thirty-five feet in height: it is found in New York and the other middle states, as well as at the north. The trunk, like the foregoing, is covered with silvery bark; the branches are slender, and generally drooping when the tree attains considerable size. The leaves are smooth on both surfaces, heart-shaped at the base, very acuminate, and doubly and irregularly toothed. The petioles are slightly twisted, and the leaves are almost as tremulous as those of the aspen. It is a beautiful small tree for ornamental plantations.

The common Black or Sweet birch. (*B. lenta.*) This is the sort most generally known by the name of the birch, and is widely diffused over the middle and southern states. In color and appearance the bark much resembles that of the cherry tree; on old trees, at the close of winter, it is frequently detached in transverse portions, in the form of hard ligneous plates six or eight inches broad. The leaves, for a fortnight after their appearance, are covered with a thick silvery down, which disappears soon after. They are about two inches long, serrate, heart-shaped at the base, acuminate at the summit, and of a pleasing tint and fine texture. The wood is of excellent quality, and Michaux recommends its introduction largely into the forests of the north of Europe.

The Yellow birch (*B. lutea*) grows most plentifully in Nova Scotia, Maine, and New Brunswick, on cool, rich soils, where it is a tree of the largest size. It is remarkable for the color and arrangement of its outer bark, which is of a brilliant golden yellow, and is frequently seen divided into fine strips rolled backwards at the end, but attached in the middle. The leaves are about three and a half inches long, two and a half broad, ovate, acuminate, and bordered with sharp and irregular teeth. It is a beautiful tree, with a trunk of nearly uniform diameter, straight, and destitute of branches for thirty or forty feet.

The Red birch (*B. rubra*) belongs chiefly to the south, being scarcely ever seen north of Virginia. It prefers the moist soil of river banks, where it reaches a noble height.

It takes its name from the cinnamon or reddish color of the outer bark on the young trees; when old it becomes rough, furrowed, and greenish. The leaves are light green on the upper surface, whitish beneath, very pointed at the end, and terminated at the base in an acute angle. The twigs are long, flexible, and pendulous; and the limbs of a brown color, spotted with white.

The European White birch. (*B. alba.*) This species, the common birch tree of Europe, is intermediate in appearance and qualities between our Canoe birch and White birch. The latter it resembles in its foliage, the former in its large size and the excellence of its wood. There is a distinct variety of this, to which we have alluded, called the Weeping birch (*Var. pendula*), which is very rapid in its growth, and highly graceful in its form. From the great beauty of our native species, this is perhaps the only European sort which it is very desirable to introduce into our collections.

The Alder Tree. *Alnus.*

Nat. Ord. Betulaceæ. *Lin. Syst.* Monœcia, Tetrandria.

The alder tree is a native of the whole of Europe, where it grows to the altitude of from thirty to sixty feet. Our common Black alder (*A. glauca*), and Hazel-leaved alder (*A. serrulata*), are low shrubs of little value or interest. This, however, is a neat tree, remarkable for its love of moist situations, and thriving best in places even too wet for the willows; although it will also flourish on dry and elevated soils. The leaves are roundish in form, wavy, and serrated in their margins, and dark green in color. The tree rapidly forms an agreeable pyramidal head of foliage, when growing in damp situations. As it is a foreign tree we shall quote from Gilpin its character in scenery. "The alder," says he, "loves a low, moist soil, and frequents the banks of rivers, and will flourish in the poorest forest swamps where nothing else will grow. It is perhaps the most picturesque of any of the aquatic tribe, except the weeping willow. He who would see the alder in perfection must follow the banks of the Mole in Surrey, through the sweet vales of Dorking and Mickleham, into the groves of Esher. The Mole, indeed, is far from being a beautiful river; it is a silent and sluggish stream, but what beauty it has it owes greatly to the alder, which everywhere fringes its meadows, and in many places forms very pleasing scenes. It is always associated in our minds with river scenery, both of that tranquil description most frequently to be met with in the vales of England, and with that wider and more stirring cast which is to be found amidst the deep glens and ravines of Scotland; and nowhere is this tree found in greater perfection than on the wild banks of the river Findhorn and its tributary streams, where scenery of the most romantic description everywhere prevails."*

Although the beauty of the alder is of a secondary kind, it is worth occasional introduction into landscapes where there is much water to be planted round, or low running streams to cover with foliage. In these damp places, like the willow, it grows very well from truncheons or large limbs, stuck in the ground, which take root and become trees speedily. There are two principal varieties, the

* Lauder's Gilpin, i. p. 136.

common alder (*A. glutinosa*), and the cut-leaved alder (*A. glutinosa laciniata*). The latter is much the handsomer tree, and is also the rarest in our nurseries.

The Maple Tree. *Acer.*

Nat. Ord. Aceraceæ.   *Lin. Syst.* Polygamia, Monœcia.

The great esteem in which the maples are held in the middle states, as ornamental trees, although they are by no means uncommon in every piece of woods of any extent, is a high proof of their superior merits for such purposes. These consist in the rapidity of their growth, the beauty of their form, the fine verdure of their foliage, and in some sorts, the elegance of their blossoms. Among all the species, both native and foreign, we consider the Scarlet-flowering maple as decidedly the most ornamental species. In the spring this tree bursts out in gay tufts of red blossoms, which enliven both its own branches and the surrounding scene long before a leaf is seen on other deciduous trees, and when the only other appearances of vegetation are a few catkins of some willows or poplars swelling into bloom. At that season of the year the Scarlet maple is certainly the most beautiful tree of our forests. Besides this, it grows well either in the very moist soil of swamps, or the dry one of upland ridges, forms a fine clustering head of foliage, and produces an ample and delightful shade; while it is also as little infected by insects of any description as any other tree. The latter advantage, the Sugar maple and our other varieties equally possess. As a handsome spreading tree, perhaps the White maple deserves most praise, its outline and surface being, in many cases, quite picturesque. There is no quality, however, for which the American maples are entitled to higher consideration as desirable objects in scenery, than for the exquisite beauty which their foliage assumes in autumn, as it fades and gradually dies off. At the first approach of cold we can just perceive a bright yellow stealing over the leaves, then a deeper golden tint, then a few faint blushes, until at length the whole mass of foliage becomes one blaze of crimson or orange.

> "Tints that the maple woods disclose
> Like opening buds or fading rose,
> Or various as those hues that dye
> The clouds that deck a sunset sky."

The contrast of coloring exhibited on many of our fine river shores in a warm dry autumn, is perhaps superior to anything of the kind in the world: and the leading and most brilliant colors, viz. orange and scarlet, are produced by maples. Even in Europe, they are highly valued for this autumnal appearance, so different from that of most of the trees of the old world. Very beautiful effects can be produced by planting the Scarlet and Sugar maples in the near neighborhood of the ash, which, as we have already noticed, assumes a fine brownish purple; of the sycamore, which is yellow, and some of the oaks, which remain green for a long time: if to these we add a few evergreens, as the White pine and hemlock, to produce depth, we shall have a kind of kaleidoscope ground, harmonious and beautiful as the rainbow.

When the maple is planted to grow singly on the lawn, or in small groups, it should never be trimmed up ten or twenty feet high, a very common practice in some places, as this destroys half its beauty; but if it be suffered to branch out quite low down, it will form a very elegant head. The maple is well suited to scenes expressive of graceful beauty, as they unite to a considerable variation of surface, a pleasing softness and roundness of outline. In bold or picturesque scenes, they can be employed to advantage by intermingling them with the more striking and majestic forms of the oak, etc., where variety and contrast is desired. The European sycamore, which is also a maple, has a coarser foliage, and more of strength in its growth and appearance: it perhaps approaches nearer in general expression and effect to the plane tree, than to our native maples.

It is unnecessary for us to recommend this tree for avenues, or for bordering the streets of cities, as its general prevalence in such places sufficiently indicates its acknowledged claims for beauty, shade, and shelter. It bears pruning remarkably well, and is easily transplanted, even when of large size, from its native woods or swamps. The finest trees, however, are produced from seed.

The Sugar maple (*Acer saccharinum*) is a very abundant tree in the northern states and the Canadas, where it sometimes forms immense forests. The bark is white; the leaves four or five inches broad, and five-lobed; varying, however, in size according to the age of the tree. The flowers are small, yellowish, and suspended by slender drooping peduncles. The seed is contained in two capsules united at the base, and terminated in a membranous wing; they are ripe in October. From certain parts of the trunks of old Sugar maples, the fine wood called *bird's-eye maple* is taken, which is so highly prized by the cabinet-makers; and the sap, which flows in abundance from holes bored in the stem of the tree early in March, produces the well-known *maple sugar*. This can be clarified, so as to equal that of the cane in flavor and appearance; and it has been demonstrated that the planting of maple orchards, for the production of sugar, would be a profitable investment.

The Scarlet-flowering maple (*A. rubrum*) is found chiefly on the borders of rivers, or in swamps; the latter place appears best suited to this tree, for it there often attains a very large size: it is frequently called the Soft maple or Swamp maple. The blossoms come out about the middle of April while the branches are yet bare of leaves, and their numerous little pendulous stamens appear like small tufts of scarlet or purple threads. The leaves somewhat resemble those of the Sugar maple, but are rather smaller, and only three or four lobed, glaucous or whitish underneath, and irregularly toothed on the margin. This tree may easily be distinguished when young from the former, by the bark of the trunk, which is grey, with large whitish spots. Its trunk, in the choicest parts, furnishes the beautiful wood known as the *curled maple*.

The White or Silver-leaved maple. (*A. eriocarpum*.) This species somewhat resembles the Scarlet-flowering maple, and they are often confounded together in the eastern and middle states, where it grows but sparingly. West of the Alleghany mountains it is seen in perfection, and is well known as the White maple. Its flowers are very pale in color, and much smaller than those of the foregoing sorts. The leaves are divided into four lobes, and have a beautiful white under surface. Michaux, speaking of this tree, says: "In no part of the United

States is it more multiplied than in the western country, and nowhere is its vegetation more luxuriant than on the banks of the Ohio. There, sometimes alone and sometimes mingled with the willow, which is found along these waters, it contributes singularly, by its magnificent foliage, to the embellishment of the scene. The brilliant white of the leaves beneath, forms a striking contrast with the bright green above; and the alternate reflection of the two surfaces in the water, heightening the beauty of this wonderful moving mirror, aids in forming an enchanting picture, which, during my long excursions in a canoe in these regions of solitude and silence, I contemplated with unwearied admiration."* There, on those fine, deep, alluvial soils, it often attains twelve or fifteen feet in circumference.

As an ornamental variety, the Silver-leaved maple is one of the most valuable. It is exceedingly rapid in its growth, often making shoots six feet long in a season; and the silvery hue of its foliage, when stirred by the wind, as well as its fine, half drooping habit, render it highly interesting to the planter. Admirable specimens of this species may be seen in the wide streets of Burlington, N. J.

The Moose wood, or Striped maple (*A. striatum*), is a small tree with beautifully striped bark. It is often seen on the mountains which border the Hudson, but abounds most profusely in the north of the continent. *Acer nigrum* is the Black sugar tree of Genesee. *A. Negundo*,† the Ash-leaved maple, has handsome pinnated foliage of a light green hue; it forms a pleasing tree of medium size. These are our principal native species ‡

---
* N. A. Sylva, i. 214.   † *Negundo fraxinifolium.*
‡ Mr. Douglas has discovered a very superb maple (*A. macrophyllum*), on the Columbia river, with very large leaves, and fine fragrant yellow blossoms.

Among the finest foreign sorts is the Norway maple (*A. platanoides*), with leaves intermediate in appearance between those of the plane tree and Sugar maple. The bark of the trunk is brown, and rougher in appearance than our maples, and the tree is more loose and spreading in its growth; it also grows more rapidly, and strongly resembles at a little distance, the button-wood in its young state. Another interesting species is the sycamore tree or Great maple (*A. pseudo-platanus*). The latter also considerably resembles the plane; but the leaves, like those of the common maple, are smoother. They are five-lobed, acute in the divisions, and are placed on much longer petioles than those of most of the species. The flowers, strung in clusters like those of the common currant, are greenish in color. It is much esteemed as a shade-tree in Scotland and some parts of the Continent, and grows with vigor, producing a large head, and widely spreading branches.

The Locust Tree. *Robinia.*

Nat. Ord. Leguminosæ.   *Lin. Syst.* Diadelphia, Decandria.

This is a well-known American tree, found growing wild in all of the states west of the Delaware River. It is a tree of secondary size, attaining generally the height of forty or fifty feet. The leaves are pinnated, bluish-green in color, and are thinly scattered over the branches. The white blossoms appear in June, and are highly fragrant and beautiful; and from them the Paris perfumers distil an *extrait* which greatly resembles orange-flower water, and is used for the same purposes.

As an ornamental tree we do not esteem the locust highly. The objections to it are, 1st, its meagreness and lightness of foliage, producing but little shade; secondly, the extreme brittleness of its branches, which are liable to be broken and disfigured by every gale of wind; and lastly, the abundance of suckers which it produces. Notwithstanding these defects, we would not entirely banish the locust from our pleasure-grounds; for its light foliage of a fresh and pleasing green may often be used to advantage in producing a variety with other trees; and its very fragrant blossoms are beautiful, when in the beginning of summer they hang in loose pendulous clusters from among its light foliage. These will always speak sufficiently in its favor to cause it to be planted more or less, where a variety of trees is desired. It should, however, be remembered that the foliage comes out at a late period in spring, and falls early in autumn, which we consider objections to any tree that is to be planted in the close vicinity of the mansion. It is valuable for its extremely rapid growth when young; as during the first ten or fifteen years of its life it exceeds in thrifty shoots almost all other forest trees: but it is comparatively short-lived, and in twenty years' time many other trees would completely overtop and outstrip it. It is easily propagated by seed, which is by far the best mode of raising it, and it prefers a deep, rich, sandy loam.*

---

* There is a great difference in the growth of this tree. In cold or indifferent soils it presents a rough and rugged aspect; but in deep, warm, sandy soils it becomes quite another tree in appearance. The highest specimens we have ever seen are now growing in such soil on the estate of J. P. Derwint, Esq., at

198   LANDSCAPE GARDENING.

As a timber tree of the very first class, the locust has but few rivals. It is found to be stronger and more durable than the best oak or Red cedar ; while it is lighter and equally durable with the Live oak of the south. Its excellency for ship-building is therefore unsurpassed ; and as much of the timber as can be procured of sufficient size, commands a high price for that purpose. Great use is also made of it in tree-nails (the wooden pins which fasten the side planks to the ship's frame), and it is now extensively substituted for the iron ones formerly used for that purpose ; a considerable quantity of the wood is now even exported to England for this purpose. For posts it is more durable than the Red cedar, and is therefore in high estimation for fencing. In France, where the tree was introduced by Jean Robin, herbalist to Henry IV. (whence the name *Robinia*), it is much cultivated for the poles used in supporting the grapes in vineyards. It has the remarkable property, says Michaux, of beginning from the third year to convert its sap into perfect wood ; which is not done by the elm, oak, beech, or chestnut, until after the tenth or fifteenth year. Hence excellent and durable timber can be obtained from this tree in a shorter period than from any other.*

Fishkill Landing, on the banks of the Hudson, New York. Some specimens there measure 90 feet, which is higher than Michaux saw on the deep alluvials in Kentucky, where they are natives. The finest single tree is one standing in front of the mansion at Clermont, on the Hudson, which is four feet in diameter.

* Cobbett, who, *en passant*, though a most remarkable man, was as great a quack in gardening as the famous pill-dealers now are in medicine, carried over from this country when he returned to England, a great quantity of seeds of the locust, which he reared and sold in immense quantities. In his " Woodlands," which appeared about that time, he praised its value and utility in the most exaggerated terms, affirming " that no man in America will pretend to say he

The locust can be cultivated to advantage as a timber tree, only upon deep, mellow, and rather rich, *sandy* soils; there, its growth is wonderfully vigorous, and an immense number may be grown upon a small area of ground. In clayey, heavy, or strong loamy soils the tree never attains much size, and is extremely liable to the attacks of the borer, which renders its wood in a great measure valueless. In particularly favorable situations its culture may be made extremely profitable.*

There are but two distinct species of locust which attain

---

ever saw a bit of it in a decayed state;" and that "its wood is *absolutely indestructible by the powers of earth, air, and water.*" "The time will come," he continues, " and it will not be very distant, when the locust tree will be more common in England than the oak; when a man would be thought mad if he used anything but locust in the construction of sills, posts, gates, joists, feet for rick stands, stocks and axletrees for wheels, hop-poles, pales, or for anything where there is liability to rot. This time will not be distant, seeing that the locust tree grows so fast. The next race of children but one, that is to say, those who will be born 60 years hence, will think the locust trees have always been the most numerous trees in England; and some curious writer of a century or two hence, will tell his readers, that wonderful as it may seem, 'the locust was hardly known in England until about the year 1823, when the nation was introduced to a knowledge of it by William Cobbett.' What he will say of me besides, I do not know; but I know he will say this of me. I enter this upon account, therefore, knowing that I am writing for centuries to come."!! For a fuller account of his locust phrensy, we refer our readers to the very complete article on Robinia, in that magnificent work, the "Arboretum Britannicum."

* There is a well known instance of the profit of this tree, which we perceive has found its way into the memoirs of the Agricultural Society of Paris. A farmer on Long Island, some sixty years ago, on the year of his marriage, planted fourteen acres of his farm with the Yellow locust. When his eldest son married at twenty-two, he cut twelve hundred dollars' worth of timber from the field, as a marriage portion, which he gave his son to buy a settlement in Lancaster County, Pennsylvania, then considered a part of the " western country." Three years after the locust grove yielded as much for a daughter; and in this way his whole family were provided for; as the rapidity with which the young suckers grew up fully repaired the breaches made in the fourteen acres.

the size of trees in this country, viz. the Yellow locust (*R. pseud-acacia*), so called from the color of its wood ; and the Honey locust (*R. viscosa*), a smaller tree, with reddish flowers, and branches covered with a viscid honey-like gum. Some pretty varieties of the former have been originated in gardens abroad, among which the Parasol locust (*Var. umbraculifera*) is decidedly the most interesting. We recollect some handsome specimens which were imported by the late M. Parmentier, and grew in his garden at Brooklyn, Long Island. They were remarkable for their unique, rounded, umbrella-like heads, when grafted ten or twelve feet high on the common locust.

There are two pretty distinct varieties of the common Yellow locust, cultivated on the Hudson. That most frequently seen is the *White* variety, which forms a tall and narrow head ; the other is the *Black* locust, with a broad and more spreading head, and larger trunk ; the latter may be seen in fine condition at Clermont. It is a much finer ornamental tree, and appears less liable to the borer than the White variety.

The Three-thorned Acacia Tree. *Gleditschia.*

Nat. Ord. Leguminosæ.   Lin. Syst. Polygamia, Diœcia.

This tree is often called the Three-thorned locust, from some resemblance to the latter tree. Its delicate, doubly pinnate leaves, however, are much more like those of the Acacias, a family of plants not hardy enough to bear our climate. It is a much finer tree in appearance than the common locust, although the flowers are greenish, and inconspicuous, instead of possessing the beauty and fra-

DECIDUOUS ORNAMENTAL TREES. 201

grance of the latter. There is, however, a peculiar elegance about its light green and beautiful foliage, which wafts so gracefully in the summer breeze, and folds up on the slightest shower, that it stands far above that tree in our estimation, for the embellishment of scenery. The branches spread out rather horizontally, in a fine, broad, and lofty head; there are none of the dead and unsightly branches so common on the locust; and the light feathery foliage, lit up in the sunshine, has an airy and transparent look, rarely seen in so large a tree, which sometimes produces very happy effects in composition with other trees. The bark is of a pleasing brown, smooth in surface the branches are studded over with curious, long, triply-pointed thorns, which also often jut out in clusters, in every direction from the trunk of the tree, to the length of four or five inches, giving it a most singular and forbidding look. In winter, these and the long seed-pods, five or six inches in length, which hang upon the boughs at that season, give the whole tree a very distinct character. These pods contain a sweetish substance, somewhat resembling honey; whence the tree has in some places obtained the name of Honey locust, which properly belongs to *Robinia viscosa.*

Another recommendation of this tree, is the variety of picturesque shapes which it assumes in growing up; sometimes forming a tall pyramidal head of 50 or 60 feet, sometimes a low horizontally branched tree, and at others it expands into a wide irregular head, quite flattened at the summit. It does not produce suckers like the locust, and may therefore be introduced into any part of the grounds. When but a limited extent is devoted to a lawn or garden, this tree should be among the first to obtain a place; as one or two Three-thorned Acacias, mingled with other larger and heavier foliage, will at once produce a charming variety.

The Three-thorned Acacia has been strongly recommended for hedges. It is too liable to become thin at the bottom, to serve well for an outer inclosure, but if kept well trimmed, it forms a capital farm fence and protection against the larger animals, growing up in much less time than the hawthorn. Like the locust, it has the disadvantage of expanding its foliage late in the spring. In the strong rich soils which it prefers, it grows very vigorously, and is easily propagated from seeds.

The Three-thorned Acacia (*G. triacanthos*) is the principal species, and is indigenous to the states west of the Alleghanies. *G. monosperma* is another kind, which is scarcely distinguishable from the Three-thorned, except in having one-seeded pods. The seedlings raised from *G. triacanthos* are often entirely destitute of thorns.

There is a fine species called the Chinese (*G. horrida*), with larger and finer foliage, and immense triple thorns, which is interesting from its great singularity. A tree of this kind which we imported, has stood our coldest winters perfectly uninjured, and promises to be beautiful and very hardy. Some noble specimens of the common Three-thorned Acacia may be seen upon the lawn at Hyde Park, the fine seat of the late Dr. Hosack.

The Judas Tree. *Cercis.*

Nat. Ord. Leguminosæ.   *Lin. Syst.* Decandria, Monogynia.

A handsome low tree, about 20 feet in height, which is

found scattered sparsely through warm sheltered valleys, along the Hudson and other rivers of the northern sections of the United States, but most abundantly on the Ohio. It is valuable as an ornamental tree, no less on account of its exceedingly neat foliage, which is exactly heart-shaped, or cordiform, and of a pleasing green tint, than for its pretty pink blossoms. These, which are pea-shaped, are produced in little clusters close to the branches, often in great profusion, early in the spring, before the leaves have expanded. From the appearance of the limbs at that period, it has in some places obtained the name of *Redbud*. It is then one of the most ornamental of trees, and, in company with the Dog-wood, serves greatly to enliven the scene, and herald the advent of the floral season. These blossoms, according to Loudon (*Encycl. of Plants*), having an agreeable poignancy, are frequently eaten in salads abroad, and pickled by the French families in Canada. The name of Judas tree appears to have been whimsically bestowed by Gerard, an old English gardener, who described it in 1596, and relates that "this is the tree whereon Judas did hange himselfe; and not upon the elder tree, as it is said."

There are two species in common cultivation; the American (*C. Canadensis*) and the European (*C. Siliquastrum*). The latter much resembles our native tree. The flowers, however, are deeper in color; the leaves darker, and less pointed at the extremity. It also produces blossoms rather more profusely than the American tree. Both species are highly worthy of a place in the garden, or near the house, where their pleasing vernal influences may be observed.

The Chestnut Tree. *Castanea.*

Nat. Ord. Corylaceæ.   *Lin. Syst.* Monœcia, Polyandria.

The chestnut, for its qualities in Landscape Gardening, ranks with that king of the forest, the oak. Like that tree, it attains an enormous size, and its longevity in some cases is almost equally remarkable. Its fine massy foliage, and sweet nuts, have rendered it a favorite tree since a very remote period. Among the ancients, the latter were a common article of food.

>   ——" Sunt nobis mitia poma,
> *Castaneæ* molles, et pressi copia lactis."
> Virg. Ecl. 1.

They appear to have been in general use, both in a raw and cooked state. In times of scarcity, they probably supplied in some measure the place of bread-stuffs, and were thence highly valued:

> " As for the thrice three angled beech nut shell,
> Or Chestnut's armed huske and hid kernell,
> No squire durst touch, the law would not afford,
> Kept for the court, and for the king's own board."
> Bp. Hall, Sat. B. III. 1.

Even to this day, in those parts of France and Italy nearest the great chestnut forests of the Appenines, these nuts form a large portion of the food which sustains the peasantry, where grain is but little cultivated, and potatoes almost unknown. There a sweet and highly nutritious flour is prepared from them, which makes a delicious bread. Large quantities of the fruit are therefore annually collected in those countries, and dried and stored away for the winter's consumption. Old Evelyn says, "the bread of the flour is exceedingly nutritive: it is a robust food, and makes women well complexioned, as I have read in a good author. They also make fritters of chestnut flour, which they wet with rose-water, and sprinkle with grated parmigans, and so fry them in fresh butter for a delicate." The fruit of the chestnut abounds in saccharine matter; and we learn from a French periodical, that experiments have been made, by which it is ascertained that the kernel yields nearly sixteen per cent. of good sugar.

As a timber tree, this is greatly inferior to the oak, being looser grained, and more liable to decay; and the American wood is more open to this objection than that produced on the opposite side of the Atlantic. It is, however, in general use among us, for posts and rails in fencing; and when the former are charred, they are found to be quite durable.

The finest natural situations for this tree appear to be the mountainous slopes of mild climates, where it attains the greatest possible perfection. Michaux informs us, that the most superb and lofty chestnuts in America are to be found in such situations, in the forests of the Carolinas. Abroad, every one will call to mind the far-famed chestnuts of Mount Etna, of wonderful age and extraordinary size. The great chestnut there, has excited the surprise of numerous travellers; at present, however, it appears to be scarcely more than a mere shell, the wreck of former greatness. When visited by M. Houel (*Arboretum Brit.*), it was in a state of decay, having lost the greater part of its branches, and its trunk was quite hollow. A house was erected in the interior, and some country people resided in it, with an oven, in which, according to the custom of the country, they dried chestnuts, filberts, and other fruits, which they wished to preserve for winter use; using as fuel, when they could find no other, pieces cut with a hatchet from the interior of the tree. In Brydone's time, in 1770, this tree measured two hundred and four feet in circumference. He says it had the appearance of five distinct trees; but he was assured that the space was once filled with solid timber, and there was no bark on the inside. This circumstance of an old trunk, hollow in the interior, becoming separated so as to have the appearance of being the remains of several distinct trees, is frequently met with in the case of very old mulberry trees in Great Britain, and olive trees in Italy. Kircher, about a century before Brydone, affirms that an entire flock of sheep might be inclosed within the Etna chestnut, as in a fold.* (*Arboretum Brit. p.* 1988.)

In considering the chestnut as highly adapted to ornament the grounds of extensive country residences, much that we have already said of the oak will apply to this tree. When young, its smooth stem, clear and bright foliage, and lively aspect, when adorned with the numerous light greenish yellow blossoms, which project beyond the mass of leaves, render it a graceful and beautiful tree.

* One of the most celebrated Chestnut trees on record, is that called the Tortworth Chestnut, in England. In 1772, Lord Ducie, the owner, had a portrait of it taken, which was accompanied by the following description: "The east view of the ancient Chestnut tree at Tortworth, in the county of Gloucester, which measures nineteen yards in circumference, and is mentioned by Sir Robert Aikins in his history of that county, as a famous tree in King John's reign: and by Mr. Evelyn in his Sylva, to have been so remarkable in the reign of King Stephen, 1135, as then to be called the great Chestnut of Tortworth; from which it may reasonably be presumed to have been standing before the Conquest, 1066." This tree is still standing.

It has long been a favorite with the poets for its grateful shade; and as the roots run deep, the soil beneath it is sufficiently rich and sheltered to afford an asylum for the minutest beauties of the woods. Tennyson sweetly says :—

> " That slope beneath the chestnut tall
> Is wooed with choicest breaths of air,
> Methinks that I could tell you all
> The cowslips and the king cups there."

When old, its huge trunk, wide-spread branches, lofty head, and irregular outline, all contribute to render it a picturesque tree of the very first class. In that state, when standing alone, with free room to develope itself on every side, like the oak, it gives a character of dignity, majesty, and grandeur, to the scene, beyond the power of most trees to confer. It is well known that the favorite tree of Salvator Rosa, and one which was most frequently introduced with a singularly happy effect into his wild and picturesque compositions, was the chestnut; sometimes a massy and bold group of its verdure, but oftener an old and storm-rifted giant, half leafless, or a barren trunk coated with a rich verdure of mosses and lichens.

The chestnut in maturity, like the oak, has a great variety of outline ; and no trees are better fitted than these for the formation of grand groups, heavy masses, or wide outlines of foliage. A higher kind of beauty, with more dignity and variety, can be formed of these two genera of trees when disposed in grand masses, than with any other forest trees of temperate climates ; perhaps we may say of any climate.

There is so little difference in the common Sweet chestnut (*Castanea vesca*) of both hemispheres, that they are generally considered the same species. Varieties have been produced in Europe, which far surpass our common chestnuts of the woods in size, though not in delicacy and richness of flavor. Those cultivated for the table in France, are known by the name of *marrons*. These improved sorts of the Spanish chestnut bear fruit nearly as large as that of the Horse-chestnut, inferior in sweetness, when raw, to our wild species, but delicious when roasted. The Spanish chestnut thrives well, and forms a large tree, south of the Highlands of the Hudson, but is rather tender north of this neighborhood. A tree in the grounds at Presque Isle, the seat of William Denning, Esq., Dutchess Co., is now 40 feet high. They may be procured from the nurseries, and we can hardly recommend to our planters more acceptable additions to our nut-bearing forest trees.

The Chinquapin, or Dwarf chestnut (*C. pumila*), is a curious low bush, from four to six feet high. The leaves are nearly the size of the ordinary chestnut, or rather smaller, and the fruit about two-thirds as large. It is indigenous to all the states south of Pennsylvania, and is often found in great abundance. It is a curious little tree, or more properly a shrub, and merits a place in the garden; or it may be advantageously planted for underwood in a group of large trees.

As the chestnut, like the oak, forms strong tap-roots, it is removed with some difficulty. The finest trees are produced from the nut, and their growth is much more rapid when young, than that of the transplanted tree. It prefers a deep sandy loam, rather moist than dry; and will not, like many forest trees, accommodate itself to wet and low situations.

The Osage Orange Tree. *Maclura.*

Nat. Ord. Urticaceæ.   *Lin. Syst.* Diœcia, Tetrandria.

This interesting tree is found growing wild on the Arkansas River, and other western tributaries of the Mississippi, south of St. Louis, where, according to Mr. Nuttall, it attains the height of 50 or 60 feet. The branches are rather light-colored, and armed with spines (produced at every joint) about an inch and a half long. The leaves are long, ovate, and acuminate, or pointed at the extremity; they are deep green, and more glossy and bright than those of the orange. The blossoms are greenish; and the fruit is about the shape and size of a large orange, but the surface much rougher than that fruit. In the south, we are told, it assumes a deep yellow color, and, at a short distance, strikingly resembles the common orange; the specimens of fruit which we have seen growing in Philadelphia, did not assume that fine color; but the appearance of the tree laden with it, is not unlike that of a large orange tree. It was first transplanted into our gardens from a village of the Osage tribe of Indians, whence the common name of Osage orange. The introduction of this tree was one of the favorable results of Lewis and Clarke's Expedition. It was named by them in honor of the late Wm. Maclure, Esq., President of the American Academy of Natural Sciences.

The wood is fine grained, yellow in color, and takes a brilliant polish. It is also very strong and elastic, and on this account the Indians of the wide district to which this tree is indigenous, employ it extensively for bows, greatly preferring it to any other timber. Hence its common name among the white inhabitants is *Bodac*, a corruption of the term *bois d'arc (bow-wood)*, of the French settlers. A fine yellow dye is extracted from the wood, similar to that of the Fustic.

As the Osage orange belongs to the monœcious class of plants, it does not perfect its fruit unless both the male and female trees are growing in the same neighborhood. Many have believed the fruit to be eatable, both from its fine appearance, and from its affinity with and resemblance to that of the bread-fruit; but all attempts to render it pleasant, either cooked or in a raw state, have hitherto failed: it is therefore probably inedible, though not injurious. Perhaps when fully ripened, some mode of preparing it by baking or otherwise, may render it palatable.

As an ornamental tree, the Osage orange is rather too loose in the disposition of its wide-spreading branches, to be called beautiful in its form. But the bright glossy hue of its foliage, and especially the unique appearance of a good sized tree when covered with the large, orange-like fruit, render it one of the most interesting of our native trees; while it has the same charm of rarity as an exotic, since it was introduced from the far west, and is yet but little planted in the United States. On a small lawn, where but few trees are needed, and where it is desirable that the species employed should all be as distinct as possible, to give the whole as much variety as can be obtained in a limited space, such trees should be selected as will not only be ornamental, but combine some other charm, association, or interest. Among such trees, we would by all means give the Osage orange a foremost place. It has the additional recommendation of being a fine shade tree, and of producing an excellent and durable wood.

The stout growth and strong thorns of this tree have been thought indicative of its usefulness for the making of hedges: a method of fencing, which sooner or later must be adopted in many parts of this country: and from the experiments which we have seen made with plants of the Osage orange, we think it likely to answer a very valuable purpose; especially in the middle and southern states. The Messrs. Landreth of Philadelphia have lately offered many thousands of them to the public at a low rate, and we hope to see the matter fairly tested in various parts of the Union.

A rich deep loam is the soil best adapted to the growth of this tree; and as it is rather tender when young (though quite hardy when it attains a considerable size) it should, as far as possible, be planted in a rather sheltered situation. A dry soil is preferable, if it must be placed in a cold aspect, as all plants not perfectly hardy are much injured by the late growth, caused by an excess of moisture and consequent upon an immature state of the wood, which is unable to resist the effects of a severe winter.

The Mulberry Tree. *Morus.*

Nat. Ord. Urticaceæ.   *Lin. Syst.* Monœcia, Tetrandria.

The three principal species of the Mulberry, are the common Red American, the European Black, and the White mulberries. None of them are truly handsome in scenery; and the two latter are generally low spreading trees, valued entirely for the excellency of the fruit, or the suitableness of the foliage for feeding silkworms. Our common mulberry, however, in free, open situations, forms a large, wide-spreading, horizontally branched, and not inelegant tree: the rough, heart-shaped leaves with which it is thickly clothed, afford a deep shade; and it groups well with the lime, the catalpa, and many other round-headed trees. We consider it, therefore, duly entitled to a place in all extensive plantations; while the pleasant flavor of its slightly acid, dark red fruit, will recommend it to those who wish to add to the delicacies of the dessert. The timber of our wild mulberry tree is of the very first quality; when fully seasoned, it takes a dull lemon-colored hue, and is scarcely less durable than the locust or Live oak. Like those trees, it is much valued by ship-builders; and at Philadelphia and Baltimore it commands a high price, for the frame-work, knees, floor-timbers, and tree-nails of vessels. The Red mulberry is much slower in its growth than the locust; but so far as we are aware it is not liable to the attacks of any insect destructive to its timber; and it would probably be found profitable to cultivate it as a timber tree. The locust, it will be remembered, grows thriftily only on peculiar soils, loose, dry, and mellow; the Red mulberry prefers deep, moist, and rich situations. No extensive experiments, so far as we can learn, have been made in its culture; but we would recommend it to the particular attention of those who have facilities for plantations of this kind.

The Black mulberry of Europe (*Morus nigra*) is a low, slow-growing tree, with rough leaves, somewhat resembling those of our Red mulberry, but more coarsely serrated, and often found divided into four or five lobes; while the leaves, which are not heart-shaped on our native species, are generally three-lobed. The European mulberry bears a fruit four or five times as large as the American, full of rich, sweet juice. It has long been a favorite in England, and is one of the most healthy and delicious fruits of the season. Glover says:

>———" There the flushing peach,
> The apple, citron, almond, pear, and date,
> Pomegranates, *purple mulberry*, and fig,
> From interlacing branches mix their hues
> And scents, the passengers' delight."
>
> LEONID. B. II.

We regret that so excellent a fruit should be so little cultivated here. It succeeds extremely well in the middle states; and as it ripens at the very period in midsummer when fruits are scarcest, there can be no more welcome addition to our pomonal treasures, than its deep purple and luscious berries. According to Loudon, it is a tree of great durability; in proof of which he quotes a specimen at Sion House, 300 years old, which is supposed to have been planted in the 16th century by the botanist Turner.

The White mulberry (*M. alba*) is the species upon the leaves of which the silkworms are fed. The fruit is insipid and tasteless, and the tree is but little cultivated to embellish ornamental plantations, though one of the most useful in the world, when its importance in the production of silk is taken into account. There are a great number of varieties of this species to be found in the different nurseries and silk plantations; among them the Chinese mulberry (*M. multicaulis*) grows rapidly, but scarcely forms more than a large shrub at the north; and its very large, tender, and soft green foliage is interesting in a large collection. The fruit is, we believe, of no importance; but it is the most valuable of all mulberries as food for the silkworm, while its growth is the most vigorous, and its leaves more easily gathered than those of any other tree of the genus.

The Paper Mulberry Tree. *Broussonetia.*

Nat. Ord. Urticaceæ.   *Lin. Syst.* Diœcia, Tetrandria.

The Paper mulberry is an exotic tree of a low growth, rarely exceeding twenty-five or thirty feet, indigenous to Japan and the South Sea Islands, but very common in our gardens. It is remarkable for the great variety of forms exhibited in its foliage; as upon young trees it is almost impossible to find two exactly alike, though the prevailing outlines are either heart-shaped, or more or less deeply cut or lobed. These leaves are considered valueless for feeding the silkworm; but in the South Seas the bark is woven into dresses worn by the females; and in China and Japan extensive use is made of it in the manufacture of a paper of the softest and most beautiful texture. This is fabricated from the inner bark of the young shoots, which is first boiled to a soft pulp, and then submitted to processes greatly similar to those performed in our paper-mills. This tree blossoms in spring and ripens its fruit in the month of August. The latter is dark scarlet, and quite singular and ornamental, though of no value. The genus is diœcious; and the reason why so few fruit-bearing trees are seen in the United States, is because we generally cultivate only one of the sexes, the female. M. Parmentier, however, who introduced the male plant from Europe, disseminated it in several parts of the country; and the beauty of the tree has thereby been augmented by the interest which it possesses when laden with its long, hairy berries.

The value of the Paper mulberry, in ornamental plantations, arises from its exotic look, as compared with other trees, from the singular diversity of its foliage, the beauty of its reddish berries, and from the rapidity of its growth. It is deficient in hardiness for a colder climate than that of New York; but further south it is considerably esteemed as a shade-tree for lining the side-walks in cities. In winter its light fawn or ash-colored bark, mottled with patches of a darker grey, contrasts agreeably with other trees. It has little picturesque beauty, and should never be planted in quantities, but only in scattered specimens, to give interest and variety to a walk in the lawn or shrubbery.

The Sweet Gum Tree. *Liquidambar.*

Nat. Ord. Platanaceæ.   *Lin. Syst.* Monœcia, Polyandria.

According to Michaux,* the Sweet gum is one of our most extensively diffused trees. On the seashore it is seen as far north as Portsmouth; and it extends as far south as the Gulf of Mexico and the Isthmus of Darien. In many of the southern states it is one of the commonest trees of the forest; it is rarely seen, however, along the banks of the Hudson (except in New Jersey), or other large streams of New York. It is not unlike the maple in general appearance, and its palmate, five-lobed leaves are in outline much

* N. A. Sylva, i. 315.

like the Sugar maple, though darker in color and firmer in texture. It may also be easily distinguished from that tree, by the curious appearance of its secondary branches, which have a peculiar roughness, owing to the bark attaching itself in plates edgewise to the trunk, instead of laterally, as in the usual manner. The fruit is globular, somewhat resembling that of the buttonwood, but much rougher, and bristling with points. The male and female catkins appear on different branches of the same tree early in spring.

This tree grows in great perfection in the forests of New Spain. It was first described by a Spanish naturalist, Dr. Hernandez, who observed that a fragrant and transparent gum issued from its trunk in that country, to which, from its appearance, he gave the name of liquid amber. This is now the common name of the tree in Europe; and the gum is at present an article of export from Mexico, being chiefly valued in medicine as a styptic, and for its healing and balsamic properties. "This substance, which in the shops is sometimes called the white balsam of Peru, or liquid storax, is, when it first issues from the tree, perfectly liquid and clear, white, with a slight tinge of yellow, quite balsamic; and having a most agreeable fragrance, resembling that of ambergris or styrax. It is stimulant and aromatic, and has long been used in France as a perfume, especially for gloves."[*] In the middle states a fragrant substance sometimes exudes from the leaves, and, by incision, small quantities of the gum may be procured from the trunk; but a warmer climate appears to be necessary to its production in considerable quantities.

We hardly know a more *beautiful* tree than the Liquid

[*] Arboretum Brit. 2051.

amber in every stage of its growth, and during every season of the year. Its outline is not picturesque or graceful, but simply beautiful, more approaching that of the maple than any other: it is, therefore, a highly pleasing, round-headed or tapering tree, which unites and harmonizes well with almost any others in composition; but the chief beauty lies in the foliage. During the whole of the summer months it preserves, unsoiled, that dark glossy freshness which is so delightful to the eye; while the singular, regularly palmate form of the leaves readily distinguishes it from the common trees of a plantation. But in autumn it assumes its gayest livery, and is decked in colors almost too bright and vivid for foliage; forming one of the most brilliant objects in American scenery at that period of the year. The prevailing tint of the foliage is then a deep purplish red, unlike any symptom of decay, and quite as rich as is commonly seen in the darker blossoms of a Dutch parterre. This is sometimes varied by a shade deeper or lighter, and occasionally an orange tint is assumed. When planted in the neighborhood of our fine maples, ashes, and other trees remarkable for their autumnal coloring, the effect, in a warm, dry autumn, is almost magical. Whoever has travelled through what are called the pine barrens of New Jersey in such a season, must have been struck with the gay tints of the numberless forest trees, which line the roads through those sandy plains, and with the conspicuous beauty of the Sweet gum, or Liquidamber.

The bark of this tree when full grown, or nearly so, is exceedingly rough and furrowed, like that of the oak. The wood is fine-grained, and takes a good polish in cabinet work; though it is not so durable, nor so much esteemed for such purposes, as that of the Black walnut and some other native trees. The average height of full grown trees is about 35 or 40 feet.

Liquidambar styraciflua is the only North American species. It grows most rapidly in moist or even wet situations, though it will accommodate itself to a drier soil.

THE WALNUT TREE. *Juglans.*

Nat. Ord. Juglandaceæ.   *Lin. Syst.* Monœcia, Polyandria.

The three trees which properly come under this head, and belong to the genus Juglans, are the Black walnut, the European walnut, and the Butternut.

The Black walnut is one of the largest trees of our native forests. In good soils it often attains a stature of **60** or **70** feet, and a diameter of three or four feet in the trunk, with a corresponding amplitude of branches. The leaves, about a foot or eighteen inches in length, are composed of six or eight pairs of opposite leaflets, terminated by an odd one. They contain a very strong aromatic odor, which is emitted plentifully when they are bruised. The large nut, always borne on the extremity of the young shoots, is round, and covered with a thick husk; which, instead of separating into pieces, and falling off like those of the hickory, rots away and decays gradually. The kernel of the Black walnut, too well known to need any description here, is highly esteemed, and is even considered by some persons to possess a finer flavor than any other walnut.

The timber of this tree is very valuable : when well seasoned it is as durable as the White oak, and is less liable to the attacks of sea-worms, etc., than almost any other; it is, therefore, highly esteemed in naval architecture for certain purposes. But its great value is in cabinet work. Its color, when exposed to the air, is a fine, rich, dark brown, beautifully veined in certain parts; and as it takes a brilliant polish, it is coming into general use in the United States for furniture, as well as for the interior finishing of houses.

The Black walnut has strong claims upon the Landscape Gardener, as it is one of the grandest and most massive trees which he can employ. When full grown it is scarcely inferior in the boldness of its ramification or the amplitude of its head to the oak or chestnut; and what it lacks in spirited outline when compared with those trees, is fully compensated, in our estimation, by its superb and heavy masses of foliage, which catch and throw off the broad lights and shadows in the finest manner. When the Black walnut stands alone on a deep fertile soil it becomes a truly majestic tree; and its lower branches often sweep the ground in a graceful curve, which gives additional beauty to its whole expression. It is admirably adapted to extensive lawns, parks, or plantations, where there is no want of room for the attainment of its full size and fair proportions. Its rapid growth and umbrageous foliage also recommend it for wide public streets and avenues.

The European walnut (*J. regia*), or, as it is generally termed here, the *Madeira nut*, is one of the most common cultivated trees of Europe, where it was introduced originally from Persia. It differs from our Black walnut (which, however, it much resembles) in the smooth, grey bark of the stem, the leaves composed of three or four pair of leaflets, and in the very thin-shelled fruit, which, though not exceeding the Black walnut in size, yet contains a much larger kernel, which is generally considered more delicate in flavor. In the interior of France orchards of the walnut are planted, and a considerable commerce is carried on in its products, consisting chiefly of the fruit, of which large quantities are consumed in all parts of Europe. The wood is greatly used in the manufacture of gun-stocks, and in cabinet-making (though it is much inferior to the American walnut for this purpose) ; and the oil extracted from the kernel is in high estimation for mixing with delicate colors used in painting and other purposes.

The European walnut is a noble tree in size, and thickly clad in foliage. It is much esteemed as a shade tree by the Dutch ; and Evelyn, who is an enthusiastic admirer of its beauties, mentions their fondness for this tree as in the highest degree praiseworthy. "The *Bergstras* [*Bergstrasse*], which extends from Heidelberg to Darmstadt, is all planted with walnuts ; for as by an ancient law the Borderers were obliged to nurse up and take care of them, and that chiefly for their ornament and shade, so as a man may ride for many miles about that country under a continual arbor or close walk,—the traveller both refreshed with the fruit and shade. How much such public plantations improve the glory and wealth of a nation! In several places betwixt Hanau and Frankfort in Germany, no young farmer is permitted to marry a wife till he bring proof that he hath planted, and is the father of a stated number of walnut trees."[*]

The nuts are imported into this country in great

---

[*] Hunter's Evelyn, p. 168.

quantities; and as they are chiefly brought from Spain and the Madeiras, they are here almost entirely known by the name of the Madeira nut. The tree is but little cultivated among us, though highly deserving more extensive favor, both on account of its value and beauty. It grows well in the climate of the middle states, and bears freely; a specimen eighteen or twenty years old, in the garden of the author, has reached thirty-five feet in height, and bears two or three bushels of fine fruit annually; from which we have already propagated several hundred individuals. It is not perfectly hardy north of this.

As an ornamental tree, Gilpin remarks, that the warm russet hue of its young foliage makes a pleasing variety among the vivid green of other trees, about the end of May; and the same variety is maintained in summer, by the contrast of its yellowish hue, when mixed in any quantity with trees of a darker tint. It stands best alone, as the early loss of its foliage is then of less consequence, and its ramification is generally beautiful.

The Butternut (*J. cathartica*) belongs to this section, and is chiefly esteemed for its fruit, which abounds in oil, and is very rich and sweet. The foliage somewhat resembles that of the Black walnut, though the leaflets are smaller and narrower. The form of the nut, however, is strikingly different, being oblong, oval, and narrowed to a point at the extremity. Unlike the walnut, the husk is covered with a sticky gum, and the surface of the nut is much rougher than any other of the walnut genus. The bark of the butternut is grey, and the tops of old trees generally have a flattened appearance. It is frequently an uncouth, ill-shapen, and ugly tree in form, though occasionally, also, quite striking and picturesque. And it is well worthy of a place for the excellence of its fruit.*

The Hickory Tree. *Carya.*

Nat. Ord. Juglandaceæ.   *Lin. Syst.* Monœcia, Polyandria.

The hickories are fine and lofty North American trees, highly valuable for their wood, and the excellent fruit borne by some of the species. The timber is extremely elastic, and very heavy, possessing great strength and tenacity. It is not much employed in architecture, as it is peculiarly liable to the attacks of worms, and decays quickly when exposed to moisture. But it is very extensively employed for all purposes requiring great elasticity and strength; as for axletrees, screws, the wooden rings used upon the rigging of vessels, whip-handles, and axe-handles; and an immense quantity of the young poles are employed in the manufacture of hoops, for which they are admirably adapted.

For fuel, no American wood is equal to this in the brilliancy with which it burns, or in the duration or amount of heat given out by it: it therefore commands the highest price in market for that purpose.

The hickories are nearly allied to the walnuts; the

---

* Loudon errs greatly in his Arboretum, in supposing the butternut to be identical with the Black walnut: no trees in the whole American forest are more easily distinguished at first sight. He also states the fruit to be rancid and of little value; but no American lad of a dozen years will accord with him in this opinion.

chief botanical distinction consisting in the covering to the nut, or husk; which in the hickories separates into four valves, or pieces, when ripe, instead of adhering in a homogeneous coat, as upon the Black walnut and butternut. In size and appearance, the hickories rank with the first class of forest trees ; most of them growing vigorously to the height of 60 or 80 feet, with fine straight trunks, well balanced and ample heads, and handsome, lively, pinnated foliage. When confined among other trees in the forest, they shoot up 50 or 60 feet without branches; but when standing singly, they expand into a fine head near the ground and produce a noble, lofty pyramid of foliage, rather rounded at the top. They have all the qualities which are necessary to constitute fine, graceful park trees, and are justly entitled to a place in every considerable plantation.

The most ornamental species are the Shellbark hickory, the Pignut, and the Pecan-nut. The former and the latter produce delicious nuts, and are highly worthy of cultivation for their fruit alone ; while all of them assume very handsome shapes during every stage of their growth, and ultimately become noble trees. Varieties of the Shellbark hickory are sometimes seen producing nuts of twice or thrice the ordinary size; and we have not the least doubt that the fruit might be so improved in size and delicacy of flavor by careful cultivation, as greatly to surpass the European walnut, for the table. This result will probably be attained by planting the nuts of the finest varieties found in our woods, in rich moist soil, kept in high cultivation; as all improved varieties of fruit have been produced in this way, and not, as many suppose, by cultivating the original species. These remarks also apply to the Pecan-nut; a western sort, which thrives well in the middle states, and which produces a nut more delicate in flavor than any other of this continent.

These trees form strong tap-roots, and are, therefore, somewhat difficult to transplant; but they are easily reared from the nut; and, for the reason stated above, this method should be adopted in preference to any other, except in particular cases.

The principal species of the hickory are the following:

The Shellbark hichory (*C. alba*), so called on account of the roughness of its bark, which is loosened from the trunk in long scales or pieces bending outwards at the extremity, and remaining attached by the middle; this takes place, however, only on trees of some size. The leaves are composed of two pair of leaflets, with an odd or terminal one. The scales which cover the buds of the Shellbark in winter, adhere only to the lower half, while the upper half of the bud is left uncovered, by which this sort is readily distinguished from the other species. The hickory nuts of our markets are the product of this tree; they are much esteemed in every part of the Union, and are exported in considerable quantities to Europe. Among many of the descendants of the original Dutch settlers of New York and New Jersey, the fruit is commonly known by the appellation of the *Kisky-tom nut*.*

The Pecan-nut (*Pacainer* of the French), (*C. olivæformis*) is found only in the western states. It abounds on the Missouri, Arkansas, Wabash, and Illinois Rivers, and

---

* In some parts, pleasant social parties which meet at stated times during the winter season, are called Kisky-toms, from the regular appearance of these nuts among the refreshments of the evening.

a portion of the Ohio: Michaux states that there is a swamp of 800 acres on the right bank of the Ohio, opposite the Cumberland river, entirely covered with it. It is a handsome, stately tree, about 60 or 70 feet in height, with leaves a foot or eighteen inches long, composed of six or seven pairs of leaflets much narrower than those of our hickories. The nuts are contained in a thin, somewhat four-sided husk; they are about an inch or an inch and a half long, smooth, cylindrical, and thin-shelled. The kernel is not, like most of the hickories, divided by partitions, and it has a very delicate and agreeable flavor. They form an object of petty commerce between Upper and Lower Louisiana. From New Orleans, they are exported to the West Indies, and to the ports of the United States.*

Besides these two most valuable species, our forests produce the Pignut hickory (*C. porcina*), a lofty tree with five to seven pairs of leaflets, so called from the comparative worthlessness of its fruit; which is very thick-shelled, and generally is left on the ground for the swine, squirrels, etc., to devour. It is easily distinguished in winter by the smaller size of its brown shoots, and its small oval buds. Its wood is considered the toughest and strongest of any of the trees of this section. The thick Shellbark hickory (*C. laciniosa*) resembles much in size and appearance the common Shellbark; but the nuts are double the size, the shell much thicker and yellowish, while that of the latter is white. It is but little known except west of the Alleghanies. The Mockernut hickory (*C. tomentosa*) is so called from the deceptive appearance of the nuts,

* N. A. Sylva, i. 168.

which are generally of large size, but contain only a very small kernel. The leaves are composed of but four pairs of sessile leaflets, with an odd one at the end. The trunk of the old trees is very rugged, and the wood is one of the best for fuel.

The Bitternut hickory (*C. amara*), sometimes called the White hickory, grows 60 feet high in New Jersey. The husk which covers the nut of this species, has four winged appendages on its upper half, and never hardens like the other sorts, but becomes soft and decays. The shell is thin, but the kernel is so bitter that even the squirrels refuse to eat it. The Water Bitternut (*C. aquatica*) is a very inferior sort, growing in the swamps and rice fields of the southern states. The leaflets are serrated, and resemble in shape the leaves of the peach tree. Both the fruit and timber are much inferior to those of all the other hickories.

The Mountain Ash Tree. *Pyrus.**

Nat. Ord. Rosaceæ.   *Lin. Syst.* Icosandria, Di-Pentagynia.

The European Mountain ash (*Pyrus aucuparia*) is an elegant tree of the medium size, with an erect stem, smooth bark, and round head. The leaves are pinnated, four or five inches in length, and slightly resemble those of the ash. The snow-white flowers are produced in large flat clusters, in the month of May, which are thickly

---

* *Sorbus* of the old Botanists.

scattered over the outer surface of the tree, and give it a lively appearance. These are succeeded by numerous bunches of berries, which in autumn turn to a brilliant scarlet, and are then highly ornamental. For the sake of these berries, this tree is a great favorite with birds; and in Germany it is called the *Vogel Beerbaum*, i. e. bird's berry tree, and is much used by bird catchers to bait their springs with.

Twenty-five feet is about the average height of the Mountain ash in this country. Abroad it grows more vigorously; and in Scotland, where it is best known by the name of the Roan or Rowan tree, it sometimes reaches the altitude of 35 or 40 feet. The lower classes throughout the whole of Britain, for a long time attributed to its branches the power of being a sovereign charm against witches; and Sir Thomas Lauder informs us that this superstition is still in existence in many parts of the Highlands, as well as in Wales. It is probable that this tree was a great favorite with the Druids; for it is often seen growing near their ancient mystical circles of stones. The dairymaid, in many parts of England, still preserves the old custom of driving her cows to pasture with a switch of the roan tree, which she believes has the power to shield them from all evil spells.* "Evelyn mentions that it is customary in Wales to plant this tree in churchyards; and Miss Kent in her Sylvan Sketches, makes the following remarks:—'In former times this tree was supposed to be possessed of the property of driving away witches and evil spirits; and this property is alluded to in one of the stanzas of a very ancient song, called the *Laidley Worm of Spindleton's Heughs*.

* Lightfoot, Flora Scotica.

'Their spells were vain; the boys return'd
To the Queen in sorrowful mood,
Crying that " witches have no power
Where there is rowan-tree wood ?"

" The last line of this stanza leads to the true reading of a stanza in Shakspeare's tragedy of Macbeth. The sailor's wife, on the witch's requesting some chestnuts, hastily answers, ' A rown-tree, witch!'—but many of the editions have it, 'aroint thee, witch!' which is nonsense, and evidently a corruption."*

The European Mountain ash is quite a favorite with cultivators here, and deservedly so. Its foliage is extremely neat, its blossoms pretty, and its blazing red berries in autumn communicate a cheerfulness to the season, and harmonize happily with the gay tints of our native forest trees. It is remarkably well calculated for small plantations or collections, as it grows in almost any soil or situation, takes but little room, and is always interesting. "In the Scottish Highlands," says Gilpin, "on some rocky mountain covered with dark pines and waving birch, which cast a solemn gloom on the lake below, a few Mountain ashes joining in a clump and mixing with them, have a fine effect. In summer the light green tint of their foliage, and in autumn the glowing berries which hang clustering upon them, contrast beautifully with the deeper green of the pines: and if they are happily blended, and not in too large a proportion, they add some of the most picturesque furniture with which the sides of those rugged mountains are invested." We have seen the Mountain ash, here, displaying itself in great beauty, mingled with a group of hemlocks, from among the deep green foliage of which, the coral

* Arboretum et Fruticetum, p. 918.

berries of the former seemed to shoot out; their color heightened by the dark back ground of evergreen boughs.

The American Mountain ash (*Pyrus Americana*) is a native of the mountains along the banks of the Hudson, and other cold and elevated situations in the north of the United States: on the Catskill we have seen some handsome specimens near the Mountain House; but generally it does not grow in so comely a shape, or form so handsome a tree as the foreign sort. In the general appearance of the leaves and blossoms, however, it so nearly resembles the European as to be thought merely a variety by some botanists. The chief difference between them appears to be in the color of the fruit, which on our native tree is copper colored or dull purplish red. It may probably assume a handsome shape when cultivated.

The Sorb or Service tree (*Pyrus Sorbus*) is an interesting species of Pyrus, a native of Europe, which is sometimes seen in our gardens, and deserves a place for its handsome foliage and its clusters of fruit; which somewhat resemble those of the Mountain ash, and are often eaten when in a state of incipient decay. The leaves are coarser than those of the Mountain ash, and the tree is larger, often attaining the height of 50 or 60 feet in its native soil.

The White Beam (*Pyrus Aria*) is another foreign species, also bearing bunches of handsome scarlet berries, and clusters of white flowers. The leaves, however, are not pinnated, but simply serrated on the margin. It grows 30 feet high, and as the foliage is dark green on the upper side, and downy white beneath, it presents an effect greatly resembling that of the Silver poplar in a slight breeze. Abroad, the timber is considered valuable; but here it is chiefly planted to produce a pleasing variety among other trees, by its peculiar foliage, and scarlet autumnal fruit.

All the foregoing trees grow naturally in the highest, most exposed, and often almost barren situations. When, however, a rapid growth is desired, they should be planted in a more moist and genial soil. They are easily propagated from the seed, and some of the sorts may be grafted on the pear or hawthorn. The seeds, in all cases, should be sown in autumn.

The Ailantus Tree. *Ailantus.*

Nat. Ord. Xanthoxylaceæ.   *Lin. Syst.* Polygamia, Monœcia.

Ailanto is the name of this tree in the Moluccas, and is said to signify Tree of Heaven; an appellation probably bestowed on account of the rapidity of its growth, and the great height which it reaches in the East Indies, its native country. When quite young it is not unlike a sumac in appearance; but the extreme rapidity of its growth and the great size of its pinnated leaves, four or five feet long, soon distinguish it from that shrub. During the first half dozen years it outstrips almost any other deciduous tree in vigor of growth, and we have measured leading stems which had grown twelve or fifteen feet in a single season. In four or five years, therefore, it forms quite a bulky head, but after that period it advances more slowly, and in 20 years would probably be overtopped by the poplar, the plane, or any other fast growing tree. There are, as yet, no specimens in this country more than 70 feet high; but the trunk shoots up in a fine column, and the head is massy and irregular in outline. In this country it is planted purely for ornament, but we learn that in Europe its wood has been applied to cabinet work; for which, from its close grain and bright satin-like lustre, it is well adapted.* The male and female flowers are borne on separate trees, and both sexes are now common, especially in New York. The male forms the finer ornamental tree, the female being rather low, and spreading in its head.

In New York and Philadelphia, the Ailantus is more generally known by the name of the *Celestial tree*, and is much planted in the streets and public squares. For such situations it is admirably adapted, as it will insinuate its strong roots into the most meagre and barren soil, where few other trees will grow, and soon produce an abundance of foliage and fine shade. It appears also to be perfectly free from insects; and the leaves, instead of dropping slowly, and for a long time, fall off almost immediately when frost commences.

The Ailantus is a picturesque tree, well adapted to produce a good effect on the lawn, either singly or grouped; as its fine long foliage catches the light well, and contrasts strikingly with that of the round-leaved trees. It has a troublesome habit of producing suckers, however, which must exclude it from every place but a heavy sward, where the surface of the ground is never stirred by cultivation.

The branches of this tree are entirely destitute of the small spray so common on most forest trees, and have a singularly naked look in winter, well calculated to fix the attention of the spectator at that dreary season.

* Annales de la Societé d'Horticulture.

The largest Ailantus trees in America are growing in Rhode Island, where it was introduced from China, under the name of the Tillou tree. It has since been rapidly propagated by suckers, and is now one of the commonest ornamental trees sold in the nurseries. The finest trees, however, are those raised from seed.

The Kentucky Coffee Tree. *Gymnocladus.*

Nat. Ord. Leguminosæ.   *Lin. Syst.* Diœcia, Decandria.

This unique tree is found in the western part of the State of New York, and as far north as Montreal, in Canada. But it is seen in the greatest perfection, in the fertile bottoms of Kentucky and Tennessee. Sixty feet is the usual height of the Coffee tree in those soils; and judging from specimens growing under our inspection, it will scarcely fall short of that altitude, in well cultivated situations, anywhere in the middle states.

When in full foliage, this is a very beautiful tree. The whole leaf, doubly compound and composed of a great number of bluish-green leaflets, is generally three feet long, and of two-thirds that width on thrifty trees; and the whole foliage hangs in a well-rounded mass, that would look almost too heavy, were it not lightened in effect by the loose, tufted appearance of each individual leaf. The flowers, which are white, are borne in loose spikes, in the beginning of summer; and are succeeded by ample brown pods, flat and somewhat curved, which contain six or seven large grey seeds, imbedded in a sweet pulpy substance. As the genus is diœcious, it is necessary that both sexes of this tree should be growing near each other, in order to produce seed.

When Kentucky was first settled by the adventurous pioneers from the Atlantic States, who commenced their career in the primeval wilderness, almost without the necessaries of life, except as produced by them from the fertile soil, they fancied that they had discovered a substitute for coffee in the seeds of this tree, and accordingly the name of Coffee tree was bestowed upon it: but when a communication was established with the seaports, they gladly relinquished their Kentucky beverage for the more grateful flavor of the Indian plant; and no use is at present made of it in that manner. It has, however, a fine, compact wood, highly useful in building or cabinet-work.

The Kentucky Coffee tree is well entitled to a place in every collection. In summer, its charming foliage and agreeable flowers render it a highly beautiful lawn tree; and in winter, it is certainly one of the most novel trees, in appearance, in our whole native sylva. Like the Ailantus, it is entirely destitute of small spray, but it also adds to this the additional singularity of thick, blunt, terminal branches, without any perceptible buds. Altogether it more resembles a dry, dead, and withered combination of sticks, than a living and thrifty tree. Although this would be highly monotonous and displeasing, were it the common appearance of our deciduous trees in winter; yet, as it is not so, but a rare and very unique exception to the usual beautiful diversity of spray and ramification, it is highly interesting to place such a tree as the present in the neighborhood of other full-sprayed species, where the curiosity which it excites will add greatly to its value as an interesting object at that period of the year.*

[Fig. 38. The Kentucky Coffee Tree.]

The seeds vegetate freely, and the tree is usually propagated in that manner. It prefers a rich, strong soil, like most trees of the western states.

THE WILLOW TREE. *Salix.*

Nat. Ord. Salicaceæ.   *Lin. Syst.* Diœcia, Diandria.

A very large genus, comprising plants of almost every

* There are some very fine specimens upon the lawn at Dr. Hosack's seat, Hyde Park, N. Y., which have fruited for a number of years. *See Fig.* 38.

stature, from minute shrubs of three or four inches in height, to lofty and wide-spreading trees of fifty or sixty feet.* They are generally remarkable for their narrow leaves, and slender, round, and flexible branches.

There are few of these willows which are adapted to add to the beauty of artificial scenery; but among them are three or four trees, which, from their peculiar character, deserve especial notice. These are the Weeping, or Babylonian willow (*Salix Babylonica*), the White, or Huntington willow (*S. alba*), the Golden willow (*S. vitellina*), the Russell willow (*S. Russelliana*), and the profuse Flowering willow (*S. caprea*).

The above are all foreign sorts, which, however (except the last), have long ago been introduced, and are now quite common in the United States. All of them except the first, have an upright or wavy, spreading growth, and form lofty trees, considerably valued abroad for their timber. The White willow and the Russell willow are very rapid in their growth, and have a pleasing light green foliage. The Golden willow is remarkable for its bright yellow bark, which renders it quite ornamental, even in winter. It is a middle sized tree, and is often seen growing along the road-sides in the eastern and middle states. *Salix caprea* is deserving a place in collections for the beauty of its abundant blossoms at an early and cheerless period in the spring. There are a number of other species found growing in different parts of the Union, which may perhaps possess sufficient interest to recommend themselves to the planter.

* Dr. Barratt of Middletown, Conn., who has paid great attention to the willow, enumerates 100 species, as growing in North America, either indigenous or introduced.

The chief, and indeed almost the only value of these willows in Landscape Gardening, is to embellish low grounds, streams of water, or margins of lakes. When mingled with other trees, they often harmonize so badly from their extremely different habits, foliage, and color, that unless very sparingly introduced, they cannot fail to have a bad effect. On the banks of streams, however, they are extremely appropriate, hanging their slender branches over the liquid element, and drawing genial nourishment from the moistened soil.

"Le saule incliné sur la rive penchante,
Balançant mollement sa tête blanchissante."

In the middle distance of a scene, also, where a stream winds partially hidden, or which might otherwise wholly escape the eye, these trees, if planted along its course, connected as they are in our minds with watery soils, will not fail to direct the attention and convey forcibly the impression of a brook or river, winding its way beneath their shade.

The Weeping willow, however, is at once one of the most elegant, graceful, and interesting trees; elegant in its light and delicate waving foliage; graceful in the soft flowing lines formed by its drooping branches; and interesting by the melancholy, poetical, and scriptural associations connected with it. Every one will call to mind the captivity of the children of Israel, as connected with this tree: "By the waters of Babylon we sat down and wept, O Zion! As for our harps, we hanged them upon the willow trees:" *Psalm* cxxxvii. And the gentle sigh of the faintest breeze through its light foliage, still recalls to the mind the plaintive murmur of those abandoned harps, which one may fancy to have bequeathed their last tones of music to its pensile branches.

Since that period, the willow appears to have been, more or less, consecrated to a tender sentiment of grief,

> "Trailing low its boughs, to hide
> The gleaming marble."

To these offices of pensive melancholy, it appears to be dedicated in almost all countries. The Chinese and other Asiatic nations, and the Turks, as well as the enlightened Europeans, universally plant it in their cemeteries and last places of repose. A French writer thus speaks of it in contrasting its merits for those purposes, with the cypress. "The cypress was long considered as the appropriate ornament of the cemetery; but its gloomy shade among the tombs, and its thick, heavy foliage of the darkest green, inspire only depressing thoughts, and present the image of death under its most appalling form. The Weeping willow, on the contrary, rather conveys a picture of grief for the loss of the departed, than of the darkness of the grave. Its light and elegant foliage flows like the dishevelled hair and graceful drapery of a sculptured mourner over a sepulchral urn; and conveys those soothing, though softly melancholy reflections which have made one of our poets exclaim, 'There is a pleasure even in grief.'"[*] On this passage, Loudon remarks: "Notwithstanding the preference thus given the willow, the shape of the cypress conveying to a fanciful mind the idea of a flame pointing upwards, has been supposed to afford an emblem of the hope of immortality; it is still planted in many church-

[*] Poiteau, "nouveau du Hamel."

yards on the continent, and alluded to in the epitaphs, under this light." *

Abroad, the willow was in ancient days worn by young girls, as a symbol of grief for one of their own sex who died young:

> " Lay a garland on my hearse,
> Of the dismal yew;
> Maidens, willow branches wear,
> Say I died true."

The poets often allude to the willow:

> " A willow garland thou did'st send
> Perfumed last day to me;
> Which did but only this portend,
> I was forsook by thee.
> Since so it is, I'll tell thee what,
> To-morrow thou shalt see
> Me wear the willow, after that
> To die upon the tree." HERRICK.

In landscapes, the Weeping willow is peculiarly expressive of grace and softness. Although a highly beautiful tree, great care must be used in its introduction, to preserve the harmony and propriety of the whole; as nothing could be more strikingly inappropriate than to intermix it frequently with trees expressive of dignity or majesty, as the oak, etc.; where the violent contrast exhibited in the near proximity of the two opposite forms, could only produce discord. The favorite place, where it is most true to nature and itself, is near water, where

> ——— " it dips
> Its pendent boughs, stooping as if to drink." COWPER.

* Arb. Brit.

There, when properly introduced, not in too great abundance, hanging over some rustic bridge, or cool jutting spring, and supported, and brought into harmony with surrounding vegetation by such other graceful and light-sprayed trees as the Birch and Weeping elm, its effect is often surpassingly beautiful and appropriate. There it is one of the first in the vernal season to burst its buds, and mirror its soft green foliage in the flood beneath, and one of the last in autumn to yield its leafy vesture to the chilling frosts, or fitful gusts of approaching winter.

We consider the Weeping willow ill calculated for a place near a mansion which has any claims to size, magnificence, or architectural beauty; as it does not in any way contribute by its form or outline to add to or strengthen such characteristics in a building. The only place where it can be happily situated in this way, is in the case of very humble or inconspicuous cottages, which we have seen much ornamented by being completely hidden, as it were, beneath the soft veil of its streaming foliage.

There is a very singular variety of the Weeping willow cultivated in our gardens, under the name of the Ringlet willow; which is so remarkable in the form of its foliage, and so different from all other trees, that it is well worth a place as a curiosity. Each leaf is curled round like a ring or hoop, and the appearance of a branch in full foliage is not unlike a thinly curled ringlet; whence its common name. It forms a neat, middle-sized tree, with drooping branches, though hardly so pendent as the Weeping willow.

The uses of the willow are extremely numerous. Abroad it is extensively cultivated in coppices, for timber and fuel, for hoops, ties, etc.; and we are informed, that in the northern parts of Europe, and throughout the Russian Empire, the twigs are employed in manufacturing domestic utensils, harness, cables, and even for the houses of the peasantry themselves. From the fibres of the bark, it is said that a durable cloth is woven by the Tartars; and the bark is used for tanning in various parts of the eastern continent.

But by far the most extensive use to which this plant is applied, is in the manufacture of baskets. From the earliest periods it has been devoted to this purpose, and large plantations, or osier-fields, as they are called, are devoted to the culture of particular kinds for this purpose, both in Europe and America. The common Basket willow, an European species (*S. viminalis*), is the sort usually grown for this purpose, but several others are also employed. For the culture of the basket willows, a deep, moist, though not inundated soil is necessary; such as is generally found on the margins of small streams, or low lands. "Ropes and baskets made from willow twigs, were probably among the very earliest manufactures, in countries where these trees abound. The Romans used the twigs for binding their vines, and tying their reeds in bundles, and made all sorts of baskets of them. A crop of willows was considered so valuable in the time of Cato, that he ranks the Salictum, or willow field, next in value to the vineyard and the garden. (Art. *Salix. Arb. Brit.*)

Among us, the European Basket willow is extensively cultivated, and very large plantations are to be seen in the low grounds of New Jersey and Pennsylvania. The wood of some of the tree willows, and particularly that of the Yellow willow, and the Shining willow (*S. lucida*), is

DECIDUOUS ORNAMENTAL TREES.   241

greatly used in making charcoal for the manufacture of gunpowder.

It is almost unnecessary to say that all the willows grow readily from slips or truncheons planted in the ground. So tenacious of life are they, that examples are known where small trees have been taken up and completely inverted, by planting the branches and leaving the roots exposed, which have nevertheless thrown out new roots from the former tops, and the roots becoming branches, the tree grew again with its ordinary vigor.

THE SASSAFRAS TREE. *Laurus.*

Nat. Ord. Lauraceæ.   *Lin. Syst.* Enneandria, Monogynia.

The Sassafras is a neat tree of the middle size, belonging to the same family as the European laurel or Sweet bay; it is found, more or less plentifully, through the whole territory of the United States. In favorable soils, along the banks of the Hudson, it often grows to 40 or 50 feet in height; but in the woods it seldom reaches that altitude. The flowers are yellow, and appear in small clusters in May, and the fruit is a small, deep blue berry, seated on a red footstalk or cup. The bark of the wood and roots has an agreeable smell and taste, and is a favorite ingredient, with the branches of the spruce, in the small beer made by the country people. Medicinally, it is considered antiscorbutic and sudorific; and is thought efficacious in purifying the blood. It was formerly in great repute with practitioners abroad, and large quantities of the bark of the roots were shipped to England; but the demand has of late greatly decreased.

16

The Sassafras is a very agreeable tree to the eye, decked as it is with its glossy, deep green, oval, or three-lobed leaves. When fully grown, it is also quite picturesque for a tree of so moderate a size; as its branches generally have an irregular, somewhat twisted look, and the head is partially flattened, and considerably varied in outline. After ten years of age, this tree always looks older than it really is, from its rough, deeply cracked, grey bark, and rather crooked stem. It often appears extremely well on the borders of a plantation, and mixes well with almost any of the heavier deciduous trees. As it is by no means so common a tree as many of those already noticed, it is generally the more valued, and may frequently be seen growing along the edges of cultivated fields and pastures, appearing to thrive well in any good mellow soil.

The Catalpa Tree. *Catalpa.*

Nat. Ord. Bignoniaceæ.   *Lin. Syst.* Diandria, Monogynia.

A native of nearly all the states south and west of Virginia, this tree has become naturalized also throughout the middle and eastern sections of the Union, where it is generally planted for ornament.

In Carolina it is called the Catawba tree, after the Catawba Indians, a tribe that formerly inhabited that country; and it is probable that the softer epithet now generally bestowed upon it in the north, is only a corruption of that original name.

The leaves of this tree are very large, often measuring six or seven inches broad; they are heart-shaped in form, smooth, and pale green on the upper side, slightly downy beneath. The blossoms are extremely beautiful, hanging, like those of the Horse-chestnut, in massy clusters beyond the outer surface of the foliage. The color is a pure and delicate white, and the inner part of the corolla is delicately sprinkled over with violet, or reddish and yellow spots; indeed, the individual beauty of the flowers is so great when viewed closely, that one almost regrets that they should be elevated on the branches of a large forest tree. When these fall, they are succeeded by bean-like capsules or seed-vessels, which grow ten or twelve inches long, become brown, and hang pendent upon the branches during the greater part of the winter.

The Catalpa never, or rarely, takes a symmetrical form when growing up; but generally forms a wide-spreading head, forty or fifty feet in diameter. Its large and abundant foliage affords a copious shade, and its growth is quite rapid, soon forming a large and bulky tree. In ornamental plantations it is much valued on account of its superb and showy flowers, and is therefore deserving a place in every lawn. It is generally seen to best advantage when standing alone, but it may also be mingled with other large round-leaved trees, as the basswood, etc., when it produces a very pleasing effect. The branches are rather brittle, like those of the locust, and are therefore somewhat liable to be broken by the wind. Accustomed to a warmer climate, the leaves expand late in the spring, and wither hastily when frost approaches; but the soft tint of their luxuriant vegetation is very grateful to the eye, and it appears to be uninjured by the hottest rays of summer. North of this place the Catalpa is rather too tender for exposed situations.

We have seen the Catalpa employed to great advantage in fixing and holding up the loose soil of river banks, where, if planted, it will soon insinuate its strong roots, and retain the soil firmly. In Ohio, experiments have been made with the timber for the posts used in fencing; and it is stated on good authority that it is but little inferior, when well seasoned, to that of the locust in durability.

Michaux mentions that he has been assured that the honey collected from the flowers is poisonous; but this we are inclined to doubt; or at least we have witnessed no ill effects from planting it in abundance in the middle States, in those neighborhoods where bees are kept in considerable numbers.

The Catalpa is very easily propagated from seeds sown in any light soil; and the growth of the young plants is extremely rapid. *C. syringafolia* is the only species.

The Persimon Tree. *Diospyros.*

Nat. Ord. Ebenaceæ.   *Lin. Syst.* Polygamia, Diœcia.

The Highlands of the Hudson, and about the same latitude on the Connecticut, may be considered the northern limits of this small tree. It generally forms a spreading loose head, of some twenty or thirty feet high, in good soils in the middle states; but we have seen a specimen of nearly eighty feet, in the old Bartram Garden at Philadelphia; and fifty feet is probably the average growth on deep fertile lands in the southern states.

The Persimon bears a small, round, dull red fruit, about an inch in diameter, containing six or seven stones; it is insufferably austere and bitter, until the autumnal frosts have mellowed it and lessened its harshness, when it becomes quite palatable. Considerable quantities of the fruit are annually brought into New York market and its vicinity, from New Jersey, and sold: the produce is very abundant, a single tree often yielding several bushels. A strong brandy has been distilled from them; and in the south they are said to enter into the composition of the country beer. For the latter purpose they are pounded up with bran, dried, and kept for use till wanted.

The foliage of the Persimon is handsome; the leaves being four or five inches long, simple, oblong, dark green, and glossy, like those of the orange. The blossoms are green and inconspicuous.

The Persimon has no importance as a tree to recommend it; but it may be admitted in all good collections for its pleasing shining foliage, and the variety which its singular fruit adds to the productions of a complete country residence. The common sort (*D. Virginiana*) grows readily from the seed.

There is an European Species (*Dyosporus Lotus*), with yellow fruit about the size of a cherry, rather less palatable than our native kind. The specimens of this tree, which we have imported, appear too tender to bear our winters unprotected, so that it will probably not prove hardy in the northern states.

The Peperidge Tree. *Nyssa.*

Nat. Ord. Santalaceæ.   *Lin. Syst.* Polygamia, Diœcia.

The Peperidge, Tupelo, or sour gum tree, as it is called in various parts of the Union, grows to a moderate size, and is generally found in moist situations, though we have seen it in New York State, thriving very well in dry upland soils. The diameter of the trunk is seldom more than eighteen inches, and the general height is about forty or fifty feet. The flowers are scarcely perceptible, but the fruit borne in pairs, is about the size of a pea, deep blue, and ripens in October.

The leaves are oval, smooth, and have a beautiful gloss on their upper surface. The branches diverge from the main trunk almost horizontally, and sometimes even bend downwards like those of some of the Pine family, which gives the tree a very marked and picturesque character.

The Peperidge when of moderate size is not difficult to transplant, and we consider it a very fine tree, both on account of its beautiful, dark green, and lustrous foliage in summer, and the brilliant fiery color which it takes when the frost touches it in autumn. In this respect it is fully equal in point of beauty to that of the Liquidambar or Sweet gum, and the maples which we have already described; and so fine a feature do we consider this autumnal beauty of foliage that we would by all means advise the introduction of such trees as the Peperidge into the landscape for that reason alone, were it not also valuable for its peculiar form and polished leaves in summer.

Besides the Peperidge there are three other Nyssas, natives of this continent, viz. the Black gum (*N. Sylvatica*), a tree of greater dimensions, and larger, more elongated leaves, whose northern boundary is the neighborhood of Philadelphia; the Large Tupelo (*N. grandidentata*), a tree of the largest size, with large, coarsely toothed foliage, and a large blue fruit, three-fourths of an inch long, which is sometimes called the wild olive; and the sour Tupelo (*N. capitata*), with long, smooth, laurel-like leaves, and light red, oval fruit, called the Wild Lime, from its abounding in a strong acid, resembling that of the latter fruit. Both the latter trees are natives of the southern states, and are little known north of Philadelphia.

The wood of all the foregoing trees is remarkable for the peculiar arrangement of its fibres; which, instead of running directly through the stem in parallel lines, are curiously twisted and interwoven together. Owing to this circumstance it is extremely difficult to split, and is therefore often used in the manufacture of wooden bowls, trays, etc. That of the Peperidge is also preferred for the same reason, and for its toughness, by the wheelwrights, in the construction of the naves of wheels, and for other similar purposes.

Michaux remarks that he is unable to give any reason why the names of Sour gum, Black gum, etc., have been bestowed upon these trees, as they spontaneously exude no sap or fluid which could give rise to such an appellation. We suspect that the term has arisen from a comparison of the autumnal tints of these trees belonging to the genus Nyssa, with those of the Sweet gum or Liquidambar, which, at a short distance, they so much resemble in the early autumn.

The Thorn Tree. *Crategus.*

Nat. Ord. Rosaceæ.   *Lin. Syst.* Icosandria, Di-pentagynia.

A tree of the smallest size; but though many of the sorts attain only the stature of ordinary shrubs, yet some of our native species, as well as the English Hawthorn (*C. oxycantha*), when standing alone, will form neat, spreading-topped trees, of twenty or thirty feet in height.

Although the thorn is not generally viewed among us as a plant at all conducive to the beauty of scenery, yet we are induced to mention it here, and to enforce its claims in that point of view, as they appear to us highly entitled to consideration. First, the foliage—deep green, shining, and often beautifully cut and diversified in form—is prettily tufted and arranged upon the branches; secondly, the snowy blossoms—often produced in such quantities as to completely whiten the whole head of the tree, and which in many sorts have a delightful perfume—present a charming appearance in the early part of the season; and thirdly, the ruddy crimson or purple haws or fruit, which give the whole plant a rich and glowing appearance in and among our fine forests, open glades, or wild thickets, in autumn.

The most ornamental and the strongest growing indigenous kinds are the Scarlet Thorn tree (*C. coccinea*), and its varieties, the Washington Thorn (*C. populifolia*), and the Cockspur Thorn (*C. crus-galli*); all of which, in good soil, will grow to the height of twenty or thirty feet, and can readily be transplanted from their native sites.

The English Hawthorn is not only a beautiful small tree, but it is connected in our minds with all the elegant, poetic, and legendary associations which belong to it in England; for scarcely any tree is richer in such than this. With the floral games of *May*, this plant, from its blooming at that period, and being the favorite of the season, has become so identified, that the blossoms are known in many parts of Britain chiefly by that name. Among the ancient Greeks and Romans, they were dedicated to Flora, whose festival began on the first of that month; and in the olden times of merry England, the May-pole, its top decked with the gayest garlands of these blossoms, was raised amid the shouts of the young and old assembled to celebrate this happy rustic festival. *Chaucer* alludes to the custom, and describes the hawthorn thus:.

> Marke the faire blooming of the Hawthorne tree,
> Which finely cloathed in a robe of white,
> Fills full the wanton eye with May's delight.
> COURT OF LOVE.

And *Herrick* has left us the following lines to "Corrina going a Maying:"

> " Come, my Corrina, come; and coming marke
> How eche field turns a street, eche street a park
> Made green, and trimmed with trees; see how
> Devotion gives eche house a bough
> Or branch; eche porch, eche doore ere this,
> An arke, a tabernacle is,
> Made up of Hawthorne, neatly interwove,
> As if here were those cooler shades of love."

The following lines descriptive of the English species, we extract from the "*Romance of Nature.*"

> " Come let us rest this hawthorn tree beneath,
> And breathe its luscious fragrance as it flies,

And watch the tiny petals as they fall,
Circling and winnowing down our sylvan hall."

The berries, or *haws*, as they are called, have a very rich and coral-like look when the tree, standing alone, is completely covered with them in October. There are some elegant varieties of this species, which highly deserve cultivation for the beauty of their flowers and foliage. Among them we may particularly notice the Double White, with beautiful blossoms like small white roses; the Pink and the Scarlet flowering, both single and double, and the Variegated-leaved hawthorn, all elegant trees; as well as the Weeping hawthorn, a rarer variety, with pendulous branches.

The Hawthorn is most agreeable to the eye in composition when it forms the undergrowth or thicket, peeping out in all its green freshness, gay blossoms, or bright fruit, from beneath and between the groups and masses of trees; where, mingled with the hazel, etc., it gives a pleasing intricacy to the whole mass of foliage. But the different species display themselves to most advantage, and grow also to a finer size, when planted singly, or two or three together, along the walks leading through the different parts of the pleasure-ground or shrubbery.

The Magnolia Tree. *Magnolia.*

Nat. Ord. Magnoliaceæ.   *Lin. Syst.* Polyandria, Polygynia.

The North American trees composing the genus Magnolia are certainly among the most splendid productions of the forests in any temperate climate; and when we consider the size and fragrance of their blossoms, or the beauty of their large and noble foliage, we may be allowed to doubt whether there is a more magnificent and showy genus of deciduous trees in the world. With the exception of a few shrubs or smaller trees, natives of China and the mountains of Central Asia, it belongs exclusively to this continent, as no individuals of this order are indigenous to Europe or Africa. The American species attracted the attention of the first botanists who came over to examine the riches of our native flora, and were transplanted to the gardens of England and France more than a hundred years ago, where they are still valued as the finest hardy trees of that hemisphere.

The Large Evergreen Magnolia (*M. grandiflora*), or Big Laurel, as it is sometimes called, is peculiarly indigenous to that portion of our country south of North Carolina, where its stately trunk, often seventy feet in height, and superb pyramid of deep green foliage, render it one of the loveliest and most majestic of trees. The leaves, which are evergreen, and somewhat resemble those of the laurel in form, are generally six or eight inches in length, thick in texture and brilliantly polished on the upper surface. The highly fragrant flowers are composed of about six petals, opening in a wide cup-like form, of the most snowy whiteness of color. Scattered among the rich foliage, their effect is exquisitely beautiful. The seeds are borne in an oval, cone-like carpel or seed-vessel, composed of a number of cells which split longitudinally, when the stony seed, covered with a bright red pulp, drops out. There are several varieties, which have been raised from the seed of this species abroad: the most beautiful is the Exmouth Magnolia, with fine foliage, rusty beneath; it produces its flowers much earlier and more abundantly than the original sort.

We regret that this tree is too tender to bear the open air north of Philadelphia, as it is one of the choicest evergreens. At the nurseries of the Messrs. Landreth, and at the Bartram Botanic Garden of Col. Carr, near that city, some good specimens of this Magnolia and its varieties are growing thriftily; but in the State of New York, and at the east, it can only be considered a greenhouse plant.

The Cucumber Magnolia (*C. acuminata*), (so called from the appearance of the young fruit, which is not unlike a green cucumber) takes the same place in the north, in point of majesty and elevation, that the Big Laurel occupies in the south. Its northern limit is Lake Erie; and it abounds along the whole range of the Alleghanies to the southward, in rich mountain acclivities, and moist sheltered valleys. There it often measures three or four feet in diameter, and eighty in height. The leaves, which are deciduous, like those of all the Magnolias except the *M. grandiflora*, are also about six inches long and four broad, acuminate at the point, of a bluish green on the upper surface. The flowers are six inches in diameter, of a pale yellow, much like those of the Tulip tree, and slightly fragrant. The fruit is about three inches long, and cylindrical in shape. Most of the inhabitants of the country bordering on the Alleghanies, says Michaux, gather these cones about midsummer, when they are half ripe, and steep them in whiskey; the liquor produced, they take as an antidote against the fevers prevalent in those districts

The Umbrella Magnolia (*M. tripetala*), though found

sometimes in the northwest of New York, is rare there, and abounds most in the south and west. It is a smaller tree than the preceding kinds, rarely growing more than thirty feet high. The leaves on the terminal shoots are disposed three or four in a tuft, which has given rise to the name of Umbrella tree. They are of fine size, eighteen inches or two feet long, and seven or eight broad, oval, pointed at both ends; the flowers are also large, white, and numerous; and the conical fruit-vessel containing the seeds, assumes a beautiful rose-color in autumn. From its fine tufted foliage, and rapid growth, this is one of the most desirable species for our pleasure-grounds.

The Large-leaved Magnolia (*M. macrophylla*) is the rarest of the genus in our forests, being only found as yet in North Carolina. The leaves grow to an enormous size when the tree is young, often measuring three feet long, and nine or ten inches broad. They are oblong, oval, and heart-shaped at the base. The flowers are also immense, opening of the size of a hat-crown, and diffusing a most agreeable odor. The tree attains only a secondary size, and is distinguished in winter by the whiteness of its bark, compared with the others. It is rather tender north of New York.

The Heart-leaved Magnolia (*M. cordata*) is a beautiful southern species, distinguished by its nearly round, heart-shaped foliage, and its yellow flowers about four inches in diameter. It blooms in the gardens very young, and very abundantly, often producing two crops in a season.

Magnolia auriculata grows about forty feet high, and is also found near the southern Alleghany range of mountains. The leaves are light green, eight or nine inches long, widest at the top, and narrower towards the base, where they are rounded into lobes. The flowers are not so fine as those of the preceding kinds, but still are handsome, pale greenish white, and about four inches in diameter.

Besides these, there is a smaller American Magnolia, which is the only sort that in the middle or eastern sections of the Union grows within 150 miles of the seashore. This is the Magnolia of the swamps of New Jersey and the South (*M. glauca*), of which so many fragrant and beautiful bouquets are gathered in the season of its inflorescence, brought to New York and Philadelphia, and exposed for sale in the markets. It is rather a large bush, than a tree; with shining, green, laurel-like leaves, four or five inches long, somewhat mealy or glaucous beneath. The blossoms, about three inches broad, are snowy white, and so fragrant that where they abound in the swamps, their perfume is often perceptible for the distance of a quarter of a mile.

The foreign sorts introduced into our gardens from China, are the Chinese purple (*M. purpurea*), which produces an abundance of large delicate purple blossoms early in the season; the Yulan or Chinese White Magnolia (*M. conspicua*), a most abundant bloomer, bearing beautiful white, fragrant flowers in April, before the leaves appear; and Soulange's Magnolia (*M. Soulangiana*), a hybrid between the two foregoing, with large flowers delicately tinted with white and purple. These succeed well in sheltered situations, in our pleasure-grounds, and add greatly to their beauty early in the season. Grafted on the cucumber tree, they form large and vigorous trees of great beauty.

The Magnolia, in order to thrive well, requires a deep,

rich soil; which in nearly all cases, to secure their luxuriance, should be improved by adding thereto some leaf mould or decayed vegetable matter from the woods. When transplanted from the nursery, they should be preferred of small or only moderate size, as their succulent roots are easily injured, and they recover slowly when large. Most of them may be propagated from seed; but they flower sooner, grow more vigorously, and are much hardier when grafted upon young stocks of the Cucumber Magnolia. This we have found to be particularly the case with the Chinese species and varieties.

All these trees are such superbly beautiful objects upon a lawn in their rich summer garniture of luxuriant foliage, and large odoriferous flowers, that they need no further recommendation from us to insure their regard and admiration from all persons who have room for their culture. If possible, situations somewhat sheltered either by buildings or other trees, should be chosen for all the species, except the Cucumber Magnolia, which thrives well in almost any aspect not directly open to violent gales of wind.

The White-wood, or Tulip Tree. *Liriodendron.*

Nat. Ord. Magnoliaceæ.   *Lin. Syst.* Polyandria, Polygynia.

The Tulip tree belongs to the same natural order as the Magnolias, and is not inferior to most of the latter in all that entitles them to rank among our very finest forest trees.

The taller Magnolias, as we have already remarked, do

not grow naturally within 100 or 150 miles of the seacoast; and the Tulip tree may be considered as in some measure supplying their place in the middle Atlantic states. West of the Connecticut river, and south of the sources of the Hudson, this fine tree may be often seen reaching in warm and deep alluvial soils 80 or 90 feet in height. But in the western states, where indeed the growth of forest trees is astonishingly vigorous, this tree far exceeds that altitude. The elder Michaux mentions several which he saw in Kentucky, that were fifteen and sixteen feet in girth; and his son confirms the measurement of one, three miles and a half from Louisville, which, at five feet from the ground, was found to be twenty-two feet and six inches in circumference, with a corresponding elevation of 130 feet.

The foliage is rich and glossy, and has a very peculiar form; being cut off, as it were, at the extremity, and slightly notched and divided into two-sided lobes. The breadth of the leaves is six or eight inches. The flowers, which are shaped like a large tulip, are composed of six thick yellow petals, mottled on the inner surface with red and green. They are borne singly on the terminal shoots, have a pleasant, slight perfume, and are very showy. The seed-vessel, which ripens in October, is formed of a number of scales surrounding the central axis in the form of a cone. It is remarkable that young trees under 30 or 35 feet high, seldom or never perfect their seeds.

Whoever has once seen the Tulip tree in a situation where the soil was favorable to its free growth, can never forget it. With a clean trunk, straight as a column, for 40 or 50 feet, surmounted by a fine, ample summit of rich green foliage, it is, in our estimation,

DECIDUOUS ORNAMENTAL TREES.   257

decidedly the most *stately* tree in North America. When standing alone, and encouraged in its lateral growth, it will indeed often produce a lower head, but its tendency is to rise, and it only exhibits itself in all its stateliness and majesty when, supported on such a noble columnar trunk, it towers far above the heads of its neighbors of the park or forest. Even when at its loftiest elevation, its large specious blossoms, which, from their form, one of our poets has likened to the chalice;

>———Through the verdant maze
>The Tulip tree
>Its golden chalice oft triumphantly displays.
>   PICKERING.

jut out from amid the tufted canopy in the month of June, and glow in richness and beauty. While the tree is less than a foot in diameter, the stem is extremely smooth, and it has almost always a refined and finished appearance. For the lawn or park, we conceive the Tulip tree eminently adapted: its tall upright stem, and handsome summit, contrasting nobly with the spreading forms of most deciduous trees. It should generally stand alone, or near the border of a mass of trees, where it may fully display itself to the eye, and exhibit all its charms from the root to the very summit; for no tree of the same grandeur and magnitude is so truly beautiful and graceful in every portion of its trunk and branches. Where there is a taste for avenues, the Tulip tree ought by all means to be employed, as it makes a most magnificent overarching canopy of verdure, supported on trunks almost architectural in their symmetry. The leaves also, from their bitterness, are but little liable to the attacks of any insect.

This tree was introduced into England about 1668 ; and is now to be found in almost every gentleman's park on the Continent of Europe, so highly is it esteemed as an ornamental tree of the first class. We hope that the fine native specimens yet standing, here and there, in farm lands along our river banks, may be sacredly preserved from the barbarous infliction of the axe, which formerly despoiled without mercy so many of the majestic denizens of our native forests.

In the western states, where this tree abounds, it is much used in building and carpentry. The timber is light and yellow, and the tree is commonly called the Yellow Poplar in those districts, from some fancied resemblance in the wood, though it is much heavier and more durable than that of the poplar.

When exposed to the weather, the wood is liable to warp, but as it is fine grained, light, and easily worked, it is extensively employed for the panels of coaches, doors, cabinet-work, and wainscots. The Indians who once inhabited these regions, hollowed out the trunks, and made their canoes of them. There are two sorts of timber known ; viz. the Yellow and the White Poplar, or Tulip tree. These, however, it is well known are the same species (*L. tulipifera*) ; but the variation is brought about by the soil, which if dry, gravelly, and elevated, produces the white, and if rich, deep, and rather moist, the yellow timber.

It is rather difficult to transplant the Tulip tree when it has attained much size, unless the roots have undergone preparation, as will hereafter be mentioned ; but it is easily propagated from seed, or obtained from the nurseries, and the growth is then strong and rapid.

The Dogwood Tree. *Cornus.*

Nat. Ord. Cornaceæ.   *Lin. Syst.* Tetrandria, Monogynia.

There are a number of small shrubs that belong to this genus, but the common Dogwood (*Cornus florida*) is the only species which has any claims to rank as a tree. In the middle states, where it abounds, as well as in most other parts of the Union, the maximum height is thirty-five feet, while its ordinary elevation is about twenty feet.

The Dogwood is quite a picturesque small tree, and owes its interest chiefly to the beauty of its numerous blossoms and fruit. The leaves are oval, about three inches long, dark green above, and paler below. In the beginning of May, while the foliage is beginning to expand rapidly, and before the tree is in full leaf, the flowers unfold, and present a beautiful spectacle, often covering the whole tree with their snowy garniture. The principal beauty of these consists in the involucrum or calyx, which, instead of being green, as is commonly the case, in the Dogwood takes a white or pale blue tint. The true flowers may be seen collected in little clusters, and are, individually, quite small, though surrounded by the involucrum, which produces all the effect of a fine white blossom.

In the early part of the season, the Dogwood is one of the gayest ornaments of our native woods. It is seen at that time to great advantage in sailing up the Hudson river. There, in the abrupt Highlands, which rise boldly many hundred feet above the level of the river, patches of the Dogwood in full bloom gleam forth in snowy whiteness from among the tender green of the surrounding young foliage, and the gloomier shades of the dark evergreens, which clothe with a rich verdure the rocks and precipices that overhang the moving flood below.

The berries which succeed these blossoms become quite red and brilliant in autumn ; and, as they are plentifully borne in little clusters, they make quite a display. When the sharp frosts have lessened their bitterness, they are the food of the robin, which, at that late season, eats them greedily.

The foliage in autumn is also highly beautiful, and must be considered as contributing to the charms of this tree. The color it assumes is a deep lake-red; and it is at that season as easily known at a distance by its fine coloring, as the Maple, the Liquidambar, and the Nyssa, of which we have already spoken. Taking into consideration all these ornamental qualities, and also the fact that it is every day becoming scarcer in our native wilds, we think the Dogwood tree should fairly come under the protection of the picturesque planter, and well deserves a place in the pleasure-ground and shrubbery.

The wood is close-grained, hard, and heavy, and takes a good polish. It is too small to enter into general use, but is often employed for the lesser utensils of the farm. The bark has been very successfully employed by physicians in Philadelphia, and elsewhere, and is found to possess nearly the same properties as the Peruvian bark. Bigelow states in his American Botany, that its use in fevers has been known and practised in many sections of the Union by the country people, for more than fifty years.

Besides this native species there is an European dogwood (*Cornus mascula*), commonly called the Cornelian cherry, which is now planted in many of our gardens, and grows to the height of twenty or thirty feet. The small yellow flowers come out close to the branches in March or April, and the whole tree is quite handsome in autumn, from the size and color of its fine oval scarlet berries. These are as large as a small cherry, transparent, and hang for a long time upon the tree. The leaves are much like those of the common Dogwood. Although the blossoms are produced when the plant is quite a bush, yet it must attain some age before the fruit sets. Altogether, the Cornelian cherry is one of the most desirable of small trees.

The Salisburia, or Ginko Tree.

Nat. Ord. Taxaceæ.   *Lin. Syst.* Monœcia, Polyandria.

This fine exotic tree, which appears to be perfectly hardy in this climate, is one of the most singular in its foliage that has ever come under our observation. The leaves are wedge-shaped, or somewhat triangular, attached to the petioles at one of the angles, and pale yellowish green in color; the ribs or veins, instead of diverging from the central mid-rib of the leaf, as is commonly the case in dicotyledonous plants, are all parallel; in short, they almost exactly resemble (except in being three or four times as large) those of the beautiful Maiden hair fern (*Adiantum*) common in our woods : being thickened at the edges and notched on the margin in a similar manner. The male flowers are yellow, sessile catkins; the female is seated in a curious kind of cup, formed by the enlargement of the summit of the peduncle. The fruit is a drupe, about an inch in length, containing a nut, which, according to Dr. Abel, is almost always to be seen for sale in the markets of China and Japan, the native country of this tree. They are eaten after having been roasted or boiled, and are considered excellent.

The Salisburia was introduced into this country by that zealous amateur of horticulture and botany, the late Mr. Hamilton, of Woodlands, near Philadelphia, who brought it from England in 1784, where it had been received from Japan about thirty years previous. There are several of these now growing at Woodlands; and the largest measures sixty feet in height, and three feet four inches in circumference. The next largest specimen which we have seen is now standing on the north side of that fine public square, the Boston Common. It originally grew in the grounds of Gardiner Green, Esq., of Boston; but though of fine size, it was, about three years since, carefully removed to its present site, which proves its capability for bearing transplanting. Its measurement is forty feet in elevation, and three in circumference. There is also a very handsome tree in the grounds of Messrs. Landreth, Philadelphia, about thirty-five feet high and very thrifty.

We have not learned that any of these trees have yet borne their blossoms; at any rate none but male blossoms have yet been produced. Abroad, the Salisburia has fruited in the South of France, and young trees have been reared from the nuts.

The bark is somewhat soft and leathery, and on the trunk and branches assumes a singular tawny yellow or greyish color. The tree grows pretty rapidly, and forms an exceedingly neat, loose, conical, or tapering head. The timber is very solid and heavy; and the tree is said to grow to enormous size in its native country. Bunge, who accompanied the mission from Russia to Pekin, states that he saw near a Pagoda, an immense Ginko tree, with a trunk nearly forty feet in circumference, and still in full vigor of vegetation.*

Although nearly related to the Pine tribe, and forming, apparently, the connecting link between the *coniferæ* and exogenous trees, yet, unlike the former tribe, the wood of the tree is perfectly free from resin.

The Ginko tree is so great a botanical curiosity, and is so singularly beautiful when clad with its fern-like foliage, that it is strikingly adapted to add ornament and interest to the pleasure ground. As the foliage is of that kind which must be viewed near by to understand its peculiarity, and as the form and outline of the tree are pleasing, and harmonize well with buildings, we would recommend that it be planted near the house, where its unique character can be readily seen and appreciated.

Salisburia adiantifolia is the only species. In the United States it appears to flourish best in a rich fertile soil, rather dry than otherwise. South of Albany it is perfectly hardy, and may therefore be considered a most valuable acquisition to our catalogue of trees of the first class. It has hitherto been propagated chiefly from layers; but cuttings of the preceding year's growth, planted early in the spring, in a fine sandy loam, and kept shaded and watered, will also root without much difficulty. When the old trees already mentioned (which have doubtless been raised from seed) begin to blossom, plants reared from them by cuttings or grafts, will, of course, produce blossoms and fruit much more speedily than when reared from the nut.

* Bull. de la Soc. d'Agr. du départ. de l'Herault. Arb. Brit.

The American Cypress Tree. *Taxodium.*

Nat. Ord. Coniferæ.   *Lin. Syst.* Monœcia, Monadelphia.

The Southern or Deciduous cypress (*Taxodium distichum*)* is one of the most majestic, useful, and beautiful trees of the southern part of North America. Naturally, it is not found growing north of Maryland, or the south part of Delaware, but below that boundary it becomes extremely multiplied. The low grounds and alluvial soils subject to inundations, are constantly covered with this tree; and on the banks of the Mississippi and other great western rivers, for more than 600 miles from its mouth, those vast marshes, caused by the periodical bursting and overflowing of their banks, are filled with huge and almost endless growths of this tree, called Cypress swamps. Beyond the boundaries of the United States its geographical range extends to Mexico; and Michaux estimates that it is found more or less abundantly, over a range of country more than 3000 miles in extent.

" In the swamps of the southern states and the Floridas, on whose deep, miry soil a new layer of vegetable mould is deposited every year by the floods, the Cypress attains its utmost development. The largest stocks are 120 feet in height, and from 25 to 40 feet in circumference above the conical base, which at the surface of the earth is always three or four times as large as the continued diameter of the trunk; in felling them, the negroes are obliged to raise themselves upon scaffolds five or six feet from the ground. The roots of the largest stocks, particularly of such as are

---
* Cupressus disticha.

most exposed to inundation, are charged with conical protuberances, commonly from eighteen to twenty-four inches, and sometimes four or five feet in thickness; these are always hollow, smooth on the surface, and covered with a reddish bark, like the roots, which they resemble also in the softness of their wood; they exhibit no sign of vegetation, and I have never succeeded in obtaining shoots by wounding their surface and covering them with the earth. No cause can be assigned for their existence: they are peculiar to the Cypress, and begin to appear when it is twenty or twenty-five feet in height; they are not made use of except by the negroes for bee-hives."

" The foliage is open, light, and of a fresh, agreeable tint; each leaf is four or five inches long, and consists of two parallel rows of leaflets, upon a common stem. The leaflets are small, fine, and somewhat arching, with the convex side outwards. In the autumn, they change from a light green to a dull red, and are shed soon after."

" The Cypress blooms in Carolina about the first of February. The male and female flowers are borne separately, by the same tree; the first in flexible pendulous aments, and the second in bunches, scarcely apparent. The cones are about as large as the thumb, hard, round, of an uneven surface, and stored with small, irregular, ligneous seeds, containing a cylindrical kernel; they are ripe in October, and retain their productive virtue for two years."[*]

Such is the account given of the Cypress in its native soils. In the middle states it is planted only as an ornamental tree; and while, in the South, its great abundance

[*] N. A. Sylva. ii. 332.

causes it to be neglected or disregarded as such, its rarity here allows us fully to appreciate its beauty. North of the 43° of latitude it will not probably stand the winter without protection; but south of that, it will attain a good size. The finest planted specimen which we have seen, and one which is probably equal in grandeur to almost any in their native swamps, is growing in the Bartram Botanic Garden, near Philadelphia. That garden was founded by the father of American botanists, John Bartram, who explored the southern and western territories, then vast wilds, at the peril of his life, to furnish the *savans* and gardens of Europe, with the productions of the new world, and who commenced the living collection, now unequalled, of American trees, in his own garden. In the lower part of it stands the *great Cypress*, a tree of noble dimensions, measuring at this time 130 feet in height and 25 in circumference. The tree was held by Bartram's son, William, while his father assisted in planting it, *ninety-nine* years ago. The elder Bartram at the time expressed to his son, the hope that the latter might live to see it a large tree. Long before he died (not many years since), it had become the prodigy of the garden, and great numbers from the neighboring city annually visit it, to admire its vast size, and recline beneath its ample shade.

The foliage of the Cypress is peculiar; for while it has a resemblance to the ·Hemlock, Yew, and other evergreen trees, its cheerful, bright green tint, and loose airy tufts of foliage, give it a character of great lightness and elegance. In young trees, the form of the head is pyramidal or pointed; but when they become old, Michaux remarks, the head becomes widely spread, and even depressed, thus assuming a remarkably picturesque aspect. This is also heightened by the deep furrows or channels in the trunk, and the singular excrescences or knobs already described, which, jutting above the surface of the ground, give a strange ruggedness to the surface beneath the shadow of its branches. A single Cypress standing alone, like that in the Bartram Garden, is a grand object, uniting with the expression of great elegance and lightness in its foliage, that of magnificence, when we perceive its extraordinary height, and huge stem and branches.

In composition, the Cypress produces the happiest effect, when it is planted with the hemlock and firs, with which it harmonizes well in the form of its foliage, while its soft light green hue is beautifully opposed to the richer and darker tints of those thickly-clad evergreens. Wherever there is a moist and rather rich soil, the Cypress may be advantageously planted: for although we have seen it thrive well on a fertile dry loam, yet to attain all its lofty proportions, it requires a soil where its thirsty roots can drink in a sufficient supply of moisture. There its growth is quite rapid; and although it may, at first, suffer a little from the cold at the north, in severe winters, yet it continues its progress, and ultimately becomes a stately tree.

In many parts of the southern states, the timber of this tree, which is of excellent quality, is extensively used in the construction of the framework and outer covering of houses. It is also esteemed for shingles; and a large trade has long been carried on from the south in Cypress shingles. Posts made of this tree are found to be very lasting; and it is also employed for water-pipes, masts of vessels, etc. In the north, its place is supplied by the Pine timber, but in many southern cities, particularly New Orleans, it will be found to enter into the composition of almost every building.

In the nurseries, the Cypress is usually propagated from the seed; and as it sends down strong roots, it should be transplanted where it is finally to grow before it attains too great a development.

The European Cypress (*Cupressus sempervirens*), a beautiful evergreen tree, shaped like a small Lombardy poplar, which is the principal ornament of the churchyards and cemeteries abroad, is unfortunately too tender to endure the winter in any of the states north of Virginia. South of that state, it may probably become naturalized, and serve to add to the catalogue of beautiful indigenous evergreen trees.

From its dark and sombre tint, and perpetual verdure, it is peculiarly the emblem of grief:

> " Binde you my brows with *mourning* Cyparesse,
> And palish twigs of deadlier poplar tree,
> Or if some sadder shades ye can devise,
> Those sadder shades vaile my light-loathing eyes."
>
> Bp. Hall.

The Larch Tree. *Larix.*

Nat. Ord. Coniferæ.   *Lin. Syst.* Monœcia, Monadelphia.

The Larch is a resinous, cone-bearing tree, belonging to the Pine family, but differing from that genus in the annual shedding of its leaves like other deciduous trees. In Europe it is a native of the coldest parts of the Alps and Appenines; and in America, is indigenous to the most

DECIDUOUS ORNAMENTAL TREES. 269

northern parts of the Union, and the Canadas. The leaves are collected in little bunches, and the branches shoot out from the main stem in a horizontal, or, more generally, in a declining position.

[Fig. 39. The European Larch.]

For picturesque beauty, the Larch is almost unrivalled. Unlike most other trees which must grow old, uncouth, and misshapen before they can attain that expression, this is singularly so, as soon almost as it begins to assume the stature of a tree. It can never be called a beautiful tree, so far as beauty consists in smooth outlines, a finely rounded head, or gracefully drooping branches. But it has what is perhaps more valuable, as being more rare,—the expression of boldness and picturesqueness peculiar to itself, and which it seems to have caught from the wild and rugged chasms, rocks, and precipices of its native mountains. There its irregular and spiry top and branches, harmonize admirably with the abrupt variation of the surrounding hills, and suit well with the gloomy grandeur of those frowning heights.

Like all highly expressive and characteristic trees, much more care is necessary in introducing the Larch into artificial scenery judiciously, than round-headed trees. If planted in abundance, it becomes monotonous, from the similitude of its form in different specimens; it should therefore be introduced sparingly, and always for some special purpose. This purpose may be either to give spirit to a group of other trees, to strengthen the already picturesque character of a scene, or to give life and variety to one naturally tame and uninteresting. All these objects can be fully effected by the Larch; and although it is by far the most suited to harmonize with and strengthen the expression of scenery naturally grand, or picturesque, with which it most readily enters into combination; yet, in the hands of taste, there can be no reason why so marked a tree should not be employed in giving additional expression to scenery of a tamer character.

The extremely rapid growth of this tree when planted upon thin, barren, and dry soils, is another great merit which it possesses as an ornamental tree; and it is also a necessary one to enable it to thrive well on those very rocky and barren soils, where it is most in character with the surrounding objects. It is highly valuable to produce effect or shelter suddenly, on portions of an estate, too thin or meagre in their soil to afford the sustenance necessary to the growth of many other deciduous trees.

The Larch is the great timber tree of Europe. Its wood is remarkably heavy, strong, and durable, exceeding in all those qualities the best English oak. To these, it is said to add the peculiarity of being almost uninflammable, and resisting the influence of heat for a long time. Vitruvius relates that when Cæsar attacked the castle of Larignum, near the Alps, whose gate was commanded by a tower built of this wood, from the top of which the besieged annoyed him with their stones and darts, he commanded his army to surround it with fagots, and set fire to the whole. When, however, all the former were consumed, he was astonished to find the Larch tower uninjured.* The wood is also recommended for the decks of vessels and the masts of ships, as it is little liable either to fly in splinters during an engagement or to catch fire readily.

In Great Britain, immense plantations of this tree are made with a view to profit; and although as yet nothing like rearing trees for timber has been attempted here, nevertheless the time must come when our attention will necessarily be turned in this direction. When such is the case, it is probable that the Larch will be found to be as much an object of profit on this side of the Atlantic as on the other. Indeed, we are much inclined to believe that thousands of acres of our sterile soils in some districts, might now be profitably planted with this tree.

In Scotland, the Larch was first introduced in the year 1738, when eleven plants were given to the Duke of Athol, who afterwards, struck by the rapidity of their growth and the excellency of their timber, planted thousands of acres with them. As a specimen of what is done in timber

* Newton's Vitruvius, p. 40.

growing abroad, and the peculiar capacity of the Larch for thriving on poor soils, we shall make some extracts from the account given of its growth in Scotland, by Sir T. D. Lauder.

The late Duke of Athol planted large districts with this tree, and thereby converted the heathy wastes into valuable forests; but this was not the whole of the improvement he thus created. The Larch being a deciduous tree, sheds upon the earth so great a shower of decayed spines every succeeding autumn, that the annual addition which is made to the soil cannot be less than from a third of an inch to half an inch, according to the magnitude of the trees. This we have had opportunities of proving by our remarks made on the surfaces of newly cleaned pleasure walks. The result of planting a moor with Larches then, is, that when the trees have grown so much as to exclude the air and moisture from the surface, the heath is soon exterminated; and the soil gradually increasing by the decomposition of the leaflets annually thrown down by the Larches, grass begins to grow as the trees rise in elevation, so as to allow greater freedom for the circulation of the air below,—and thus, land which was not worth one shilling an acre becomes most valuable pasture; and we can say that our own experience amply bears out the fact. The Duke of Athol found that the value of the pasture in oak copses was worth five or six shillings (sterling) per acre for eight years only in twenty-four, when the copse is cut down again. Under a Scotch fir plantation it is not worth sixpence more per acre than it was before it was planted; under Beech and Spruce, it is worth less than it was before. But under Larch, where the ground was not worth one shilling per acre, before it was planted, the pasture becomes worth from eight to ten shillings per acre, after the first thirty years, when all the thinnings have been completed, and the trees left for naval purposes, at the rate of four hundred to the acre, and twelve feet apart.

The Larch is a very quick grower. Between 1740 and 1744, eleven trees were planted at Blair, the girths of which, at growths from seventy-three to seventy-six years, ranged from eight feet two inches to ten feet. This lot was calculated to average one hundred feet each, in the whole one thousand two hundred feet. The total measurement of this lot of twenty-two trees, therefore, is two thousand six hundred and forty-five feet, which, at the moderate value of two shillings per foot, would give the sum of £264 10s. ($1174) for twenty-two Larch trees, of something under eighty years' old. We find by the Duke of Athol's tables of measurement, that trees planted by him in 1743 were nine feet three inches in circumference, when measured at four feet from the ground, in 1795.

The plantations of Larch made by James Duke of Athol, between 1733 and 1759, amounted to one thousand nine hundred and twenty-eight trees. Of these, eight hundred and seventy-three were cut down between 1809 and 1816. The Duke of Athol had the satisfaction to behold a British frigate built in 1819 and 1820 at Woolwich yard, out of timber planted at Blair and Dunkeld, by himself and the Duke his predecessor. And the extensive and increasing Larch forests of those districts may yet be called upon largely to supply both naval and mercantile dock-yards. Mankind are prone to cherish and embalm the memory of individuals whose claims to notoriety have originated in their wide-spread destruction of the human race; but they are too apt to forget those who have been the benefactors of mankind. That a vessel formed from trees of his introduction and planting should have waved the British flag over the ocean, is likely to be all the reward contemporaneous or posthumous, which will ever adhere to the noble Duke, for the great good he has done to his country, and for the blessed legacy he has left to his descendants, by the plantation of about fifteen thousand five hundred and seventy-three English acres of ground, which consumed above twenty-seven millions, four hundred and thirty-one thousand, and six hundred trees.

The following is the probable supply of Larch timber from Athol, beginning twelve years from 1817.

				Loads annually.	Scotch acres about.
12 years before cutting, or in			1829		
12 years before cutting,		. .	1841	4,250	
10 do.	do.	. .	1851	8,000	
8 do.	do.	. .	1859	18,000	2000
8 do.	do.	. .	1867	30,000	
16 do.	do.	. .	1883	52,000	3000
3 do.	do.	. .	1886	120,000	
69 years calculated to finish 3 plants marked out.			1889	130,000	1500
72 years.					Scotch acres, 7000

The Larch is unquestionably the most enduring timber that we have. It is remarkable, that whilst the red wood or heart wood is not formed at all in the other resinous trees, till they have lived for a good many years, the Larch, on the contrary, begins to make it soon after it is planted; and while you may fell a Scotch fir of thirty years old, and find no red wood in it, you can hardly cut down a young Larch large enough to be a walking stick, without finding just such a proportion of red wood compared to its diameter as a tree, as you will find in the largest Larch tree in the forest, compared to its diameter. To prove the value of the Larch as a timber tree, several experiments were made in the river Thames. Posts of equal thickness and strength, some of Larch and others of oak, were driven down facing the river wall, where they were alternately covered with water by the effect of the tide, and then left dry by its fall. This species of alternation is the most trying of all circumstances for the endurance of timber; and accordingly the oaken posts decayed, and were twice renewed in the course of a very few years, while those that were made of the Larch remained altogether unchanged.

Besides the foregoing species (*Larix Europea*) we have two native sorts much resembling it; which are chiefly found in the states of Maine, Vermont, and New Hampshire. These are known by the names of the Red Larch (*L. Microcarpa*) and the Black Larch (*L. pendula*), which latter is often called *Hackmatack*. In the coldest parts of the Union, these often grow to 80 and 100 feet high; but in the middle states they are only seen in the swamps, and appear not to thrive so well except in such situations. For this reason the European Larch is of course greatly preferable when plantations are to be made, either for profit or ornament. The latter is generally increased from seed in the nurseries.

The American Larches are well worthy a place where sufficient moisture can be commanded, as their peculiar forms are striking, though not so finely picturesque as that of the European species.

In the upper part of Massachusetts, we have observed them in their native soils growing 70 or 80 feet high, and assuming a highly pleasing appearance. Their foliage is bluish-green, and more delicate; yet altogether the American Larch appears to be more stiff and formal (except far north) than the foreign tree.

THE VIRGILIA TREE. *Virgilia.**

Nat. Ord. Leguminaceæ.   *Lin. Syst.* Decandria, Monogynia.

This fine American tree, still very rare in our ornamental plantations, is a native of West Tennessee, and the banks of the Kentucky river, and in its wild localities seems confined to rather narrow limits. It was named, when first discovered, after the poet Virgil, whose agreeable *Georgics* have endeared him to all lovers of nature and a country life.

The Virgilia is certainly one of the most beautiful of all that class of trees bearing papilionaceous, or pea-shaped flowers, and pinnate leaves, of which the common locust may serve as a familiar example. It grows to a fine, rather broad head, about 30 or 40 feet high, with dense and luxuriant foliage—much more massy and finely tufted than that of most other pinnated-leaved trees. Each leaf is composed of seven or eight leaflets, three or four inches long, and half that breadth, the whole leaf being more than a foot in length. These expand rather late in the spring, and are, about the middle of May, followed by numerous terminal racemes, or clusters, of the most delicate and charming pea-shaped blossoms, of a pure white. These clusters are six or eight inches in length, and quite broad, the flowers daintily formed, and arranged in a much more graceful, loose, and easy manner, than those of the locust.

* Cladeastris tinctoria. *Torrey and Gray.*

They have a very agreeable, slight perfume, especially in the evening, and the whole effect of the tree, when standing singly on a lawn and filled with blossoms, is highly elegant.

When the blossoms disappear, they are followed by the pods, about the fourth of an inch wide, and three or four inches long, containing a few seeds. These ripen in July or August.

This tree is frequently called the *Yellow-wood* in its native haunts—its heart wood abounding in a fine yellow coloring matter, which, however, is said to be rather difficult to fix, or render permanent. The bark is beautifully smooth, and of a greenish grey color. In autumn, the leaves, when they die off, take a lively yellow tint.

This tree grows pretty rapidly, and is very agreeable in its form and foliage, even while young. It commences flowering when about ten or fifteen feet high, and we can recommend it with confidence to the amateur of choice trees as worthy of a conspicuous place in the smallest collection.

The only species known is *Virgilia lutea*. It was first described by Michaux, and was sent to England about the year 1812. Quite the finest planted specimens within our knowledge are growing in some of the old seats in the northern suburbs of Philadelphia, where there are several thirty or forty feet in height, and exceedingly beautiful, both in their form and blossoms. A small specimen on our lawn, eighteen feet high, blossoms now very profusely.

The Paulownia Tree. *Paulownia.*

Nat. Ord. Scrophulariaceæ.   *Lin. Syst.* ——— ———

The Paulownia is an entirely new ornamental tree, very lately introduced into our gardens and pleasure-grounds from Japan, and is likely to prove hardy here, wherever the Ailantus stands the winter, being naturally from the same soil and climate as that tree. It has already attained a great notoriety in the gardening world of the other continent; and from a cost of four or five guineas a plant, it is now reduced to as many shillings, being very readily propagated. In the north of France it is perfectly hardy, and will no doubt prove equally so here, south of the latitude of Boston. With our own plants being newly received, we have not yet had the opportunity of testing this point.

The Paulownia is remarkable for the large size of its foliage, and the great rapidity of its growth. The largest leaves are more than two feet in diameter, slightly rough or hairy, and serrated on the edges. They are heart-shaped, and have been likened to those of the Catalpa, but they perhaps more nearly resemble those of the common sun-flower.

In its growth, this tree, while young, equals or exceeds the *Ailantus*. In rich soils, near Paris, it has produced shoots, in a single season, 12 or 14 feet in length. After being two or three years planted, it commences yielding its blossoms in panicled clusters. These are bluish lilac, of an open mouthed, tubular form, are very abundantly distributed, and, together with the large foliage, and the robust habit of growth, give this tree a gay and striking appearance. Its flower buds open during the last of April, or early in May, and have a slight, syringa-like perfume.

The Paulownia, though yet very rare, is easy of propagation by cuttings; and even pieces of the roots grow freely. Should it prove as hardy as (from our fine dry summers for ripening its wood) we confidently anticipate, it will be worthy of a prominent place in every arrangement of choice ornamental trees.

SECTION V.

EVERGREEN ORNAMENTAL TREES.

The History and Description of all the finest Hardy Evergreen Trees. REMARKS on THEIR EFFECTS in LANDSCAPE GARDENING, INDIVIDUALLY AND IN COMPOSITION. Their Cultivation, etc. The Pines. The Firs. The Cedar of Lebanon, and the Deodar Cedar. The Red Cedar. The Arbor Vitæ. The Holly. The Yew, etc.

> Beneath the forest's skirt I rest,
>   Whose branching Pines rise dark and high,
> And hear the breezes of the West
>   Among the threaded foliage sigh.
> BRYANT.

THE PINE TREE. *Pinus.*

Nat. Ord. Coniferæ.   *Lin. Syst.* Monœcia, Monadelphia.

H E Pines compose by far the most important genus of evergreen trees. In either continent they form the densest and most extensive forests known, and their wood in civil and naval architecture, and for various other purposes, is more generally used than any other. In the United States and the Canadas, there are ten species; in the territory west of the Mississippi to the Pacific, including Mexico, there are fourteen; in Europe fourteen; in Asia, eight, and in Africa, two species. All the colder parts of the old world

—the mountains of Switzerland and the Alps, the shores of the Baltic, vast tracts in Norway, Sweden, Germany, Poland, and Russia, as well as millions of acres in our own country, abound with immense and interminable forests of Pine. Capable of enduring extreme cold, growing on thin soils, and flourishing in an atmosphere, the mean temperature of which is not greater than 37° or 38° *Fahrenheit*, they are found as far north as latitude 68° in Lapland; while on mountains they grow at a greater elevation than any other arborescent plant. On Mount Blanc, the Pines grow within 2,800 feet of the line of perpetual snow.* In Mexico, also, Humboldt found them higher than any other tree; and Lieut. Glennie describes them as growing in thick forests on the mountain of Popocatapetl, as high as 12,693 feet, beyond which altitude vegetation ceases entirely.†

The Pines are, most of them, trees of considerable magnitude and lofty growth, varying from 40 to 150 or even 200 feet in height in favorable situations, rising with a perpendicular trunk, which is rarely divided into branches bearing much proportionate size to the main stem, as in most deciduous trees. The branches are much more horizontal than those of the latter class (excepting the Larch). The leaves are linear or needle-shaped, and are always found arranged in little parcels of from two to six, the number varying in the different species. The blossoms are produced in spring, and the seeds, borne in cones, are not ripened, in many sorts, until the following autumn. Every part of the stem abounds in a resinous juice, which is extracted, and forms in the

* Edinburgh Phil. Journ.
† Proc. Geological Soc. Lond.   Arb. Brit.

various shapes of tar, pitch, rosin, turpentine, balsam, etc., a considerable article of trade and export.

As ornamental trees, the Pines are peculiarly valuable for the deep verdure of their foliage, which, unchanged by the severity of the seasons, is beautiful at all periods, and especially so in winter; for the picturesque forms which many of them assume when fully grown; and for the effectual shelter and protection which they afford in cold, bleak, and exposed situations. We shall here particularize those species, natives of either hemisphere, that are most valuable to the planter, and are also capable of enduring the open air of the middle states.

The White Pine (*P. strobus*), called also Sapling Pine, and Apple Pine, in various parts of this country, and Weymouth Pine abroad, is undoubtedly the most beautiful North American tree of the genus. The foliage is much lighter in color, more delicate in texture, and the whole tufting of the leaves more airy and pleasing than that of the other species. It is also beautiful in every stage of its growth, from a plant to a stately tree of 150 feet. When it grows in strong soil, it becomes thick and compact in its head; but its most beautiful form is displayed when it stands in a dry and gravelly site; there it shoots up with a majestic and stately shaft, studded every six or eight feet with horizontal tiers of branches and foliage. The hue of the leaves is much paler and less sombre than that of the other native sorts; and being less stiffly set upon the branches, is more easily put in motion by the wind; the murmuring of the wind among the Pine tops is, poetically, thought to give out rather a melancholy sound:—

> " The pines of Mœnalus were heard to mourn,
> And sounds of woe along the grove were borne,"

says Virgil, speaking of the European Pine. But the murmur of the slight breeze among the foliage of the White Pine gives out a remarkably soothing and agreeable sound, which agrees better with the description of Leigh Hunt:

> " And then there fled by me a rush of air
> That stirr'd up all the other foliage there,
> Filling the solitude with panting tongues,
> At which the Pines woke up into their songs,
> Shaking their *choral locks*."

Pickering, one of our own poets, thus characterizes the melody:

> " The overshadowing pines alone, through which I roam,
> Their verdure keep, although it darker looks ;
> And hark ! as it comes sighing through the grove,
> The exhausted gale, a spirit there awakens,
> That wild and melancholy music makes."

This species—the White Pine—seldom becomes flattened or rounded on the summit in old age, like many other sorts, but preserves its graceful and tapering form entire. From its pleasing growth and color, we consider it by far the most desirable kind for planting in the proximity of buildings, and its growth for an evergreen is also quite rapid.

The leaves of the White Pine are thickly disposed on the branches, in little bundles or parcels of five. The cones are about five inches long: they hang, when nearly ripe, in a pendulous manner from the branches, and open, to shed their seeds, about the first of October. The bark on trees less than twenty years old is remarkably smooth, but becomes cracked and rough, like that of the other

Pines, when they grow old, although it never splits and separates itself from the trunk in scales, as in other species.

The great forests of White Pine lie in the northern parts of the Union; and the geographical range of this tree is comprised chiefly between New York and the 47th degree of north latitude, it being neither capable of resisting the fierce heat of the south, nor the intense cold of the extreme northern regions. In Maine, New Hampshire, and Vermont, the White Pine abounds in various situations, adapting itself to every variety of soil, from dry, gravelly upland, to swamps constantly wet. Michaux measured two trunks near the river Kennebec, one of which was 154 feet long, and 54 inches in diameter; the other 144 feet long, and 44 inches in diameter, at three feet from the ground. Dr. Dwight also mentions a specimen on the Kattskill 249 feet long, and several on the Unadilla 200 feet long, and three in diameter.* These, though they are remarkable specimens, show the stately altitude which this fine species sometimes attains, equalling in majesty the grandest specimens of the old world:

> ———The rougher rinded Pine,
> The great Argoan ship's brave ornament,
> Which, coveting with his high top's extent
> To make the mountains touch the stars divine,
> Decks all the forest with embellishment.
>
> SPENSER.

The Yellow Pine (*P. mitis*) is a fine evergreen, usually reaching a stature of 50 or 60 feet, with a nearly uniform diameter of about 18 inches for two-thirds of its length. The branches generally take a handsome conical shape, and the whole head considerably resembles that of the spruce,

---

* Dwight's Travels, Vol. iv. p. 21—26.

whence it is sometimes called the *Spruce Pine*. The term Yellow Pine arises from the color of the wood as contrasted with that of the foregoing sort, which is white. The leaves of this species are long and flexible, arranged in pairs upon the branches, and have a fine dark green color. The cones are very small, scarcely measuring an inch and a half in length, and are clothed on the exterior with short spines. The growth is quite slow.

The Yellow Pine is rarely found above Albany to the northward, but it extends as far south as the Floridas. It grows in the greatest abundance in New Jersey, Maryland, and Virginia, and sometimes measures five or six feet in circumference. In plantations, it has the valuable property to recommend it, of growing on the very poorest lands.

The Pitch Pine (*P. rigida*) is a very distinct sort, common in the whole of the United States east of the Alleghanies. It is very stiff and formal in its growth when young, but as it approaches maturity, it becomes one of the most picturesque trees of the genus. The branches, which shoot out horizontally, bend downwards at the extremities, and the top of the tree, when old, takes a flattened shape. The whole air and expression of the tree is wild and romantic, and is harmonious with portions of scenery where these characters predominate. The leaves are collected in threes, and the color of the foliage is a dark green. The cones are pyramidal, from one to three inches long, and armed with short spines.

The bark of this kind of Pine is remarkably rough, black, and furrowed, even upon young trees; and the wood is filled with resinous sap, from which pitch and tar are copiously supplied. The trees grow in various parts of the country, both on the most meagre soils and in moist swamps, with almost equal facility. In the latter situations they are, however, comparatively destitute of resin, but the stems often rise to 80 feet in elevation.

The foregoing are the finest and most important species of the north. The Red Pine (*Pinus rubra*) and the Grey Pine are species of small or secondary size, chiefly indigenous to British America. The Jersey Pine (*P. inops*) is a dwarfish species, often called the Scrub Pine, which seldom grows more than 25 feet high.

There are some splendid species that are confined to the southern states, where they grow in great luxuriance. Among the most interesting of these is the Long-leaved Pine (*P. Australis*), a tree of 70 feet elevation, with superb wandlike foliage, borne in threes, often nearly a foot in length. The cones are also seven or eight inches long, containing a kernel or seed of agreeable flavor. As this tree grows as far north as Norfolk in Virginia, we are strongly inclined to believe that it might be naturalized in the climate of the middle states, and think it would become one of the most valuable additions to our catalogue of evergreen trees. The Loblolly Pine (*P. Tœda*) of Virginia has also fine foliage, six inches or more in length, and grows to 80 feet in height. Besides these already named, the southern states produce the Pond Pine (*P. Serotina*), which resembles considerably the Pitch Pine, with, however, longer leaves, and the Table Mountain Pine (*P. Pungens*), which grows 40 or 50 feet high, and is found exclusively upon that part of the Alleghany range.

We must not forget in this enumeration of the Pines of North America, the magnificent species of California and the North-West coast. The most splendid of these was discovered in Northern California, and named the ***Pinus***

Lambertiana, in honor of that distinguished botanist, A. B. Lambert, Esq., of London, the author of a superb work on this genus of trees. It is undoubtedly one of the finest evergreens in the world, averaging from 100 to 200 feet in height. Its discoverer, Mr. Douglass, the indefatigable collector of the Horticultural Society of London, measured one of these trees that had blown down, which was two hundred and fifteen feet in length, and fifty-seven feet nine inches in circumference, at three feet from the root; while at one hundred and thirty-four feet from the root, it was seventeen feet five inches in girth. This, it is stated, is by no means the maximum height of the species. The cones of the Lambert Pine measure sixteen inches in length; and the seeds are eaten by the natives of those regions, either roasted or made into cakes, after being pounded. The other species found by Mr. Douglass grow naturally in the mountain valleys of the western coast, and several of them, as the *Pinus grandis* and *nobilis*, are almost as lofty as the foregoing sort; while *Pinus monticola* and *P. Sabiniana* are highly beautiful in their forms and elegant in foliage. The seeds of nearly all these sorts were first sent to the garden of the London Horticultural Society, where many of the young trees are now growing; and we hope that they will soon be introduced into our plantations, which they are so admirably calculated, by their elegant foliage and stupendous magnitude, to adorn.

The European Pines next deserve our attention. The most common species in the north of Europe is the Scotch Pine (*P. sylvestris*), a dark, tall, evergreen tree, with bluish foliage, of 80 feet in height, which furnishes most of the deal timber of Europe. It is one of the most rapid of all the Pines in its growth, even on poor soils, and is therefore valuable in new places. The Stone Pine (*P. pinea*) is a native of the South of Europe, where it is decidedly the most picturesque evergreen tree of that continent. It belongs peculiarly to Italy, and its "vast canopy, supported on a naked column of great height, forms one of the chief and peculiar beauties in Italian scenery, and in the living landscapes of Claude." We regret that it is too tender to bear our winters, but its place may in a great measure be supplied by the Pinaster or Cluster Pine (*P. pinaster*), which is quite hardy, and succeeds well in the United States. This has much of the same picturesque expression, depressed or rounded head, and tall columnar stem, which mark the Stone Pine; while its thickly massed foliage, clustering cones, and rough bark, render it distinct and strikingly interesting.

The Corsican Pine (*P. larica*) is a handsome, regular shaped, pyramidal tree, with the branches disposed in tiers like those of the White Pine. It grows to a large size, and is valued for its extremely dark green foliage, thickly spread upon the branches. It is also one of the most rapid growers among the foreign sorts, and has been found to grow remarkably well upon the barren chalk downs of England. *Pinus cembra* is a very slow growing, though valuable kind, indigenous to Switzerland, and hardy here.

These are the principal European species that deserve notice here for their ornamental qualities. Some splendid additions have been made to this genus, by the discovery of new species on the Himalaya mountains of Asia; and from the great elevation at which they are found growing wild, we have reason to hope that they will become naturalized in our climate.

We must not leave this extensive family of trees without

adverting to their numerous and important uses. In the United States, full four-fifths of all the houses built are constructed of the White and Yellow Pine, chiefly of the former. Soft, easily worked, light and fine in texture, it is almost universally employed in carpentry, and for all the purposes of civil architecture; while the tall stately trunks furnish masts and spars, not only for our own vessels, but many of those of England. A great commerce is therefore carried on in the timber of this tree, and vast quantities of the boards, etc., are annually exported to Europe. The Yellow and Pitch Pine furnish much of the enormous supplies of fuel consumed by the great number of steamboats employed in navigating our numerous inland rivers. The Long-leaved Pine is the great timber tree of the southern states; and when we take into account all its various products, we must admit it to be the most valuable tree of the whole family. The consumption of the wood of this tree in building, in the southern states, is immense; and its sap furnishes nearly all the turpentine, tar, pitch, and rosin, used in this country, or exported to Europe. The *turpentine* flows from large incisions made in the trunk (into boxes fastened to the side of the trees for that purpose) during the whole of the spring and summer. *Spirit of turpentine* is obtained from this by distillation. *Tar* is procured by burning the dead wood in kilns, when it flows out in a current from a conduit made in the bottom. *Pitch* is prepared by boiling tar until it is about one half diminished in bulk; and *rosin* is the residuum of the distillation when spirit of turpentine is made. The Carolinas produce all these in the greatest abundance, and so long ago as 1807, the exportation of them to England alone amounted to nearly $800,000 in that single year.

The Fir Trees. *Abies.*

Nat. Ord. Coniferæ.   *Lin. Syst.* Monœcia, Monadelphia.

The Fir trees differ from the Pines, to which they are nearly related, in having much shorter leaves, which are placed *singly* upon the branches, instead of being collected in little bundles or parcels of two, three, or five, as is the case in all Pines. They generally grow in a more conical manner than the latter, and in ornamental plantations owe their beauty in most cases more to their symmetrical regularity of growth than to picturesque expression.

The Balsam, or Balm of Gilead Fir (*A. balsamea*), sometimes also called the American Silver Fir, is one of the most ornamental of our native evergreens. It is found most abundantly in Maine and Nova Scotia, but is scattered more or less on the mountain tops, and in cold swamps, through various other parts of the Union. At Pine Orchard, near the Catskill Mountain-house, it flourishes well, though never seen below the elevation of 1,800 feet. When standing singly, it forms a perfect pyramid of fine dark green foliage, 30 or 40 feet high, regularly clothed from the bottom to the top. The leaves, about half or three-fourths of an inch long, are silvery white on the under surface, though dark green above; and are inserted both on the sides and tops of the branches. It is one of the most beautiful evergreens for planting in grounds near the house, and is perhaps more cultivated for that purpose than any other in the Union. The cones, which are four or five inches long, like those of the European Silver Fir point upwards. However small the plants of this Fir may be, they are still interesting, as they display the same symmetry as full grown trees. The deep green color of the verdure of the Balm of Gilead Fir is retained unchanged in all its beauty through the severest winters, which causes it to contrast agreeably with the paler tints of the Spruces. On the trunks of trees of this species are found small vesicles or blisters, filled with a liquid resin, which is extracted and sold under the name of Balm of Gilead,* for its medicinal virtues.

The European Silver Fir (*A. picea*) strongly resembles, when young, the Balsam Fir. But its leaves are longer and coarser, and the cones are much larger, while it also attains twice or three times the size of the latter. In the forests of Germany it sometimes rises over 100 feet; and it always becomes a large tree in a favorable soil. It grows slowly during the first twenty years, but afterwards advances with much more rapidity. It thrives well, and is quite hardy in this country.

The Norway Spruce Fir (*A. communis*†) is by far the handsomest of that division of the Firs called the Spruces. It generally rises with a perfectly straight trunk to the height of from 80 to 150 feet. It is a native, as its name denotes, of the colder parts of Europe, and consequently grows well in the northern states. The branches hang down with a fine graceful curve or sweep; and although the leaves are much paler than those of the foregoing kinds, yet the thick fringe-like tufts of foliage which clothe the branches, give the whole tree a rich, dark appearance. The large cones, too, always nearly six inches long, are

---

* The true Balm of Gilead is an Asiatic herb, *Amyris gileadensis*.

† *Abies excelsa*.

beautifully pendent, and greatly increase the beauty of an old tree of this kind.

The Norway Spruce is the great tree of the Alps; and as a park tree, to stand alone, we scarcely know a more beautiful one. It then generally branches not quite down to the ground; and its fine, sweeping, feathery branches hang down in the most graceful and pleasing manner. There are some superb specimens of this species in various gardens of the middle states, 80 or 100 feet high.

The Black, or Double Spruce (*A. nigra*), sometimes also called the Red Spruce, is very common in the north; and, according to Michaux, forms a third part of the forests of Vermont, Maine, New Hampshire, as well as New Brunswick and Lower Canada. The leaves are quite short and stiff, and clothe the young branches around the whole surface; and the whole tree, where it much abounds, has rather a gloomy aspect. In the favorable humid black soils of those countries, the Black Spruce grows 70 feet high, forming a fine tall pyramid of verdure. But it is rarely found in abundance further south, except in swamps, where its growth is much less strong and vigorous. Mingled with other evergreens, it adds to the variety, and the peculiar coloring of its foliage gives value to the livelier tints of other species of Pine and Fir.

The White or Single Spruce (*A. alba*) is a smaller and less common tree than the foregoing, though it is often found in the same situations. The leaves are more thinly arranged on the young shoots, and they are longer and project more from the branches. The color, however, is a distinguishing characteristic between the two sorts; for while in the Black Spruce it is very dark, in this species it is of a light bluish green tint. The cones are also much larger on the White Spruce tree.

The Hemlock Spruce, or, as it is more commonly called, the *Hemlock* (*A. canadensis*), is one of the finest and most distinct of this tribe of trees. It is most abundantly multiplied in the extreme northern portions of the Union; and abounds more or less, in scattered groups and thickets, throughout all the middle states, while at the south it is confined chiefly to the mountains.

It prefers a soil, which, though slightly moist, is less humid than that where the Black Spruce succeeds best; and it thrives well in the deep cool shades of mountain valleys. In the Highlands of the Hudson it grows in great luxuriance; and in one locality, the sides of a valley near Crow's nest, the surface is covered with the most superb growths of this tree, reaching up from the water's edge to the very summit of the hill, 1,400 feet high, like a rich and shadowy mantle, sprinkled here and there only with the lighter and more delicate foliage of deciduous trees.

The average height of the Hemlock in good soils is about 70 or 80 feet; and when standing alone, or in very small groups, it is one of the most beautiful coniferous trees. The leaves are disposed in two rows on each side of the branches, and considerably resemble those of the Yew, though looser in texture, and livelier in color. The foliage, when the tree has grown to some height, hangs from the branches in loose pendulous tufts, which give it a peculiarly graceful appearance. When young, the form of the head is regularly pyramidal; but when the tree attains more age, it often assumes very irregular and picturesque forms.

Sometimes it grows up in a thick, dense, dark mass of foliage, only varied by the pendulous branches, which project beyond the grand mass of the tree; at others it forms a loose, airy, and graceful top, permeable to the slightest breeze, and waving its loose tufts of leaves to every passing breath of air. In almost all cases, it is extremely ornamental, and we regret that it is not more generally employed in decorating the grounds of our residences. It should be transplanted (like all of this class of trees) quite early in the spring, the roots being preserved as nearly entire as possible, and not suffered to become the least dried, before they are replaced in the soil.

The uses of the Fir tree are important. The Norway Spruce Fir furnishes the white deal timber so extensively employed in Europe for all the various purposes of building; and its tall, tapering stems afford fine masts for vessels. The Black Spruce timber is also highly valuable, and is thought by many persons to surpass in excellence that of the Norway Spruce. The young shoots also enter into the composition of the celebrated *Spruce beer* of this country, a delightful and very healthful beverage. And the Hemlock not only furnishes a vast quantity of the joists used in building frame-houses, but supplies the tanners with an abundance of bark, which, when mixed with that of the oak, is highly esteemed in the preparation of leather.

We regret that the fine evergreen trees both of this country and Europe, which compose the Pine and Fir tribes, have not hitherto received more of the attention of planters. It is inexpressible how much they add to the beauty of a country residence in winter. At that season, when, during three or four months the landscape is bleak and covered with snow, these noble trees, properly intermingled with the groups in view from the window, or those surrounding the house, give an appearance of verdure and life to the scene which cheats winter of half its dreariness. In exposed quarters, also, and in all windy and bleak situations, groups of evergreens form the most effectual shelter at all seasons of the year, while many of them have the great additional recommendation of growing upon the most meagre soils.

In fine country residences abroad, it is becoming customary to select some extensive and suitable locality, where all the species of Pines and Firs are collected together, and allowed to develope themselves in their full beauty of proportion. Such a spot is called a *Pinetum ;* and the effect of all the different species growing in the same assemblage, and contrasting their various forms, heights, and peculiarities, cannot but be strikingly elegant. One of the largest and oldest collections of this kind is the Pinetum of Lord Grenville, at Dropmore, near Windsor, England. This contains nearly 100 kinds, comprising all the sorts known to English botanists, that will endure the open air of their mild climate. The great advantage of these Pinetums is, that many of the more delicate species, which if exposed singly would perish, thrive well, and become quite naturalized under the shelter of the more hardy and vigorous sorts.

The Cedar of Lebanon Tree. *Cedrus.*

Nat. Ord. Coniferæ.   *Lin. Syst.* Monœcia, Monadelphia.

The Cedar of Lebanon is universally admitted by European authors to be the noblest evergreen tree of the old world. Its native sites are the elevated valleys and ridges of Mount Lebanon and the neighboring heights of the lofty groups of Asia Minor. There it once covered immense forests, but it is supposed these have never recovered from the inroads made upon them by the forty score thousand hewers employed by Solomon to procure the timber for the erection of the Temple. Modern travellers speak of them as greatly diminished in number, though there are still specimens measuring thirty-six feet in circumference. Mount Lebanon is inhabited by numerous Maronite Christians, who hold annually a celebration of the Transfiguration under the shade of the existing trees, which they call the "*Feast of Cedars.*"

The Cedar of Lebanon is nearly related to the Larch, having its leaves collected in parcels like that tree, but differs widely in the circumstance of its foliage being *evergreen.* It is remarkable for the wide extension of its branches, and the immense surface covered by its overshadowing canopy of foliage. In the sacred writings it is often alluded to as an emblem of great strength, beauty, and duration. "Behold the Assyrian was a Cedar in Lebanon, with fair branches, and with a shadowing shroud, and of an high stature ; and his top was among the thick boughs. His boughs were multiplied, and his branches became long. The fir trees were not like his boughs, nor the chestnut trees like his branches, nor any tree in the garden of God like unto him in beauty."*

In England the Cedar of Lebanon appears to have become quite naturalized. There it is considered by far the most ornamental of all the Pine tribe,—possessing, when full grown, an air of dignity and grandeur beyond any other tree. To attain the fullest beauty of development, it should always stand alone, so that its far-spreading horizontal branches can have full room to stretch out and expand themselves on every side. Loudon, in his Arboretum, gives a representation of a superb specimen now growing at Sion House, the seat of the Duke of Northumberland, which is 72 feet high, 24 in circumference, and covers an area, with its huge depending branches, of 117 feet. There are many other Cedars in England almost equal to this in grandeur. Sir T. D. Lauder gives an account of one at Whitton, which blew down in 1779: it then measured 70 feet in height, 16 feet in circumference, and covered an area of 100 feet in diameter. To show the rapidity of the growth of this tree, he quotes three Cedars of Lebanon, which were planted at Hopetoun House, Scotland, in the year 1748. The measurement is the circumference of the trunks, and shows the rapid increase *after* they have attained a large size.

	1801.	1820.	1825.	1833.	Increase in 32 years.
	ft. in.	ft. in.	ft. in.	ft. in.	ft. in.
First Cedar,	10 0	13 1½	14 0	15 1	5 1
Second do.	8 6	10 9½	11 4	12 3	3 9
Third do.	7 10	9 9½	10 8	11 6	3 8
A Chestnut measured at the same periods, only increased				2	7

Ezekiel xxxi.

From the above table, it will be seen how congenial even the cold climate of Scotland is to the growth of this tree. Indeed in its native soils, the tops of the surrounding hills are almost perpetually covered with snow, and it is, therefore, one of the very hardiest of the evergreens of the old world. There is no reason why it should not succeed admirably in many parts of the United States; and when we consider its great size, fine dark green foliage, and wide spreading limbs which

> " ———Overarching, frame
> Most solemn domes within,"
>
> SHELLEY.

as well as the many interesting associations connected with it, we cannot but think it better worth our early attention, and extensive introduction, than almost any other foreign tree. Evergreens are comparatively difficult to import, and as we have made the experiment of importing Cedars of Lebanon from the English nurseries with but indifferent success, we would advise that persons attempting its cultivation should procure the cones containing the seeds from England, when they may be reared directly in our own soil, which will of course be an additional advantage to the future growth of the tree.*

The situations found to be most favorable to this Cedar, in the parks and gardens of Europe, are sandy or gravelly soils, either with a moist subsoil underneath, or in the neighborhood of springs, or bodies of water. In such places it is found to advance with a rapidity equal to the Larch,

---

* The finest Cedar of Lebanon in the Union, is growing in the grounds of T. Ash, Esq., of Westchester Co., N. Y., being 50 feet high and of corresponding breadth. It stands near a Purple-leaved Beech, equally large and beautiful.

one of the fastest growing timber trees, as we have already noticed.

The Deodara, or Indian Cedar (*Cedrus Deodara*), is a magnificent species of this tree, recently introduced from the high mountains of Nepal and Indo-Tartary. It stands the climate of Scotland, and appears likely to succeed here wherever the Cedar of Lebanon will flourish. In its native country it is described as being a lofty and majestic tree, frequently attaining the height of 150 feet, with a trunk 30 feet in circumference. The leaves are larger than those of the Cedar of Lebanon, of a deeper bluish green, covered with a silvery bloom; the cones, borne in pairs, are of a reddish brown color, and are both longer and broader than those of the latter species. In some parts of Upper India it is considered a sacred tree (*Deodara*—tree of God), and is only used to burn as incense in days of high ceremony; but in others it is held in the highest esteem as a timber tree, having all the good qualities of the Cedar of Lebanon —its great durability being attested by its sound state in the roofs of temples of that country, which cannot have been built less than 200 years.

We have but just introduced the Deodara into the United States, and can therefore say little of its growth or beauty here, though we have little doubt that it will prove one of the noblest evergreen trees for our pleasure grounds. Loudon says, "the specimens in England are yet small; but the feathery lightness of its spreading branches, and the beautiful glaucous hue of its leaves, render it, even when young, one of the most ornamental of the coniferous trees; and all the travellers who have seen it full grown, agree that it unites an extraordinary degree of majesty and grandeur with its beauty. The tree thrives in every part of

Great Britain where it has been tried, even as far north as Aberdeen, where, as in many other places, it is found hardier than the Cedar of Lebanon. It is readily propagated by seeds, which preserve their vitality when imported in the cones. It also grows freely by cuttings, which appear to make as handsome free-growing plants as those raised from seed." The soil and culture for this tree are precisely those for the Cedar of Lebanon.

The Red Cedar Tree. *Juniperus.*

Nat. Ord. Coniferæ.   *Lin. Syst.* Diœcia, Monadelphia.

The Red Cedar is a very common tree, indigenous to this country, and growing in considerable abundance from Maine to Florida; but thriving with the greatest luxuriance in the sea-board states. When fully grown, the Red Cedar is about 40 feet in height, and little more than a foot in diameter. The leaves are very small, composed of minute scales, and lie pretty close to the branches. Small blue berries, borne thickly upon the branches of the female trees in autumn and winter, contain the seeds. These are covered with a whitish exudation, and are sometimes used, like those of the foreign juniper, in the manufacture of gin.

The Red Cedar has less to recommend it to the eye than most of the evergreens which we have already described. The color of the foliage is dull and dingy at many seasons, and the form of the young tree is too compactly conical to please generally. When old, however, we have seen it throw off this formality, and become an interesting, and indeed a picturesque tree. Then its branches shooting out in a horizontal direction, clad with looser and more pendent foliage, give the whole tree quite another character. The twisted stems, too, when they become aged, have a singular, dried-looking, whitish bark, which is quite unique and peculiar. There is a very fine natural avenue of Red Cedars near Fishkill landing, in Duchess Co., composed of two rows of noble trees 35 or 40 feet high, which is a very agreeable walk in winter and early spring. This has given the name of *Cedar Grove* to the country seat in question, where the Red Cedar grows spontaneously upon a slate subsoil with great luxuriance. There the trees are disseminated widely by the birds, which feed with avidity upon the berries.

The Red Cedar is well known to every person as one of our very best timber trees. It takes its name from the reddish hue of the perfect wood. This has a fragrant odor, and is not only light, fine-grained, and close in texture, but extremely durable. It is therefore much employed (though of late it is becoming scarcer) in conjunction with Live oak, which is too heavy alone, in ship-building. It is also valued for its great durability as posts for fencing; and is exported to Europe, to be used in the manufacture of pencils, and other useful purposes.

The Arbor Vitæ Tree. *Thuja.*

Nat. Ord. Coniferæ.   *Lin. Syst.* Monœcia, Monadelphia.

The Arbor Vitæ (*Thuja occidentalis*), sometimes also called Flat Cedar, or White Cedar, is distinguished from most evergreens by its flat foliage, composed of a great number of scales closely imbricated, or overlaying each other, which give the whole a compressed appearance. The seeds are borne in a small cone, usually not more than half an inch in length.

This tree is extremely formal and regular in outline in almost every stage of growth; generally assuming the shape of an exact cone or pyramid of close foliage, of considerable extent at the base, close to the ground, and narrowing upwards to a sharp point. So regular is their outline in many cases, when they are growing upon favorable soils, that at a short distance they look as if they had been subjected to the clipping-shears. The sameness of its form precludes the employment of this evergreen in so extensive a manner as most others; that is, in intermingling it promiscuously with other trees of less artificial forms. But the Arbor Vitæ, from this very regularity, is well suited to support and accompany scenery when objects of an avowedly artificial character predominate, as buildings, etc., where it may be used with a very happy effect. There is also no evergreen tree indigenous or introduced, which will make a more effectual, close, and impervious screen than this: and as it thrives well in almost every soil, moist, dry, rich, or poor, we strongly recommend it whenever such thickets are desirable. We have ourselves tried the experiment with a hedge of it about 200 feet long, which was transplanted about five or six feet high from the native *habitats* of the young trees, and which fully answers our expectations respecting it, forming a perfectly thick screen, and an excellent shelter on the north of a range of buildings at all seasons of the year, growing perfectly thick without trimming, from the very ground upwards.

The only fault of this tree as an evergreen, is the comparatively dingy green hue of its foliage in winter. But to compensate for this, it is remarkably fresh looking in its spring, summer, and autumn tints, comparing well at those seasons even with the bright verdure of deciduous trees.

The Arbor Vitæ is very abundant in New Brunswick, Vermont, and Maine. In New York, the shores of the Hudson, at Hampton landing, 70 miles above the city of New York, are lined on both sides with beautiful specimens of this tree, many of them being perfect cones in outline; and it is here much more symmetrical and perfect in its growth than we have seen it. Forty feet is about the maximum altitude of the Arbor Vitæ, and the stem rarely measures more than ten or twelve inches in diameter.

The wood is very light, soft, and fine-grained, but is reputed to be equally durable with the Red Cedar. It is consequently employed for various purposes in building and fencing, where, in the northern districts, it grows in sufficient abundance, and of suitable size.

The Chinese Arbor Vitæ (*T. orientalis*) is a tree of much smaller and more feeble growth. It cannot, therefore, as an ornamental tree, be put in competition with our native species. Bnt it is a beautiful evergreen for the garden and shrubbery, where it finds a more suitable and sheltered site, being rather tender north of New York.

The White Cedar (*Thuja spheroida**), which belongs to the same genus as the Arbor Vitæ, is a much loftier

---

* *Cupressus thuyoides* of the old botanists.

tree, often growing 80 feet high. It can hardly be considered a tree capable of being introduced into cultivated situations, as it is found only in thick swamps and wet grounds. The foliage considerably resembles that of the common Arbor Vitæ, though rather narrower, and more delicate in texture. The cones are small and rugged, and change from green to a blue or brown tint in autumn. In the south it is often called the Juniper.

The White Cedar furnishes excellent shingles, much more durable than those made of either Pine or Cypress; in Philadelphia the wood is much esteemed and greatly used in cooperage. "Charcoal," according to Michaux, "highly esteemed in the manufacture of gunpowder, is made of young stocks, about an inch and a half in diameter, deprived of their bark; and the seasoned wood affords beautiful lamp-black, lighter and more intensely colored than that obtained from the Pine."

The American Holly Tree. *Ilex.*

Nat. Ord. Aquifoliaceæ.   *Lin. Syst.* Diœcia, Tetrandria.

The European Holly is certainly one of the *evergreen glories* of the English gardens. There its deep green, glossy foliage, and bright coral berries, which hang on for a long time, are seen enlivening the pleasure-grounds and shrubberies throughout the whole of that leafless and inactive period in vegetation—winter. It is also, in our mother tongue, inseparably connected with the delightful associations of merry Christmas gambols and feastings, when both the churches and the dwelling-houses are decorated with its boughs. We have much to regret, therefore, in the severity of our winters, which will not permit the European Holly to flourish in the middle or eastern states, as a hardy tree. South of Philadelphia, it may become acclimated; but it appears to suffer greatly further north.

A beautiful succedaneum, however, may, we believe, be found in the American Holly (*Ilex opaca*), which indeed very closely resembles the foreign species in almost every particular. The leaves are waved or irregular in surface and outline, though not so much so as those of the latter, and their color is a much lighter shade of green. Like those of the foreign plant, they are armed on the edges with thorny prickles, and the surface is brilliant and polished. The American Holly is seen in the greatest perfection on the eastern shore of Maryland and Virginia, and the lower part of New Jersey. There it thrives best upon loose, dry, and gravelly soils. Michaux says it is also common through all the extreme southern states, and in West Tennessee, in which latter places it abounds on the margins of shady swamps, where the soil is cool and fertile. In such spots it often reaches forty feet in height, and twelve or fifteen inches in diameter.

Although the growth of the Holly is slow, yet it is *always* beautiful; and we regret that the American sort, which may be easily brought into cultivation, is so very rarely seen in our gardens or grounds. The seeds are easily procured, and if scalded and sowed in autumn, immediately after being gathered, they vegetate freely. For hedges the Holly is altogether unrivalled ; and it was also one of the favorite plants for *verdant sculpture*, in the ancient style of gardening. Evelyn, in the edition of his

Sylva, published in London in 1664, thus bursts out in eloquent praise of it: "Above all natural greens which enrich our home-born store, there is none certainly to be compared to the Holly; insomuch that I have often wondered at our curiosity after foreign plants and expensive difficulties, to the neglect of the culture of this vulgar but incomparable tree,—whether we will propagate it for use and defence, or for sight and ornament. Is there under heaven a more glorious and refreshing object of the kind, than an impregnable hedge of one hundred and sixty-five feet in length, seven high, and five in diameter, which I can show in my poor gardens, at any time of the year, glittering with its armed and varnished leaves? The taller standards at orderly distances blushing with their natural coral. It mocks the rudest assaults of the weather, beasts, or hedge-breaker:—

'Et illum nemo impune lacessit.'"

The Yew Tree. *Taxus*.

Nat. Ord. Taxaceæ.   *Lin. Syst.* Monœcia, Monadelphia.

The European Yew is a slow-growing, evergreen tree, which often, when full grown, measures forty feet in height, and a third more in the diameter of its branches. The foliage is flat, linear, and is placed in two rows, like that of the Hemlock tree, though much darker in color. The flowers are brown or greenish, and inconspicuous, but they are succeeded by beautiful scarlet berries, about half or three-fourths of an inch in diameter, which are open at the end, where a small nut or seed is deposited. These berries have an exquisitely delicate, waxen appearance, and contribute highly to the beauty of the tree.

The growth of this tree, even in its native soil, is by no means rapid. In twenty years, says Loudon, it will attain the height of fifteen or eighteen feet, and it will continue growing for one hundred years; after which it becomes comparatively stationary, but will live many centuries.

When young, the Yew is rather compact and bushy in its form; but as it grows old, the foliage spreads out in fine horizontal masses, the outline of the tree is irregularly varied, and the whole ultimately becomes highly venerable and picturesque. When standing alone, it generally shoots out into branches at some three or four feet above the surface of the ground, and is ramified into a great number of close branches.

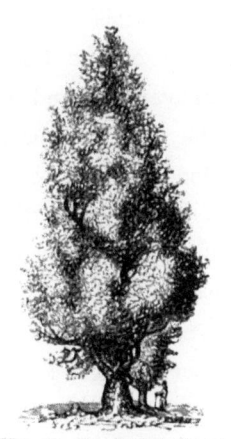

[Fig. 40. The English Yew.]

In England, it has been customary, since the earliest settlement of that island by the Britons, to plant the Yew in churchyards; and it is therefore as decidedly consecrated to this purpose there, as the Cypress is in the south of Europe. For the decoration of places of burial it is well adapted, from the deep and perpetual verdure of its foliage, which, conjointly with its great longevity, may be considered as emblematical of immortality. The custom still exists, in a few places in Ireland and Wales, of carrying twigs of this and other evergreen trees in funerals, and throwing them into the grave, with the corpse.*

> "—— Yet strew
> Upon my dismall grave
> Such offerings as ye have,
> Forsaken Cypresse and *Yewe;*
> For kinder flowers can have no birth
> Or growth from such unhappy earth."
> STANLY.

There is a mournful yet sweet and pensive pleasure, in thus adorning these last places of repose with such beautiful, unfading memorials of grief. They rob the graveyard or cemetery of its horrors, and by their perpetual garlands of verdure and freshness, inevitably lead the mind from the ideas of death which an ordinary barren churchyard alone inspires, to reflections of a purer and loftier cast; the immortality which awaits the soul when disenthralled of clay. Among the old English poets, we find much of these feelings in favor of decorating the precincts of the grave, and surrounding them with what may be called the *poetry of grief.* Herrick, one of the sweetest of the number, in some lines addressed to the Cypress and Yew, says:

> " Bothe of ye have
> Relation to the grave;
> And where
> The funeral trump sounds, you are there.

* Encyclopædia of Plants, 849.

I shall be made
Ere longe a fleeting shade;
Pray come,
And do some honor to my tomb."

Some of the old Yews in the churchyards and gardens of England have attained a wonderful period of longevity. Gilpin mentions one in the churchyard of Tisbury in Dorsetshire, now standing and in fine foliage, though the trunk is quite hollow, which measures thirty-seven feet in circumference, and the limbs are proportionately large. The tree is entered by a rustic gate; and seventeen persons lately breakfasted in its interior. It is said to have been planted many generations ago by the Arundel family. The famous Yew at Arkenwyke House, which Henry VIII. made his place of meeting with Anna Boleyn when she was there, is supposed to be upwards of a thousand years old; it is forty-nine feet high, twenty-seven in circumference, and the branches extend over an area of two hundred and seven feet. There are, besides these, a great number of other celebrated Yews in England, of immense size and age, which are preserved with the greatest care and veneration.

It is a common saying of the inhabitants of the New Forest in England, says Gilpin, that "a post of Yew will outlast a post of iron. The wood is extremely durable, and being hard and very fine-grained, as well as beautifully variegated with reddish or orange veins, it is much prized for inlaying, veneering, and other similar purposes by the cabinet-makers abroad. Tables made of it are said to be more beautiful than those of mahogany; and the wood of the root to vie in beauty with that of the Citron.

It is also remarkably elastic, and is therefore much valued for bows. In ancient times, when bows and arrows were the chief weapons of destruction in war, the bows made of the Yew tree were valued by the ancient Britons above all others. According to the Arboretum Britannicum, in Switzerland, where this tree was scarce, it was formerly forbidden, under heavy penalties, to cut down the Yew for any other purpose than to make bows of the wood. The Swiss mountaineers call it "William's tree," in memory of William Tell.

The Yew, like the Holly, makes an excellent evergreen hedge—close, dark green, and beautiful when clad in the rich scarlet berries. We desire, however, rather to see this tree naturalized in our gardens and lawns as an evergreen tree of the first class, than in any other form. Judging from specimens which we have growing in our own grounds, we should consider it quite hardy anywhere south of the 41° of latitude. And although it is somewhat slow in its growth, yet, like many other evergreens, it is as beautiful when a small bush or a thrifty young tree, as it is venerable and picturesque when ages or even centuries have witnessed its never failing verdure. It appears to grow most vigorously and thrive best on a rich and heavy soil, and in situations rather shaded than exposed to a burning sun.

There are several beautiful varieties of the Yew (*Taxus baccata*) cultivated in the nurseries; the Irish Yew (*T. b. fastigiata*), remarkable for its dark green foliage, and very handsome, upright growth, and the Yellow berried Yew (*T. b. fructo-flava*), are the most ornamental.

The North American Yew (*T. canadensis*) is a low

trailing shrub, scarcely rising above the height of four or six feet, though the branches extend to a considerable distance. In foliage, berries, etc., it so strongly resembles the European plant, that many botanists consider it only a dwarf variety. The leaves are nevertheless shorter and narrower, and the male flowers always solitary. It is found in shady, rocky places, in the Highlands, and various other localities from Canada to Virginia.

SECTION VI.

VINES AND CLIMBING PLANTS.

Value of this kind of Vegetation. Fine natural effects. The European Ivy. The Virginia Creeper. The Wild Grape Vine. The Bittersweet. The Trumpet Creeper. The Pipe Vine, and the Clematis. The Wistarias. The Honeysuckles and Woodbines. The Jasmine and the Periploca. Remarks on the proper mode of introducing vines. Beautiful effects of climbing plants in connexion with buildings.

> Quite over-canopied with lush woodbine,
> With sweet musk roses, and with eglantine.
>
> SHAKSPEARE.

VINES and climbing plants are objects full of interest for the Landscape Gardener, for they seem endowed with the characteristics of the graceful, the beautiful, and the picturesque, in their luxuriant and ever-varying forms. When judiciously introduced, therefore, nothing can so easily give a spirited or graceful air to a fine or even an ordinary scene, as the various plants which compose this group of the vegetable kingdom. We refer particularly now to those which have woody and perennial stems, as all annual or herbaceous stemmed plants are too short-lived to afford any lasting or permanent addition to the beauty of the lawn or pleasure-ground.

Climbing plants may be classed among the *adventitious beauties of trees*. Who has not often witnessed with delight in our native forests, the striking beauty of a noble tree, the old trunk and fantastic branches of which were enwreathed with the luxuriant and pliant shoots and rich foliage of some beautiful vine, clothing even its decayed limbs with verdure, and hanging down in gay festoons or loose negligent masses, waving to and fro in the air. The European Ivy (*Hedera Helix*) is certainly one of the finest, if not the very finest climbing plant (or more properly, creeping vine, for by means of its little fibres or rootlets on the stems, it will attach itself to trees, walks, or any other substance), with which we are acquainted. It possesses not only very fine dark green palmated foliage in great abundance, but the foliage has that agreeable property of being evergreen,—which, while it enhances its value tenfold, is at the same time so rare among vines. The yellow flowers of the Ivy are great favorites with bees, from their honied sweetness; they open in autumn, and the berries ripen in the spring. When planted at the root of a tree, it will often, if the head is not too thickly clad with branches, ascend to the very topmost limbs; and its dark green foliage, wreathing itself about the old and furrowed trunk, and hanging in careless drapery from the lower branches, adds greatly to the elegance of even the most admirable tree. Spenser describes the appearance of the Ivy growing to the tops of the trees,

> " Emongst the rest, the clamb'ring Ivie grew,
> Knitting his wanton arms with grasping hold,
> Lest that the poplar happely should renew
> Her brother's strokes, whose boughs she doth enfold
> With her lythe twigs, till they the top survew,
> And paint with pallid green her buds of gold."

The fine contrasts between the dark coloring of the leaves of the Ivy, and the vernal and autumnal tints of the foliage of deciduous trees, are also highly pleasing. Indeed this fine climbing plant may be turned to advantage in another way ; in reclothing dead trees with verdure. Sir T. D. Lauder says, that "trees often die from causes which we cannot divine, and there is no one who is master of extensive woods, who does not meet with many such instances of unexpected and unaccountable mortality. Of such dead individuals we have often availed ourselves, and by planting Ivy at their roots, we have converted them into more beautiful objects than they were when arrayed in their own natural foliage."

The Ivy is not only ornamental upon trees, but it is also remarkably well adapted to ornament cottages, and even large mansions, when allowed to grow upon the walls, to which it will attach itself so firmly by the little rootlets sent out from the branches, that it is almost impossible to tear it off. On wooden buildings, it may perhaps be injurious, by causing them to decay; but on stone buildings, it fastens itself firmly, and holds both stone and mortar together like a coat of cement. The thick garniture of foliage with which it covers the surface, excludes stormy weather, and has, therefore, a tendency to preserve the walls, rather than accelerate their decay. This vine is the inseparable accompaniment of the old feudal castles and crumbling towers of Europe, and borrows a great additional interest from the romance and historical recollections connected with such spots. Indeed half the interest, picturesque as well as poetical, of those time-worn buildings, is conferred by this plant, which seeks to bind together and adorn with something of their former richness, the crumbling fragments that are fast tottering to decay:—

> "The Ivy, that staunchest and firmest friend,
> That hastens its succoring arm to lend
> To the ruined fane where in youth it sprung,
> And its pliant tendrils in sport were flung.
> When the sinking buttress and mouldering tower
> Seem only the spectres of former power,
> Then the Ivy clusters round the wall,
> And for tapestry hangs in the moss-grown hall,
> Striving in beauty and youth to dress
> The desolate place in its loneliness."
> ROMANCE OF NATURE.

The Ivy lives to a great age, if we may judge from the specimens that overrun some of the oldest edifices of Europe, which are said to have been covered with it for centuries, and where the main stems are seen nearly as large as the trunk of a middle sized tree.

> "Whole ages have fled, and their works decayed,
> And nations have scattered been;
> But the stout old Ivy shall never fade
> From its hale and hearty green;
> The brave old plant in its lonely days,
> Shall fatten upon the past;
> For the stateliest building man can raise,
> Is the Ivy's food at last."

The Ivy is not a native of America; nor is it by any means a very common plant in our gardens, though we know of no apology for the apparent neglect of so beautiful a climber. It is hardy south of the latitude of 42°, and we have seen it thriving in great luxuriance as far north as Hyde Park, on the Hudson, eighty miles above New York. One of the most beautiful growths of this plant, which has ever met our eyes, is that upon the old mansion in the Botanic Garden at Philadelphia, built by the elder Bartram. That picturesque and quaint stone building is beautifully overrun by the most superb mantle of Ivy, that no one who has once seen can fail to remember with admiration. The dark grey of the stone-work is finely opposed by the rich verdure of the plant, which falls away in openings here and there, around the windows, and elsewhere. It never thrives well if suffered to ramble along the ground, but needs the support of a tree, a frame, or a wall, to which it attaches itself firmly, and grows with vigorous shoots. Bare walls or fences may thus be clothed with verdure and beauty equal to the living hedge, in a very short period of time, by planting young Ivy roots at the base.

The most desirable varieties of the common Ivy are: the Irish Ivy, with much larger foliage than the common sort, and more rapid in its growth; the Silver-striped and the Gold-striped leaved Ivy, both of which, though less vigorous, are much admired for the singular color of their leaves. The common English Ivy is more hardy than the others in our climate.

Although, as we have said, the Ivy is not a native of this country, yet we have an indigenous vine, which, at least in summer, is not inferior to it. We refer to the Virginia Creeper (*Ampelopsis hederacea*), which is often called the American Ivy. The leaves are as large as the hand, deeply divided into five lobes, and the blossoms are succeeded by handsome, dark blue berries. The Virginia Creeper is a most luxuriant grower, and we have seen it climbing to the extremities of trees 70 or 80 feet in height. Like the Ivy it attaches itself to whatever it can lay hold of, by the little rootlets which spring out of the branches;

and its foliage, when it clothes thickly a high wall, or folds itself in clustering wreaths around the trunk and branches of an open tree, is extremely handsome and showy. Although the leaves are not evergreen, like those of the Ivy, yet in autumn they far surpass those of that plant in the rich and gorgeous coloring which they then assume. Numberless trees may be seen in the country by the roadside, and in the woods, thus decked in autumn in the borrowed glories of the Virginia Creeper; but we particularly remember two as being remarkably striking objects; one, a wide-spread elm—the trunk and graceful diverging branches completely clad in scarlet by this beautiful vine, with which its own leaves harmonized well in their fine deep yellow dress; the other, a tall and dense Cedar, through whose dark green boughs gleamed the rich coloring of the Virginia Creeper, like a half-concealed, though glowing fire.

In the American forests nothing adds more to the beauty of an occasional tree, than the tall canopy of verdure with which it is often crowned by the wild Grape vine. There its tall stems wind themselves about until they reach the very summit of the tree, where they cluster it over, and bask their broad bright green foliage in the sunbeams. As if not content with this, they often completely overhang the head of the tree, falling like ample drapery around on every side, until they sweep the ground. We have seen very beautiful effects produced in this way by the grape in its wild state, and it may easily be imitated. The delicious fragrance of these wild grape vines when in blossom, is unsurpassed in delicacy ; and we can compare it to nothing but the delightful perfume which exhales from a huge bed of Mignonette in full bloom. The Bittersweet (*Celastrus scandens*) is another well known climber, which ornaments our wild trees. Its foliage is very bright and shining, and the orange-colored seed-vessels which burst open, and display the crimson seeds in winter, are quite ornamental. It winds itself very closely around the stem, however, and we have known it to strangle or compress the bodies of young trees so tightly as to put an end to their growth.

The Trumpet Creeper (*Bignonia radicans*) is a very picturesque climbing plant. The stem is quite woody, and often attains considerable size; the branches, like those of the Ivy and Virginia Creeper, fasten themselves by the roots thrown out. The leaves are pinnated, and the flowers, which are borne in terminal clusters on the ends of the young shoots about midsummer, are exceedingly showy. They are tubes five or six inches long, shaped like a trumpet, opening at the extremity, of a fine scarlet color on the outside, and orange within. The Trumpet Creeper is a native of Virginia, Carolina, and the states further south, where it climbs up the loftiest trees. It is a great favorite in the northern states as a climbing plant, and very beautiful effects are sometimes produced by planting it at the foot of a tall-stemmed tree, which it will completely surround with a pillar of verdure, and render very ornamental by its little shoots, studded with noble blossoms.

One of the most singular and picturesque climbing shrubs or plants which we cultivate, is the Pipe-vine, or Birthwort (*Aristolochia sipho*). It is a native of the Alleghany mountains, and is one of the tallest of twining plants, growing on the trees there to the height of 90 or 100 feet, though in gardens it is often kept down to a frame of four or five feet high. The leaves are of a noble size, being eight or nine inches broad, and heart-shaped in outline. The flowers, about an inch or a little more in length, are very singular. They are dark yellow, spotted with brown, in shape like a bent siphon-like tube, which opens at the extremity, the whole flower resembling, as close as possible, a very small *Dutchman's pipe*, whence the vine is frequently so called by the country people. It flowers in the beginning of summer, and the foliage, during the whole growing season, has a very rich and luxuriant appearance. *Aristolochia tomentosa* is a smaller species, with leaves and flowers of less size, the former downy or hairy on the under surface.

The various kinds of Clematis, though generally kept within the precincts of the garden, are capable of adding to the interest of the pleasure ground, when they are planted so as to support themselves on the branches of trees. The common White Clematis or Virgin's Bower (*C. virginica*) is one of the strongest growing kinds, often embellishing with its pale white blossoms, the whole interior and even the very tops of our forest trees in the middle states. After these have fallen, they are succeeded by large tufts of brown, hairy-like plumes, appendages to the clusters of seeds, which give the whole a very unique and interesting look. The Wild *Atragene*, with large purple flowers, which blossom early, has much the same habit as the Clematis, to which, indeed, it is nearly related. Among the finest foreign species of this genus are, the Single and Double-flowered purple Clematis (*C. viticella* and its varieties), which, though slender in their stems, run to considerable height, are very pretty, and blossom profusely. The sweet scented and the Japan Clematis (*C. flammula* and *C. florida*), the former very fragrant, and the latter beautiful, are perhaps too tender, except for the garden, where they are highly prized.

The Glycine or Wistaria (*Wistaria pubescens*) is a very beautiful climbing plant, and adds much to the gracefulness of trees, when trained so as to hang from their lower branches. The leaves are pinnate, and the light purple flowers, which bloom in loose clusters like those of the Locust, are universally admired. The Chinese Wistaria (*W. sinensis*) is a very elegant species of this plant, which appears to be quite hardy here; and when loaded with its numerous large clusters of pendent blossoms, is highly ornamental. It grows rapidly, and, with but little care, will mount to a great height. These vines with pinnated foliage, would be remarkably appropriate when climbing up, and hanging from the branches of such light airy trees as the Three-thorned Acacia, the Locust, etc.

We must not forget to enumerate here the charming family of the Honeysuckles; some of them are natives of the old world, some of our own continent; and all of them are common in our gardens, where they are universally prized for their beauty and fragrance. In their native localities they grow upon trees, and trail along the rocks. The species which ascends to the greatest height, is the common European Woodbine,* which twines around the stems, and hangs from the ends of the longest branches of trees:

> " As Woodbine weds the plant within her reach,
> Rough Elm, or smooth-grained Ash, or glossy Beech,
> In spiral rings ascends the trunk, and lays
> Her golden tassels on the leafy sprays."
>
> Cowper.

---

* *Woodbind* is the original name, derived from the habit of the plant of winding itself around trees, and binding the branches together.

The Woodbine (*Lonicera periclymenum*) has separate, opposite leaves, and buff-colored or paler yellow and red blossoms. There is a variety, the common monthly Woodbine, which produces its flowers all summer, and is much the most valuable plant. Another (*L. p. belgicum*), the Dutch Honeysuckle, blossoms quite early in spring; and a third (*L. p. quercifolium*) has leaves shaped like those of the oak tree.

The finest of our native sorts are the Red and Yellow trumpet Honeysuckle (*L. sempervirens* and *L. flava*), which have the terminal leaves on each branch joined together at the base, or perfoliate, making a single leaf. They blossom in the greatest profusion during the whole summer and autumn, and their rich blossom tubes, sprinkled in numerous clusters over the exterior of the foliage, as well as an abundance of scarlet berries in autumn, entitle them to high regard. There is also a very strong and vigorous species, called the Orange pubescent Honeysuckle (*L. pubescens*), with large, hairy, ciliate leaves, and fine large tawny or orange-colored flowers. It is a very luxuriant plant in its habit, and a very distinct species to the eye. All these native sorts have but very slight fragrance.

The Chinese twining Honeysuckle (*L. flexuosa*) is certainly one of the finest of the genus. In the form of the leaf it much resembles the common Woodbine; but the foliage is much darker colored, and is also sub-evergreen, hanging on half the winter, and in sheltered spots, even till spring. It blossoms when the plant is old, several times during the summer, bearing an abundance of beautiful flowers, open at the mouth, red outside, and striped with red, white, or yellow within. It grows remarkably fast, climbing to the very summit of trees in a short time; and the flowers, which first appear in June, are deliciously fragrant. In all its varieties the Honeysuckle is a charming plant, either to adorn the porch of the cottage, the latticed bower of the garden—to both of which spots they are especially dedicated—or to climb the stem of the old forest tree, where—

> " With clasping tendrils it invests the branch,
> Else unadorn'd, with many a gay festoon,
> And fragrant chaplet; recompensing well
> The strength it borrows with the grace it lends."

There it diffuses through the air a delicious breath, that renders a walk beneath the shade of the tall trees doubly delightful, while its flowers give a gaiety and brightness to the park, which forest trees, producing usually but inconspicuous blossoms. could not alone impart.

Some of the climbing Roses are very lovely objects in the pleasure-grounds. Many of them, at the north, as the Multifloras, Noisettes, etc., require some covering in the winter, and are therefore better fitted for the garden. At the south, where they are quite hardy, they are, however, most luxuriant and splendid objects. But there are two classes of Roses that are perfectly hardy climbers, and may therefore be employed with great advantage by the Landscape Gardener—the Michigan and the Boursalt trees. The single Michigan is a most compact and vigorous grower, and often, in its wild haunts in the west, clambers over the tops of tall forest trees, and decks them with its abundant clusters of pale purple flowers. There are now in our gardens several beautiful double varieties of this, and among them, one, called *Beauty of the Prairies*, is most admired for its large rich buds and blossoms of a deep rose color.

The Boursalt roses are remarkable for their profusion of flowers, and for their shining, reddish stems, with few thorns. The common Purple or Crimson Boursalt is quite a wonder of beauty in the latter part of May, when trained on the wall of a cottage, being then literally covered with blossoms; and it is so hardy that scarcely a branch is ever injured by the cold of winter. The Blush and the Elegans are still richer and finer varieties of this class of roses, all of which are well worthy of attention.

We have to regret that the inclemency of our winters will not permit us to cultivate the White European Jasmine (*Jasminum officinale*) out of the garden, as even there it requires a slight protection in winter. Below the latitude of Philadelphia, however, it will probably succeed well. In the southern states they have a most lovely plant, the Carolina Jasmine (*Gelseminum*), which hangs its beautiful yellow flowers on the very tree tops, and the woods there in spring are redolent with their perfume.

The connoisseur in vines will not forget the curious *Periploca*, which grows very rapidly to the height of 40 or 50 feet, and bears numerous branches of very curious brown or purple flowers in summer; or the Double-blossoming Brambles, both pink and white, which often make shoots of 20 or 30 feet long in a season, and bear pretty clusters of double flowers in June. All these fine climbers, and several others to be found in the catalogues, may, in the hands of a person of taste, be made to contribute in a wonderful degree to the variety, elegance, and beauty of a country residence; and to neglect to introduce them would be to refuse the aid of some of the most beautiful accessories that are capable of being combined with trees, as well as with buildings, gardens, and fences.

Some persons object to the growth of climbing plants upon trees, that, by compressing the stems and tightening themselves around the limbs of trees, they gradually check their growth, and finally by preventing the expansion of the trunk, put an end to the life of the tree. This, we have no doubt, has been the case when *young* trees in the full vigor of growth have been completely encompassed and wound about with the strong growing woody creepers; but it so rarely happens (scarcely ever in the case of middle-sized trees, on which vines are more generally planted), that we consider the objection of no moment. Indeed, were all this true, the management of the growth of any vine, however luxuriant, is so completely within the power of the cultivator, that by a very trifling annual attention, he can entirely prevent the possibility of any such injurious effects.

The reader must not imagine, from the remarks which we have here made on the beauty and charms of climbing plants, that we would desire to see every tree in an extensive park wreathed about, and overhung with fantastic vines and creepers. Such is by no means our intention. We should consider such a proceeding something in the worst possible taste. There are some trees whose rugged and ungraceful forms would refuse all such accompaniment; and others from whose dignity and majesty it would be improper to detract even by adding the gracefulness of the loveliest vine. Such, too, is never the case in nature, as for one tree decked in this manner we see a hundred which are not, and the very rarity of the example imparts additional beauty and interest to it when it appears. This should be the case in all artificial plantations; and he who has a true and lively feeling for the beautiful and picturesque, will easily understand at a glance where these expressions will be strengthened or weakened by the addition of more grace and elegance. A few scattered trees here and there, with whose forms the plans adopted harmonize, draped and festooned with the most appropriate climbing plants, will be all that can be properly introduced in any scene, unless it be of a very artificial character; but even these additional accessories, simple as they may seem, often produce an effect singularly beautiful, which shows how much in real landscape, as well as in painting, depends upon a few finishing touches to the scene.

Although we are not now writing of buildings, it is not inappropriate here to remark how much may be done in the country, and indeed even in town, by using vines and creepers to decorate buildings. The cottage in this country too rarely conveys the idea of comfort and happiness which we wish to attach to such a habitation, and chiefly because so often it stands bleak, solitary, and exposed to every ray of our summer sun, with a scanty robe of foliage to shelter it. How different such edifices, however humble, become when the porch is overhung with climbing plants,—when the blushing rose-buds peep in at the window sill, or the ripe purple clusters of the grape hang down about the eaves, those who have seen the better cottages of England well know. Very little care and trifling expense will procure all the additional beauty; and it is truly wonderful how much so little once done, adds to the happiness of the inmates. Every man feels prouder of his home when it is a pleasant spot for the eye to rest upon, than when it is situated in a desert, or overgrown with weeds. Besides this, tasteful embellishment has a tendency to refine the feelings of every member of the family; and every leisure hour spent in rendering more lovely and agreeable even the humblest cottage, is infinitely better employed than in lounging about in idle and useless dissipation.

SECTION VII.

TREATMENT OF GROUND.—FORMATION OF WALKS.

Nature of operations on Ground. Treatment of flowing and irregular surfaces to heighten their expression; flats, or level surfaces. Rocks, as materials in Landscape. Laying out Roads and Walks: Directions for the Approach: Rules by Repton. The Drive, and minor walks. The introduction of fences and verdant hedges.

———" Strength may wield the ponderous spade,
May turn the clod and wheel the compost home;
But elegance, chief grace the garden shows,
And most attractive, is the fair result
Of thought, the creature of a polished mind."

Cowper.

ROUND is undoubtedly the most unwieldy and ponderous material that comes under the care of the Landscape Gardener. It is not only difficult to remove, the operations of the leveller rarely extending below two or three feet of the surface; but the effect produced by a given quantity of labor expended upon it, is generally much less than when the same has been bestowed in the formation of plantations, or the erection of buildings. The achievements of art upon ground appear so trifling, too, when we behold the apparent facility with which nature has arranged it in such a variety of forms, that the former sink into insignificance when compared with the latter.

For these reasons, the operations to be performed

upon ground in this country, will generally be limited to the neighborhood of the house, or the scenery directly under the eye. Here, by judicious levelling and smoothing in some cases, or by raising gentle eminences with interposing hollows in others, much may be done at a moderate expense, to improve the beauty of the surrounding landscape.

It is, however, fortunately the case, that in the modern style of landscape improvement, extensive and costly operations upon ground are very seldom needed By the aid of plantations arranged as we have already suggested, much may be done to soften too great inequality of surface, as well as to heighten the apparent magnitude of gentle undulations. The art of the improver, when employed upon this material, will, therefore, be directed to the production of negative, rather than positive effects,—to the removal of existing faults or blemishes, rather than to the creation of an entirely new and artificial surface.

To pursue this method with success, it is necessary that he should refer constantly to the principle which we suggested in the commencement of our remarks: *the preservation of the natural character of the scene*, or, we may here add, the heightening of the *character intended* for the form of the surface. We have already remarked that scenes abounding in natural beauty were chiefly characterized by gentle undulations of surface, and smooth easy transitions from the level plain to the softly swelling hill or flowing hollow; and that, on the contrary, highly picturesque scenes exhibited a more irregular and broken surface, abounding with abrupt transitions, and more strongly marked elevations and depressions.

In a scene expressive of simple or graceful beauty, where the surface is more or less undulating, the first proceeding of the improver will be to remove any accidental or natural deformity which may interfere with that expression. Such are unsightly ridges of earth, small lumpish hills, the ragged elevations where old fences have been removed, or deep furrows created by the former action of the plough. If there are any uncouth pits or ugly hollows, such must be either filled up, or concealed by plantations, and all excrescences that interfere with the prevailing expression of the whole should be removed.

In the next place, the improver will examine the formation of the ground, as it appears naturally. If too rugged,—the sweeps and undulations sometimes easy and beautiful, but at others hard and disconnected,—he will endeavor to soften and remove this inequality. This will be easily executed if some of the eminences are broken into too high, sudden, and abrupt hills, by carefully lowering them into more graceful elevations, and placing the superfluous earth in the adjacent hollows: proper regard being paid to portions of the scene already pleasing, by producing such a surface as will *connect* itself naturally with the same, when the improvements shall be entirely completed.

Should the surface, on the contrary, be somewhat broken or undulating, but not distinctly so, appearing rather heavy and undecided between a level and finely varied ground, the operations must be directed in such a manner as to increase the boldness of the whole. The ground of a country residence is often brought into such a state by the continued action of the plough at some former period, which has gradually levelled down the gentle eminences and filled up the hollows, till in some places it appears scarcely struggling out of a level. The course is then obvious; the superfluous earth which chokes up the valleys, must be removed again to the neighboring hills, where it belongs, when the natural beauty of the ground will be restored. This is effected with comparative facility, as every foot of surface taken from the depression, adds by removal two feet to the height of the adjoining elevation.

The improvement of picturesque surfaces must proceed in a similar manner. When a surface is naturally and truly picturesque, art will add little or nothing to its effect. It will rather therefore endeavor to produce a perfect whole, and a connexion between the various parts, than to disturb the existing features. In the vicinity of the house, the artist will soften down that boldness and inequality which, if too great, might interfere both with convenience and the *beauty of utility*, which must there be constantly kept in view. Otherwise, the beauty of picturesque surfaces may be often heightened by various means within our reach; such as increasing the abruptness of surface by taking away a few feet of earth, or by adding other picturesque irregularities, which by connexion may strengthen the expression of the whole.

Mr. Price has remarked, that "the ugliest ground is that which has neither the beauty of smoothness, verdure, and gentle undulation, nor the picturesqueness of bold and sudden breaks, and varied tints of soil: of such kind, is ground that has been disturbed and left in that unfinished state: as in a rough ploughed field run to sward."* Such ground it is often difficult to restore to a picturesque state, even when that was its previous expression. But it is not

---

* Essay on the Picturesque, i. 193.

impossible to do so, for it must be remembered that it is not by *forming* the surface alone that nature renders it picturesque, but also by the *accessories* and *accompaniments* which she liberally bestows upon the surface when once formed. These are, vegetation, trees, rocks, etc., which, with the influence of time, will often render many a scene, that, stripped of its enriching drapery, would be positively harsh and ugly, extremely picturesque, or strikingly beautiful. Proofs of this will occur to every one who will contrast in his mind the appearance of a steep clayey river bank, or even pit, when bare, raw, and verdureless, and the same objects when nature or art has clothed them with a luxuriant and diversified garniture of trees, shrubs, and plants. In the former case, all was positively ugly and displeasing to the eye of taste; in the latter, all is picturesque and harmonious.

A perfect flat, or level surface, is often the most difficult to improve of any description of ground. In some cases, as in the example of a very large park, with an immense building, a level surface may be in excellent keeping, giving an air of grandeur to the whole scene: for both the simplicity and the wide extent of a level plain in such a situation, would be highly expressive of grandeur when united to a fine pile of building. But ordinarily, a flat surface is extremely dull and uninteresting. One unbroken plain of green is spread before the eye, varied by none of those changing lights and shadows that belong to a finely undulating lawn. It is true that this affects the mind differently in certain situations, as a broad plain is a delightful contrast and source of repose in a mountainous country. But we here speak of the greater part of the surface of the United States, where country seats are located, and where it will be found that a diversified surface is greatly to be preferred to a dead level.

Where such a level exists, in some situations, it is almost impossible to improve it much. When, for illustration, the whole surrounding country is equally tame and flat, the creation by artificial means, of undulations, hills, or hollows in a park, would be in such evident contradiction to the natural formation, that the eye would at once detect it as a deception, harmonizing badly with general nature. The best that can be done in such cases, is, perhaps, to produce the greatest possible beauty by plantations and buildings, and not to attempt any alterations of surface, which would be insignificant and absurd.

When, however, this is not the case, but the grounds themselves, though nearly level, are surrounded by more bold and spirited variations of surface, a great deal may be effected. In those portions of the grounds nearest the surrounding inequalities, the latter may be apparently carried into the former, and the artificial sweeps, breaks, or undulations in the park may be so connected with each other, and with the neighboring irregularities, as to produce the effect of accordant art joined to the charm of natural expression.

The error into which inexperienced improvers are constantly liable to fall, is a *want of breadth* and extent in their designs; which latter, when executed, are so feeble as to be full of *littleness*, out of keeping with the magnitude of the surrounding scene. Their designs, like the sketches of a novice in drawing, are cramped and meagre. This is exemplified in ground by their producing, instead of easy undulations, nothing but a succession of short sweeps and hillocks like waves in the ocean. Now the most beautiful variation in ground is undoubtedly that of gradually varying lines and *insensible transitions* of surface, and these should correspond in magnitude and breadth to the size and style of the place. Such surfaces are full of the flowing lines and rounded smoothness which Burke considers characteristic of beauty, or the long undulations exhibit the outlines of Hogarth's favorite line of grace.

In places of large extent there may be scenes in different portions of the park of totally different character; one simply beautiful, abounding with graceful and flowing lines, and another highly picturesque, and full of spirited breaks and variations. Such often form very pleasing and striking contrasts to each other, and should therefore, by all means, be preserved: but they should also be rendered distinct by their own surrounding plantations, else much of their effect as a whole, when separately considered, will be lost upon the spectator. For it should be remembered the mind is incapable of appreciating or doing justice to two distinct and dissimilar expressions at the same time. Whatever be the scene to be improved, therefore, it should be taken by itself and considered as a whole, if the eye command that scene alone. Then the improver can proceed on the principle that every piece of ground is distinguished by certain properties: it is either tame or bold, graceful or rude, continued or broken; and if any variety inconsistent with these expressions be obtruded, it has no other effect than to *weaken one idea without raising another.* " The insipidity of a flat is not taken away by a few scattered hillocks; a continuation of uneven ground can alone give the idea of irregularity. A large, deep, abrupt break, among easy swells and falls, seems at best but a piece left unfinished, and which ought to have been softened; it is not more natural because it is more rude. On the other hand, a fine small polished form, in the midst of rough, mis-shapen ground, though more elegant than all about it, is generally no better than a patch, itself disgraced and disfiguring the scene. A thousand instances might be added to show that the prevailing idea ought to pervade every part, so far at least indispensably, as to exclude whatever distracts it, and as much further as possible to accommodate the character of the ground to that of the scene to which it belongs."*

Rocks, either in detached fragments or large masses, enter into the composition of many scenes, and sometimes have an excellent effect. Indeed much of the spirit of picturesque scenery is often owing to the bold projections made by rocks in various forms. An overhanging cliff, or steep precipice, a moss-covered rocky bank, or even a group of *rocks* on a ledge, from which springs a tuft of trees and shrubs—all these give strength to a picturesque scene. Their effect may often be rendered more striking by art; sometimes by removing the earth or loose stones from the bottom of the precipice, so as greatly to increase its apparent height—for the perpendicular position is the finest in which rocks can be viewed. At other times the effect of a continuous range of rocks may be much improved by planting the summit, and making occasional breaks of verdure in the front surface.

Rocks which are too apparent, and which cannot be

* Mr. Whately has given such minute and excellent details in relation to this subject, in his *Observations on Modern Gardening*, that we gladly refer the reader who desires to pursue this subject further, to that work: which indeed is so unexceptionable in style and good taste, that *Alison* has frequently quoted it in illustration of his admirable Essay on Taste.

removed, may be concealed with trees and vegetation, or partially covered with vines and creepers. The latter often have a beautiful effect in picturesque scenery, and we have seen very charming pictures formed of over-arching cliffs and groups of rock, upon which hung and rambled in luxuriant profusion, a rich mixture of climbing plants. Where rocks thus accidentally occur in *beautiful* scenes, to which they, if left bare, would be inimical, they may be wonderfully softened and brought into keeping by a covering of the honeysuckle, the Ivy, the Virginia creeper, and other species of the gayest and most luxuriant flowering vines.

Loose and detached fragments of rocks can never be permitted to lie scattered about the lawn in any style. In a scene expressive of graceful beauty, of course they would be entirely out of place: and in a picturesque scene, they should only be suffered to remain in spots where they have some evident connexion with larger masses. If they were allowed to lie loosely around, they would only give an air of confused wildness, opposed to everything like the elegance of tasteful art or the comfort of a country residence; but if only seen in particular spots where they evidently belong, they will, by contrast, give force and spirit to the whole. We do not now speak of large rounded *boulders* or smooth stones, such as are seen lying about the soil in some of our valley tracts, as such are void of interest, and, unless they are large, or in some degree remarkable, they ought to be at once removed out of the way. Characteristic and picturesque rocks, are those with firm, rugged, and distinct outlines, externally covered with a coating of weather stains, dark lichens, or mosses, and which meet the eye with a mellow and softened tone of color.

Roads and walks are so directly connected with operations on the surface of the ground, and with the disposition of plantations, which we have already made familiar to the reader, that we shall introduce in this place a few remarks relative to their direction and formation. A French writer has remarked of them that they are "*les rubans qui attachent le bouquet,*" and they certainly serve as the connecting medium between the different parts of the estate, as well as the means of displaying its various beauties, peculiarities, and finest points of prospect.

The *Approach* is by far the most important of these routes. It is the private road, leading from the public highway, directly to the house itself. It should therefore bear a proportionate breadth and size, and exhibit marks of good keeping, in accordance with the dignity of the mansion.

In the ancient style of gardening, the Approach was so formed as to enter directly in front of the house, affording a full view of that portion of the edifice, and no other. A line drawn as directly as possible, and evenly bordered on each side with a tall avenue of trees, was the whole expenditure of art necessary in its formation. It is true, the simplicity of design was often more than counterbalanced by the difficulty of levelling, grading, and altering the surface, necessary to please the geometric eye; but the rules were as plain and unchangeable, as the lines were parallel and undeviating.

In the present more advanced state of Landscape Gardening, the formation of the Approach has become equally a matter of artistical skill with other details of the art. The house is generally so approached, that the eye shall first meet it in an angular direction, displaying not only the beauty of the architectural *façade* but also one of the *end* elevations, thus giving a more complete idea of the size, character, or elegance of the building: and instead of leading in a direct line from the gate to the house, it curves in easy lines through certain portions of the park or lawn, until it reaches that object.

If the point where the Approach is to start from the highway be not already determined past alteration, it should be so chosen as to afford a sufficient drive through the grounds before arriving at the house, to give the stranger some idea of the extent of the whole property: to allow an agreeable *diversity* of surface over which to lead it: and lastly in such a manner as not to interfere with the convenience of ready access to and from the mansion.

This point being decided, and the other being the mansion and adjacent buildings, it remains to lay out the road in such gradual curves as will appear easy and graceful, without verging into rapid turns or formal stiffness. Since the modern style has become partially known and adopted here, some persons appear to have supposed that nature "has a horror of straight lines," and consequently, believing that they could not possibly err, they immediately ran into the other extreme, filling their grounds with zigzag and regularly serpentine roads, still more horrible: which can only be compared to the contortions of a wounded snake dragging its way slowly over the earth.

There are two guiding principles which have been laid down for the formation of Approach roads. The first, that the curves should never be so great, or lead over surfaces so unequal, as to make it disagreeable to drive upon them; and the second, that the road should *never curve without some reason*, either real or apparent.

The most natural method of forming a winding Approach where the ground is gently undulating, is to follow, in some degree, the depressions of surface, and to curve round the eminences. This is an excellent method, so long as it does not lead us in too circuitous a direction, nor, as we before hinted, make the road itself too uneven. When either of these happens, the easy, gradual flow of the curve in the proper direction, must be maintained by levelling or grading, to produce the proper surface.

Nothing can be more unmeaning than to see an Approach, or any description of road, winding hither and thither, through an extensive level lawn, towards the house, without the least apparent reason for the curves. Happily, we are not, therefore, obliged to return to the straight line; but gradual curves may always be so arranged as to appear necessarily to wind round the *groups of trees*, which otherwise would stand in the way. Wherever a bend in the road is intended, a cluster or group of greater or less size and breadth, proportionate to the curve, should be placed in the projection formed. These trees, as soon as they attain some size, if they are properly arranged, we may suppose to have originally stood there, and the road naturally to have curved, to avoid destroying them.

This arrangement of trees bordering an extended Approach road, in connexion with the various other groups, masses, and single trees, in the adjacent lawn, will in most cases have the effect of concealing the house from the spectator approaching it, except, perhaps, from one or two points. It has, therefore, been considered a matter worthy of consideration, at what point or points the *first view* of the house shall be obtained. If seen at too great a distance, as in the case of a large estate, it may appear more diminutive and of less magnitude than it should; or, if first viewed at some other position, it may strike the eye of a stranger, at that point, unfavorably. The best, and indeed the only way to decide the matter, is to go over the whole ground covered by the Approach route carefully, and select a spot or spots sufficiently near to give the most favorable and striking view of the house itself. This, if openings are to be made, can only be done in winter; but when the ground is to be newly planted, it may be prosecuted at any season.

The late Mr. Repton, who was one of the most celebrated English practical landscape gardeners, has laid down in one of his works, the following rules on the subject, which we quote, not as applying in all cases, but to show what are generally thought the principal requisites of this road in the modern style.

First. It ought to be a road to the house, and to that principally.

Secondly. If it be not naturally the nearest road possible, it ought artificially to be made to appear so.

Thirdly. The artificial obstacles which make this road the nearest, ought to appear natural.

Fourthly. Where an approach quits the high road, it ought not to break from it at right angles, or in such a manner as to rob the entrance of importance, but rather at some bend of the public road, from which a lodge or gate may be more conspicuous: and where the high road may appear to branch from the approach, rather than the approach from the high road.

Fifthly. After the approach enters the park, it should avoid skirting along its boundary, which betrays the want of extent or unity of property.

Sixthly. The house, unless very large and magnificent, should not be seen at so great a distance as to make it appear much less than it really is.

Seventhly. The first view of the house should be from the most pleasing point of sight.

Eighthly. As soon as the house is visible from the approach, there should be no temptation to quit it (which will ever be the case if the road be at all circuitous), unless sufficient obstacles, such as water or inaccessible ground, appear to justify its course.*

Although there are many situations where these rules must be greatly modified in practice, yet the improver will do well to bear them in mind, as it is infinitely more easy to make occasional deviations from general rules, than to carry out a tasteful improvement without any guiding principles.

There are many fine country residences on the banks of the Hudson, Connecticut, and other rivers, where the proprietors are often much perplexed and puzzled by the *situation* of their houses; the building presenting really *two fronts*, while they appear to desire only one. Such is the case when the estate is situated between the public road on one side, and the river on the other; and we have often seen the Approach artificially tortured into a long circuitous route, in order finally to arrive at what the proprietor considers the true front, viz. the side nearest the river. When a building is so situated, much the most

---

* Repton's Inquiry into the Changes of Taste in Landscape Gardening, p. 109.

elegant effect is produced by having two fronts: one, the *entrance front*, with the porch or portico nearest the road, and the other, the *river front*, facing the water. The beauty of the whole is often surprisingly enhanced by this arrangement, for the visitor, after passing by the Approach through a considerable portion of the grounds, with perhaps but slight and partial glimpses of the river, is most agreeably surprised on entering the house, and looking from the drawing-room windows of the other front, to behold another beautiful scene totally different from the last, enriched and ennobled by the wide-spread sheet of water before him. Much of the effect produced by this agreeable surprise from the interior, it will readily be seen, would be lost, if the stranger had already driven round and alighted on the river front.

The *Drive* is a variety of road rarely seen among us, yet which may be made a very agreeable feature in some of our country residences, at a small expense. It is intended for exercise more secluded than that upon the public road, and to show the interesting portions of the place from the carriage, or on horseback. Of course it can only be formed upon places of considerable extent; but it enhances the enjoyment of such places very highly, in the estimation of those who are fond of equestrian exercises. It generally commences where the approach terminates, viz. near the house: and from thence, proceeds in the same easy curvilinear manner through various parts of the grounds, farm, or estate. Sometimes it sweeps through the pleasure grounds, and returns along the very beach of the river, beneath the fine overhanging foliage of its projecting bank; sometimes it proceeds towards some favorite point of view, or interesting spot on the landscape; or at others it leaves the lawn and traverses the farm, giving the proprietor an opportunity to examine his crops, or exhibit his agricultural resources to his friends.

Walks are laid out for purposes similar to Drives, but are much more common, and may be introduced into every scene, however limited. They are intended solely for promenades or exercise on foot, and should therefore be dry and firm, if possible, at all seasons when it is desirable to use them. Some may be open to the south, sheltered with evergreens, and made dry and hard for a warm promenade in winter; others formed of closely mown turf, and thickly shaded by a leafy canopy of verdure, for a cool retreat in the midst of summer. Others again may lead to some sequestered spot, and terminate in a secluded rustic seat, or conduct to some shaded dell or rugged eminence, where an extensive prospect can be enjoyed. Indeed, the genius of the place must suggest the direction, length, and number of the walks to be laid out, as no fixed rules can be imposed in a subject so everchanging and different. It should, however, never be forgotten, that the walk ought always to correspond to the scene it traverses, being rough where the latter is wild and picturesque, sometimes scarcely differing from a common footpath, and more polished as the surrounding objects show evidences of culture and high keeping. In *direction*, like the approach, it should take easy flowing curves, though it may often turn more abruptly at the interposition of an obstacle. The chief beauty of curved and bending lines in walks, lies in the new scenes which by means of them are opened to the eye. In the straight walk of half a mile the whole is seen at a glance, and there is too often but little to excite the spectator to pursue the search; but in the modern style, at

The Ravine Walk at Blithewood.

every few rods, a new turn in the walk opens a new prospect to the beholder, and "leads the eye," as Hogarth graphically expressed it, "a kind of wanton chase," continually affording new refreshment and variety.

Fences are often among the most unsightly and offensive objects in our country seats. Some persons appear to have a passion for subdividing their grounds into a great number of fields; a process which is scarcely ever advisable even in common farms, but for which there can be no apology in elegant residences. The close proximity of fences to the house gives the whole place a confined and mean character. "The mind," says Repton, "feels a certain disgust under a sense of confinement in any situation, however beautiful." A wide-spread lawn, on the contrary, where no boundaries are conspicuous, conveys an impression of ample extent and space for enjoyment. It is frequently the case that, on that side of the house nearest the outbuildings, fences are, for convenience, brought in its close neighborhood, and here they are easily concealed by plantations; but on the other sides, open and unobstructed views should be preserved, by removing all barriers not absolutely necessary.

Nothing is more common, in the places of cockneys who become inhabitants of the country, than a display immediately around the dwelling of a spruce paling of carpentry, neatly made, and painted white or green; an abomination among the fresh fields, of which no person of taste could be guilty. To fence off a small plot around a fine house, in the midst of a lawn of fifty acres, is a perversity which we could never reconcile, with even the lowest perception of beauty. An old stone wall covered with creepers and climbing plants, may become a picturesque barrier a thousand times superior to such a fence. But there is never one instance in a thousand where any barrier is necessary. Where it is desirable to separate the house from the level grass of the lawn, let it be done by an architectural terrace of stone, or a raised platform of gravel supported by turf, which will confer importance and dignity upon the building, instead of giving it a petty and trifling expression.

Verdant hedges are elegant substitutes for stone or wooden fences, and we are surprised that their use has not been hitherto more general. We have ourselves been making experiments for the last ten years with various hedge-plants, and have succeeded in obtaining some hedges which are now highly admired. Five or six years will, in this climate, under proper care, be sufficient to produce hedges of great beauty, capable of withstanding the attacks of every kind of cattle ; barriers, too, which will outlast many generations. The common *Arbor Vitæ* (or flat Cedar), which grows in great abundance in many districts, forms one of the most superb hedges, without the least care in trimming; the foliage growing thickly down to the very ground, and being evergreen, the hedge remains clothed the whole year. Our common Thorns, and in particular those known in the nurseries as the Newcastle and Washington thorns, form hedges of great strength and beauty. They are indeed much better adapted to this climate than the English Hawthorn, which often suffers from the unclouded radiance of our midsummer sun. In autumn, too, it loses its foliage much sooner than our native sorts, some of which assume a brilliant scarlet when the foliage is fading in autumn. In New England, the Buckthorn is preferred from its rapid and luxuriant growth ;* and in the middle states, the Maclura, or Osage Orange, is becoming a favorite for its glossy and polished foliage. The Privet, or Prim, is a rapid growing shrub, well fitted for interior divisions. Picturesque hedges are easily formed by intermingling a variety of flowering shrubs, sweet briers, etc., and allowing the whole to grow together in rich masses. For this purpose the Michigan rose is admirably adapted at the north, and the Cherokee rose at the south. In all cases where hedges are employed in the natural style of landscape (and not in close connexion with highly artificial objects, buildings, etc.), a more agreeable effect will be produced by allowing the hedge to grow somewhat irregular in form, or varying it by planting near it other small trees and shrubs to break the outline, than by clipping it in even and formal lines. Hedges may be obtained in a single season, by planting long shoots of the osier willow, or any other tree which throws out roots easily from cuttings.

A simple and pleasing barrier, in good keeping with cottage residences, may be formed of *rustic work*, as it is termed. For this purpose, stout rods of any of our native forest trees are chosen (Cedar being preferable) with the bark on, six to ten feet in length ; these are sharpened and driven into the ground in the form of a lattice, or wrought into any figures of trellis that the fancy may suggest. When covered with luxuriant vines and climbing plants, such a barrier is often admirable for its richness and variety.

* The Buckthorn is perhaps the best plant where a thick screen is very speedily desired. It is not liable to the attack of insects; grows very thickly at the bottom, at once: and will make an efficient screen sooner than almost any other plant.

The sunken fence, fosse, or *ha-ha*, is an English invention, used in separating that portion of the lawn near the house, from the part grazed by deer or cattle, and is only a ditch sufficiently wide and deep to render communication difficult on opposite sides. When the ground slopes from the house, such a sunk fence is invisible to a person near the latter, and answers the purpose of a barrier without being in the least obtrusive.

In a succeeding section we shall refer to terraces with their parapets, which are by far the most elegant barriers for a highly decorated flower garden, or for the purpose of maintaining a proper connexion between the house and the grounds, a subject which is scarcely at all attended to, or its importance even recognised as yet among us.

SECTION VIII.

TREATMENT OF WATER.

Beautiful effects of this element in nature. In what cases it is desirable to attempt the formation of artificial pieces of water. Regular forms unpleasing. Directions for the formation of ponds or lakes in the irregular manner. Study of natural lakes. Islands. Planting the margin. Treatment of natural brooks and rivulets. Cascades and waterfalls. Legitimate sphere of the art in this department.

——The dale
With woods o'erhung, and shagg'd with mossy rocks,
Whence on each hand the gushing waters play,
And down the rough cascade white-dashing fall,
Or gleam in lengthened vista through the trees.
THOMSON.

THE delightful and captivating effects of water in landscapes of every description, are universally known and admitted. The boundless sea, the broad full river, the dashing noisy brook, and the limpid meandering rivulet, are all possessed of their peculiar charms; and when combined with scenes otherwise finely disposed and well wooded, they add a hundred fold to their beauty. The soft and trembling shadows of the surrounding trees and hills, as they fall upon a placid sheet of water—the brilliant light which the crystal surface reflects in pure sunshine, mirroring, too, at times in its resplendent bosom, all the cerulean depth and snowy whiteness of the overhanging sky, give it an almost magical effect in a beautiful landscape. The murmur of the babbling brook, that

<blockquote>" In linked sweetness long drawn out,"</blockquote>

falls upon the ear in some quiet secluded spot, is inexpressibly soothing and delightful to the mind; and the deeper sound of the cascade that rushes, with an almost musical dash, over its bed of moss-covered rock, is one of the most fascinating of the many elements of enjoyment in a fine country seat. The simplest or the most monotonous view may be enlivened by the presence of water in any considerable quantity; and the most picturesque and striking landscape will, by its addition, receive a new charm, inexpressibly enhancing all its former interest. In short, as no place can be considered perfectly complete without either a water view or water upon its own grounds, wherever it does not so exist and can be *easily* formed by artificial means, no man will neglect to take advantage of so fine a source of embellishment as is this element in some of its varied forms.

<blockquote>
"——— Fleuves, ruisseaux, beaux lacs, claires fontaines,
Venez, portez partout la vie et la fraîcheur?
Ah! qui peut remplacer votre aspect enchanteur?
De près il nous amuse, et de loin nous invite:
C'est le premier qu'on cherche, et le dernier qu'on quitte.
Vous fécondez les champs; vous répétez les cieux;
Vous enchantez l'oreille, et vous charmez les yeux."
</blockquote>

In this country, where the progress of gardening and improvements of this nature, is rather shown in a simple and moderate embellishment of a large number of villas and country seats, than by a lavish and profuse expenditure on a few entailed places, as in the residences of the English nobility, the formation of large pieces of water at great cost and extreme labor, would be considered both absurd and uncalled for. Indeed, when nature has so abundantly spread before us such an endless variety of superb lakes, rivers, and streams of every size and description, the efforts of man to rival her great works by *mere imitation*, would, in most cases, only become ludicrous by contrast.

When, however, a number of perpetual springs cluster together, or a rill, rivulet, or brook, runs through an estate in such a manner as *easily* to be improved or developed into an elegant expanse of water in any part of the grounds, we should not hesitate to take advantage of so fortunate a circumstance. Besides the additional beauty conferred upon the whole place by such an improvement, the proprietor may also derive an inducement from its utility; for the possession of a small lake, well stocked with carp, trout, pickerel, or any other of the excellent pond fish, which thrive and propagate extremely well in clear fresh water, is a real advantage which no one will undervalue.

There is no department of Landscape Gardening which appears to have been less understood in this country than the management of water. Although there have not been many attempts made in this way, yet the occasional efforts that have been put forth in various parts of the country, in the shape of square, circular, and oblong pools of water, indicate a state of knowledge extremely meagre, in the art of Landscape Gardening. The highest scale to which these pieces of water rise in our estimation is that of respectable horse-ponds;—beautiful objects they certainly are not. They are generally round or square, with perfectly smooth, flat banks on every side, and resemble in tameness and insipidity, a huge basin set down in the middle of a green lawn. They are even, in most cases, denied the advantage of shade, except perhaps occasionally a few straggling trees can be said to fulfil that purpose; for richly tufted margins, and thickets of overhanging shrubs, are accompaniments rare indeed.*

* Simple and easy as would appear the artificial imitation of these variations of nature, yet to an unpractised hand and a tasteless mind, nothing is really more difficult. To produce meagre right lines and geometrical forms is extremely easy in any of the fine arts, but to give the grace, spirit, and variety of nature, requires both tasteful perception and some practice; hence, in the infancy of any art, the productions are characterized by extreme meagreness and simplicity;—of which the first efforts to draw the human figure or to form artificial pieces of water, are good examples.

Brown, who was one of the early practitioners of the modern style abroad, and who just saw far enough to lay aside the ancient formal method, without appreciating nature sufficiently to be willing to take her for his model, once disgraced half of the finest places in England with his tame, bald pieces of artificial water, and round, formal clumps of trees. Mr. Knight, in his elegant poem, "The Landscape," spiritedly rebuked this practice in the following lines:—

"Shaved to the brink our brooks are taught to flow
Where no obtruding leaves or branches grow:
While clumps of shrubs bespot each winding vale
Open alike to every gleam and gale:
Each secret haunt and deep recess display'd,
And intricacy banished with its shade.

Hence, hence! thou haggard fiend, however call'd,
The meagre genius of the bare and bald;
Thy spade and mattock here at length lay down,
And follow to the tomb, thy favorite, Brown;
Thy favorite Brown, whose innovating hand
First dealt thy curses o'er this fertile land;
First taught the walk in spiral forms to move,
And from their haunts the secret Dryads drove;
With clumps bespotted o'er the mountain's side,
And bade the stream 'twixt banks close-shaven glide;
Banish'd the thickets of high tow'ring wood
Which hung reflected o'er the glassy flood."

Lakes or ponds are the most beautiful forms in which water can be displayed in the grounds of a country residence.* They invariably produce their most pleasing effects when they are below the level of the house; as, if above, they are lost to the view, and if placed on a level with the eye, they are seen to much less advantage. We conceive that they should never be introduced where they do not naturally exist, except with the concurrence of the following circumstances. First, a sufficient quantity of running water to maintain *at all times* an overflow, for nothing can be more unpleasant than a stagnant pool, as nothing is more delightful than pure, clear, limpid water; and secondly, some natural formation of ground, in which the proposed water can be expanded, that will not only make it appear natural, but diminish, a hundred fold, the expense of formation.

The finest and most appropriate place to form a lake, is in the bottom of a small valley, rather broad in proportion to its length. The soil there will probably be found rather clayey and retentive of moisture; and the rill or brook, if not already running through it, could doubtless be easily diverted thither. There, by damming up the lower part of the valley with a head of greater or less height, the water may be thrown back so as to form the whole body of the lake.

The first subject which will demand the attention, after the spot has been selected for the lake or pond, and the

---

* Owing to the immense scale upon which nature displays this fine element in North America, every sheet of water of moderate or small size is almost universally called a *pond*. And many a beautiful, limpid, natural expanse, which in England would be thought a charming lake, is here simply a pond. The term may be equally correct, but it is by no means as elegant.

height of the head and consequent depth of water determined upon, is the proposed *form* or *outline* of the whole. And, as we have already rejected all regular and geometric forms, in scenes where either natural or picturesque beauty is supposed to predominate, we must turn our attention to examples for imitation in another direction.

If, then, the improver will recur to the most beautiful small *natural* lake within his reach, he will have a subject to study and an example to copy well worthy of imitation. If he examine minutely and carefully such a body of water, with all its accompaniments, he will find that it is not only delightfully wooded and overshadowed by a variety of vegetation of all heights, from the low sedge that grows on its margin, to the tall tree that bends its branches over its limpid wave; but he will also perceive a striking peculiarity in its *irregular outline*. This, he will observe, is neither round, square, oblong, nor any modification of these regular figures, but full of bays and projections, sinuosities, and recesses of various forms and sizes, sometimes bold, and reaching a considerable way out into the body of the lake, at others, smaller and more varied in shape and connexion. In the heights of the banks, too, he will probably observe considerable variety. At some places, the shore will steal gently and gradually away from the level of the water, while at others it will rise suddenly and abruptly, in banks more or less steep, irregular, and rugged. Rocks and stones covered with mosses, will here and there jut out from the banks, or lie along the margin of the water, and the whole scene will be full of interest from the variety, intricacy, and beauty of the various parts. If he will accurately note in his mind all these varied forms—their separate outlines, the way in which they blend into one another, and connect themselves together, and the effect which, surrounding the water, they produce as a whole, he will have some tolerably correct ideas of the way in which an artificial lake ought to be formed.

Let him go still further now, in imagination, and suppose the banks of this natural lake, without being otherwise altered, entirely denuded of grass, shrubs, trees, and verdure of every description, remaining characterized only by their original form and outline; this will give him a more complete view of the method in which his labors must commence; for uncouth and apparently mis-shapen as those banks are and must be, when raw and unclothed, to exhibit all their variety and play of light and shadow when verdant and complete, so also must the original form of the banks and margin of the piece of artificial water, in order finally to assume the beautiful or picturesque, be made to assume outlines equally rough and harsh in their raw and incomplete state.

It occasionally happens, though rarely, that around the hollow or valley where it is proposed to form the piece of water, the ground rises in such irregular form, and is so undulating, receding, and projecting in various parts, that when the water is dammed up by the head below, the natural outline formed by the banks already existing, is sufficiently varied to produce a pleasing effect without much further preparatory labor. This, when it occurs, is exceedingly fortunate; but the examples are so unfrequent, that we must here make our suggestions upon a different supposition.

When, therefore, it is found that the form of the intended lake would not be such as is desirable, it must be made so by digging. In order to do this with any exactness the improver should take his stand at that part of the ground where the dam or head is to be formed, and raising his levelling instrument to the exact height to which the intended lake will rise, sweep round with his eye upon the surrounding sides of the valley, and indicate by placing marks there, the precise line to which the water will reach. This can easily be done throughout the whole circumference by a few changes of position.

When the outline is ascertained in this way, and marked out, the improver can, with the occasional aid of the leveller, easily determine where and how he can make alterations and improvements. He will then excavate along the new margin, until he makes the water line (as shown by the instrument) penetrate to all the various bays, inlets, and curves of the proposed lake. In making these irregular variations, sometimes bold and striking, at others fainter and less perceptible, he can be guided, as we have already suggested, by no fixed rules, but such as he may deduce from the operations of nature on the same materials, or by imbuing his mind with the beauty of forms in graceful and refined art. In highly polished scenery, elegant curves and graceful sweeps should enter into the composition of the outline; but in wilder or more picturesque situations, more irregular and abrupt variations will be found most suitable and appropriate.

The intended water outline once fully traced and understood, the workmen can now proceed to form the banks. All this time the improver will keep in mind the supposed appearance of the bank of a natural lake stripped of its vegetation, etc., which will greatly assist him in his progress. In some places the banks will rise but little from the water, at others one or two feet, and at others perhaps three, four, or six times as much. This they will do, not in the same manner in all portions of the outline, sloping away with a like gradual rise on both sides, for this would inevitably produce tameness and monotony, but in an irregular and varied manner; sometimes falling back gradually, sometimes starting up perpendicularly, and again overhanging the bed of the lake itself.

All this can be easily effected while the excavations of those portions of the bed which require deepening are going on. And the better portions of the soil obtained from the latter, will serve to raise the banks when they are too low.

It is of but little consequence how roughly and irregularly the projections, elevations, etc., of the banks and outlines are at first made, so that some general form and connexion is preserved. The danger lies on the other side, viz. in producing a whole too tame and insipid ; for we have found by experience, how difficult it is to make the best workmen understand how to operate in any other way than in regular curves and straight lines. Besides, newly moved earth, by settling and the influence of rains, etc., tends, for some time, towards greater evenness and equality of surface.

Mr. Price, in his unrivalled instructions for the creation of pieces of artificial water, has suggested another excellent method by which the outlines and banks of lakes may be varied. This is, first, by cutting down the banks, in some places nearest the water, perpendicularly, and then undermining them. This will produce a gradual variation in some parts, which, falling to pieces, will produce new and irregular accidental outlines. When, by the action of rain and frost, added to that of the water itself, large fragments of mould tumble from the hollowed banks of rivers or lakes, these fragments, by the accumulation of other mould, often lose their rude and broken form, are covered with the freshest grass, and enriched with tufts of natural flowers; and though detached from the bank, and upon a lower level, still appear connected with it, and vary its outline in the softest and most pleasing manner. As fragments of the same kind will always be detached from ground that is undermined, so by their means the same effects may designedly be produced; and they will suggest numberless intricacies and varieties of a soft and pleasing, as well as of a broken kind.

It will of course be well understood that we have here *not* supposed our proposed lake to be located in a valley that must be filled to the brim, or in a tame flat when the water would rise to the same level as the adjacent ground. In such situations there could be but little room for the display of a high degree of picturesque beauty. On the contrary, when the surrounding ground in many places rises gradually, or is naturally higher than the proposed level of the water, there is room for all the variety of banks of various heights, form, and outline, which so spring out of the neighboring undulations and eminences, and connect themselves with them, as to appear perfectly natural and in proper keeping.

In arranging these outlines and banks, we should study the effect at the points from which they will generally be viewed. Some pieces of water in valleys, are looked down upon from other and higher parts of the demesne; others (and this is most generally the case) are only seen from the adjoining walk, at some point or points where the latter approaches the lake. They are most generally seen from one, and seldom from more than two sides. When a lake is viewed from above, its contour should be studied as a whole ; but when it is only seen from one or more sides or points, the beauty of the *coup d'œil* from those positions can often be greatly increased by some trifling alterations in arrangement. A piece of water which is long and comparatively narrow, appears extremely different in opposite points of view; if seen lengthwise from either extremity, its apparent breadth and extent is much increased ; while, if the spectator be placed on one side and look across, it will seem narrow and insignificant. Now, although the form of an artificial lake of moderate size should never be much less in breadth than in length, yet the contrary is sometimes unavoidably the case ; and being so, we should by all means avail ourselves of those well known laws in perspective, which will place them in the best possible position, relative to the spectator.

If the improver desire to render his banks still more picturesque, resembling the choicest *morçeaux* of natural banks, he should go a step further in arranging his materials before he introduces the water, or clothes the margin with vegetation. In analysing the finest portions of natural banks, it will be observed that their peculiar characteristics often depend on other objects besides the mere ground of the surrounding banks, and the trees and verdure with which they are clothed. These are, rocks of various size, forms, and colors, often projecting out of or holding up the bank in various places ; stones sometimes imbedded in the soil, sometimes lying loosely along the shore ; and lastly, old stumps of trees with gnarled roots, whose decaying hues are often extremely mellow and agreeable to the eye. All these have much to do with the expression of a truly picturesque bank, and cannot be excluded or taken away from it without detracting largely from its character. There is no reason, therefore, in an imitation of nature, why we should not make use of all her materials to produce a similar effect; and although in the raw and rude state of the banks at first, they may have a singular and rather *outré* aspect, stuck round and decorated here and there with large rocks, smaller stones, and old stumps of trees; yet it must be remembered that this is only the chaotic state, from which the new creation is to emerge more perfectly formed and completed; and also that the appearance of these rocks and stumps, when covered with mosses, and partially overgrown with a profusion of luxuriant vegetation and climbing plants, will be as beautifully picturesque after a little time has elapsed, as it is now uncouth and uninviting.

Islands generally contribute greatly to the beauty of a piece of water. They serve, still further, to increase the variety of outline, and to break up the wide expanse of liquid into secondary portions, without injuring the effect of the whole. The striking contrast, too, between their verdure, the color of their margins, composed of variously tinted soils and stones, and the still, smooth water around them,—softened and blended as this contrast is, by their shadows reflected back from the limpid element, gives additional richness to the picture.

The distribution of islands in a lake or pond requires some judgment. They will always appear most natural when sufficiently near the shore, on either side, to maintain in appearance some connexion with it. Although islands do sometimes occur near the middle of natural lakes, yet the effect is by no means good, as it not only breaks and distracts the effects of the whole expanse by dividing it into two distinct parts, but it always indicates a shallowness or want of depth where the water should be deepest.

There are two situations where it is universally admitted that islands may be happily introduced. These are, at the inlet and the exit of the body of water. In many cases where the stream which supplies the lake is not remarkable for size, and will add nothing to the appearance of the whole view from the usual points of sight, it may be concealed by an island or small group of islands, placed at some little distance in front of it. The head or dam of a lake, too, is often necessarily so formal and abrupt, that it is difficult to make it appear natural and in good keeping with the rest of the margin. The introduction of an island or two, placed near the main shore, on either side, and projecting as far as possible before the dam, will greatly diminish this disagreeable formality, particularly if well clothed with a rich tuft of shrubs and overhanging bushes.

Except in these two instances, islands should be generally placed *opposite the salient points* of the banks, or near those places where small breaks or promontories run out into the water. In such situations, they will increase the irregularity of the outline, and lend it additional spirit and animation. Should they, on the other hand, be seated in or near the marginal curve and indentations, they will only serve to clog up these recesses; and while their own figures are lost in these little bays where they are hidden, by lessening the already existing irregularities, they will render the whole outline tame and spiritless.

On one or two of these small islands, little rustic habitations, if it coincide with the taste of the proprietor, may be made for different aquatic birds or water fowl, which will much enliven the scene by their fine plumage. Among these the *swan* is pre-eminent, for its beauty and gracefulness. Abroad, they are the almost constant accompaniments of water in the ground of country residences; and it cannot be denied that, floating about in the limpid wave, with their snow-white plumage and superbly curved necks, they are extremely elegant objects.

After having arranged the banks, reared up the islands, and completely formed the bed of the proposed lake, the improver will next proceed, at the proper period, to finish his labors by clothing the newly formed ground, in various parts, with vegetation. This may be done immediately, if it be desirable; or if the season be not favorable, it may be deferred until the banks, and all the newly formed earth, have had time to settle and assume their final forms, after the dam has been closed, and the whole basin filled to its intended height.

Planting the margins of pieces of water, if they should be of much extent, must evidently proceed upon the same leading principle that we have already laid down for ornamental plantations in other situations. That is, there must be trees of different heights and sizes, and underwood and shrubs of lower growth, disposed sometimes singly, at others in masses, groups, and thickets: in all of which forms, *connexion* must be preserved, and the whole must be made to blend well together, while the different sizes and contours will prevent any sameness and confusion. On the retreating dry banks, the taller and more sturdy deciduous and evergreen trees, as the oak, ash, etc., may be planted, and nearer by, the different willows, the elm, the alder, and other trees that love a moister situation, will thrive well. It is indispensably necessary in order to produce breadth of effect and strong rich contrasts, that *underwood* should be employed to clothe many parts of the banks. Without it, the stems of trees will appear loose and straggling, and the screen will be so imperfect as to allow a free passage for the vision in every direction. For this purpose, we have in all our woods, swamps, and along our brooks, an abundance of hazels, hawthorns, alders, spice woods, winter berries, azaleas, spireas, and a hundred other fine low shrubs, growing wild, which are by nature extremely well fitted for such sites, and will produce immediate effect on being transplanted. These may be intermingled, here and there, with the swamp button-bush (*Cephalanthus*), which bears handsome white globular heads of blossoms, and the swamp magnolia, which is highly beautiful and fragrant. On cool north banks, among shelves of proper soil upheld by projecting ledges of rock, our native Kalmias and Rhododendrons, the common and mountain laurels, may be made to flourish. The Virginia Creeper, and other beautiful wild vines, may be planted at the roots of some of the trees to clamber up their stems, and the wild Clematis so placed that its luxuriant festoons shall hang gracefully from the projecting boughs of some of the overarching trees. Along the lower banks and closer margins, the growth of smaller plants will be encouraged, and various kinds of wild ferns may be so planted as partially to conceal, overrun, and hide the rocks and stumps of trees, while trailing plants, as the periwinkle and moneywort (*Lysamachia nummularia*), will still further increase the intricacy and richness of such portions. In this way, the borders of the lake will resemble the finest portions of the banks of picturesque and beautiful natural dells and pieces of water, and the effect of the whole when time has given it the benefit of its softening touches, if it has been thus properly executed, will not be much inferior to those matchless bits of fine landscape. A more striking and artistical effect will be produced by substituting for native trees and shrubs, common on the banks of streams and lakes in the country, only rare *foreign* shrubs, vines, and aquatic plants of hardy growth, suitable for such situations. While these are arranged in the same manner as the former, from their comparative novelty, especially in such sites, they will at once convey the idea of refined and elegant art.

If any person will take the trouble to compare a piece of water so formed, when complete, with the square or circular sheets or ponds now in vogue among us, he must indeed be little gifted with an appreciation of the beautiful, if he do not at once perceive the surpassing merit of the natural style. In the old method, the banks, level, or rising on all sides, without any or but few surrounding trees, carefully gravelled along the edge of the water, or what is still worse, walled up, slope away in a tame, dull, uninteresting grass field. In the natural method, the outline is varied, sometimes receding from the eye, at others stealing out, and inviting the gaze—the banks here slope off gently with a gravelly beach, and there rise abruptly in different heights, abounding with hollows, projections, and eminences, showing various colored rocks and soils, intermingled with a luxuriant vegetation of all sizes and forms, corresponding to the different situations. Instead of allowing the sun to pour down in one blaze of light, without any objects to soften it with their shade, the thick overhanging groups and masses of trees cast, here and there, deep cool shadows. Stealing through the leaves and branches, the sun-beams quiver. and play upon the surface of the flood, and are reflected back in dancing light, while their full glow upon the broader and more open portions of the lake is relieved, and brought into harmony by the cooler and softer tints mirrored in the water from the surrounding hues and tints of banks, rocks, and vegetation.

Natural brooks and rivulets may often be improved greatly by a few trifling alterations and additions, when they chance to come within the bounds of a country residence. Occasionally, they may be diverted from their original beds when they run through distant and unfrequented parts of the demesne, and brought through nearer portions of the pleasure grounds or lawn. This, however, can only be done with propriety when there is a natural indication in the grounds through which it is proposed to divert it—as a succession of hollows, etc., to form the future channel. Sometimes, a brisk little brook can be divided into smaller ones for some distance, again uniting at a point below, creating additional diversity by its varying form.*

Brooks, rivulets, and even rills may frequently be greatly improved by altering the form of their beds in various places. Often by merely removing a few trifling obstructions, loose stones, branches, etc., or hollowing away the

---

* The Abbe Délille has given us a fine image of a brook thus divided, in the following lines:—

" Plus loin, il se sépare en deux ruisseaux agiles,
Qui, se suivant l'un l'autre avec rapidité,
Disputent de vitesse et de limpidité ;
Puis, rejoignant tous deux le lit qui les rassemble,
Murmurent enchantés de voyager ensemble.
Ainsi, toujours errant de détour en détour,
Muet, bruyant, paisible, inquiet tour à tour,
Sous mille aspects divers son cours se renouvelle."

adjoining bank for a short distance, fine little expanses or pools of still water may be formed, which are happily contrasted with the more rugged course of the rest of the stream. Such improvements of these minor water courses are much preferable to widening them into flat, insipid, tame canals or rivers, which, though they present greater surface to the eye, are a thousand times inferior in the impetuosity of motion, and musical, "babbling sound," so delightful in rapid brooks and rivulets.*

Cascades and water-falls are the most charming features of natural brooks and rivulets. Whatever may be their size they are always greatly admired, and in no way is the peculiar stillness of the air, peculiar to the country, more pleasingly broken, than by the melody of falling water. Even the gurgling and mellow sound of a small rill, leaping over a few fantastic stones, has a kind of lulling fascination for the ear, and when this sound can be brought so near as to be distinctly heard at the residence itself, it is peculiarly delightful.† Now any one who examines a small cascade at all attentively, in a natural brook, will see that it is often formed in the simplest manner by the interposition of a few large projecting stones, which partially dam up the current and prevent the ready flow of the water. Such little cascades are easily imitated, by following exactly the same

---

* The most successful improvement of a natural brook that we have ever witnessed, has been effected in the grounds of Henry Sheldon, Esq., of Tarrytown, N. Y. The great variety and beauty displayed in about a fourth of a mile of the course of this stream, its pretty cascades, rustic bridges, rockwork, etc., reflect the highest credit on the taste of that gentleman.

† The fine stream which forms the south boundary of Blithewood, on the Hudson, the seat of R. Donaldson, Esq., affords two of the finest natural cataracts that we have seen in the grounds of any private residence. Fig. 41 is a view of the larger cascade which falls about 60 feet over a bold, rocky bed.

Fig. 41 The Cataract at Blithewood.

course, and damming up the little brook artificially; studiously avoiding, however, any formal and artificial disposition of the stones or rocks employed.

Larger water-falls and cascades cannot usually be made without some regular head or breastwork, to oppose more firmly the force of the current. Such heads may be formed of stout plank and well prepared clay;* or, which is greatly preferable, of good masonry laid in water cement. After a head is thus formed it must be concealed entirely from the eye by covering it both upon the top and sides with natural rocks and stones of various sizes, so ingeniously disposed, as to appear fully to account for, or be the cause of the water-fall.

The axe of the original backwoodsman appears to have left such a mania for *clearing* behind it, even in those portions of the Atlantic states where such labor should be for ever silenced, that some of our finest places in the country will be found much desecrated and mutilated by its careless and unpardonable use; and not only are fine plantations often destroyed, but the banks of some of our finest streams and prettiest rivulets partially laid bare by the aid of this instrument, guided by some tasteless hand. Wherever fine brooks or water courses are thus mutilated, one of the most necessary and obvious improvements is to reclothe them with plantations of trees and underwood. In planting their banks anew, much beauty and variety can often be produced by employing different growths, and arranging them as we have directed for the margins

---

* It is found that strong loam or any tenacious earth well prepared by *puddling* or beating in water is equally impervious to water as clay; and may therefore be used for lining the sides or dams of bodies of made water when such materials are required.

of lakes and ponds. In some places where easy, beautiful slopes and undulations of ground border the streams, gravel, soft turf, and a few simple groups of trees, will be the most natural accompaniments; in others where the borders of the stream are broken into rougher, more rocky, and precipitous ridges, all the rich wildness and intricacy of low shrubs, ferns, creeping and climbing plants, may be brought in to advantage. Where the extent to be thus improved is considerable, the trouble may be lessened by planting the larger growth, and sowing the seeds of the smaller plants mingled together. Prepare the materials, and time and nature, with but little occasional 'assistance, will mature, and soften, and blend together the whole, in their own matchless and inimitable manner.

From all that we have suggested in these limited remarks, it will be seen that we would only attempt in our operations with water, the graceful or picturesque imitations of natural lakes or ponds, and brooks, rivulets, and streams. Such are the only forms in which this unrivalled element can be displayed so as to harmonize agreeably with natural and picturesque scenery. In the latter, there can be no apology made for the introduction of straight canals, round or oblong pieces of water, and all the regular forms of the geometric mode; because they would evidently be in violent opposition to the whole character and expression of natural landscape. In architectural, or flower gardens (on which we shall hereafter have occasion to offer some remarks), where a different and highly artificial arrangement prevails, all these regular forms, with various jets, fountains, etc., may be employed with good taste, and will combine well with the other accessories of such places. But in the grounds of a residence in the modern style, *nature*, if possible, still more purified, as in the great *chefs-d'œuvre* of art, by an ideal standard, should be the great aim of the Landscape Gardener. And with water especially, only beautiful when allowed to take its own flowing forms and graceful motions, more than with any other of our materials, all appearance of constraint and formality should be avoided. If art be at all manifest, it should discover itself only, as in the admirably painted landscape, in the reproduction of nature in her choicest developments. Indeed, many of the most celebrated authors who have treated of this subject, appear to agree that the productions of the artist in this branch are most perfect as they approach most nearly to fac-similes of nature herself: and though art should have formed the whole, its employment must be nowhere discovered by the spectator; or as *Tasso* has more elegantly expressed the idea:

"L'ARTE CHE TUTTO FA, NULLA SI SCOPRE."

SECTION IX.

LANDSCAPE OR RURAL ARCHITECTURE.

Difference between a city and a country house. The characteristic features of a country house. Examination of the leading principles in Rural Architecture. The different styles. The Grecian style, its merits and defects, and its associations. The Roman and Italian styles. The Pointed or Gothic style. The Tudor Mansion. The English Cottage, or Rural Gothic style. These styles considered in relation to situation or scenery. Individual tastes. Entrance Lodges.

> "A house amid the quiet country's shades,
> With length'ning vistas, ever sunny glades;
> Beauty and fragrance clustering o'er the wall,
> A porch inviting, and an ample hall."

ARCHITECTURE, either practically considered or viewed as an art of taste, is a subject so important and comprehensive in itself, that volumes would be requisite to do it justice. Buildings of every description, from the humble cottage to the lofty temple, are objects of such constant recurrence in every habitable part of the globe, and are so strikingly indicative of the intelligence, character, and taste of the inhabitants, that they possess in themselves a great and peculiar interest for the mind. To have a "local habitation,"—a permanent dwelling, that we can give the impress of our own mind, and identify with our own existence,—appears to be the ardent wish, sooner or later felt, of every man: excepting only those wandering sons of Ishmael, who pitch their tents with the same indifference, and as little desire to remain fixed, in the flowery plains of Persia, as in the sandy deserts of Zahara or Arabia.

In a city or town, or in its immediate vicinity, where space is limited, where buildings stand crowded together, and depend for their attractions entirely upon the style and manner of their construction, mere architectural effect, after convenience and fitness are consulted, is of course the only point to be kept in view. There, the façade, which meets the eye of the spectator from the public street, is enriched and made attractive by the display of architectural style and decoration, commensurate to the magnitude or importance of the edifice; and the whole, so far as the effect of the building is concerned, comes directly within the province of the architect alone.

With respect to this class of dwellings we have little complaint to make, for many of our town residences are highly elegant and beautiful. But how shall we designate that singular perversity of taste, or rather that total want of it, which prompts the man, who, under the name of a villa residence, piles up in the free open country, amid the green fields, and beside the wanton gracefulness of luxuriant nature, a stiff modern "three story brick," which, like a well bred cockney with a true horror of the country, doggedly seems to refuse to enter into harmonious combination with any other object in the scene, but only serves to call up the exclamation,

> Avaunt, stiff pile! why didst thou stray
> From blocks congenial in Broadway!

Yet almost daily we see built up in the country huge

combinations of boards and shingles, without the least attempt at adaptation to situation; and square masses of brick start up here and there, in the verdant slopes of our village suburbs, appearing as if they had been transplanted, by some unlucky incantation, from the close-packed neighborhood of city residence, and left accidentally in the country, or, as Sir Walter Scott has remarked, "had strayed out to the country for an airing."

What then are the proper characteristics of a rural residence? The answer to this, in a few words, is, such a dwelling, as from its various accommodations, not only gives ample .space for all the comforts and conveniences of a country life, but by its varied and picturesque form and outline, its porches, verandas, etc., also appears to have some reasonable connexion, or be in perfect keeping, with surrounding nature. *Architectural beauty* must be considered conjointly with the *beauty of the landscape* or situation. Buildings of almost every description, and particularly those for the habitation of man, will be considered by the mind of taste, not only as architectural objects of greater or less merit, but as component parts of the general scene; united with the surrounding lawn, embosomed in tufts of trees and shrubs, if properly designed and constructed, they will even serve to impress a character upon the surrounding landscape. Their effect will frequently be good or bad, not merely as they are excellent or indifferent examples of a certain style of building, but as they are happily or unhappily combined with the adjacent scenery. The intelligent observer will readily appreciate the truth of this, and acknowledge the value, as well as necessity, of something besides architectural knowledge. And he will perceive how much more likely to be successful are the efforts of him, who, in composing and constructing a rural residence, calls in to the aid of architecture, the genius of the landscape;—whose mind is imbued with a taste for beautiful scenery, and who so elegantly and ingeniously engrafts art upon nature, as to heighten her beauties; while by the harmonious union he throws a borrowed charm around his own creation.

The English, above all other people, are celebrated for their skill in what we consider *rural adaptation*. Their residences seem to be a part of the scenes where they are situated; for their exquisite taste and nice perception of the beauties of Landscape Gardening and rural scenery, lead them to erect those picturesque edifices, which, by their varied outlines, seem in exquisite keeping with nature; while by the numberless climbing plants, shrubs, and fine ornamental trees with which they surround them, they form beautiful pictures of rural beauty. Even the various offices connected with the dwelling, partially concealed by groups of foliage, and contributing to the expression of domestic comfort, while they extend out, and give importance to the main edifice, also serve to connect it, in a less abrupt manner, with the grounds.

The leading principles which should be our guide in Landscape or Rural Architecture, have been condensed by an able writer in the following heads. "1st, As a useful art, in FITNESS FOR THE END IN VIEW: 2d, as an art of design in EXPRESSION OF PURPOSE: 3d, as an art of taste, in EXPRESSION OF SOME PARTICULAR ARCHITECTURAL STYLE."

The most enduring and permanent source of satisfaction in houses is, undoubtedly, utility. In a country residence, therefore, of whatever character, the comfort and convenience of the various members of the family being the first and most important consideration, the quality of *fitness* is universally appreciated and placed in the first rank. In many of those articles of furniture or apparel which luxury or fashion has brought into use, fitness or convenience often gives way to beauty of form or texture: but in a habitation intended to shelter us from the heat and cold, as well as to give us an opportunity to dispense the elegant hospitalities of refined life—the neglect of the various indispensable conveniences and comforts which an advanced state of civilization requires, would be but poorly compensated for by a fanciful exterior or a highly ornate style of building. Further than this, *fitness* will extend to the choice of situation; selecting a sheltered site, neither too high, as upon the exposed summit of bleak hills, nor too low, as in the lowest bottoms of damp valleys; but preferring those middle grounds which, while they afford a free circulation of air, and a fine prospect, are not detrimental to the health or enjoyment of the occupants. A proper exposure is another subject, worthy of the attention of either the architect or proprietor, as there are stormy and pleasant aspects or exposures in all climates.

However much the principle of *fitness* may be appreciated and acted upon in the United States, we have certainly great need of apology for the flagrant and almost constant violation of the second principle, viz. *the expression of purpose.* By the expression of purpose in buildings, is meant that architectural character, or *ensemble*, which distinctly points out the particular use or destination for which the edifice is intended. In a dwelling-house, the expression of purpose is conveyed by the chimney-tops, the porch or veranda, and those various appendages indicative of domestic enjoyment, which are needless, and therefore misplaced, in a public building. In a church, the spire or the dome, when present, at once stamps the building with the expression of purpose; and the few openings and plain exterior, with the absence of chimneys, are the suitable and easily recognised characteristics of the barn. Were any one to commit so violent an outrage upon the principle of the expression of purpose as to surmount his barns with the tall church spire, our feelings would at once cry out against the want of propriety. Yet how often do we meet in the northern states, with stables built after the models of Greek temples, and barns with elegant Venetian shutters—to say nothing of mansions with none but concealed chimney-tops, and without porches or appendages of any kind, to give the least hint to the mind of the doubting spectator, whether the edifice is a chapel, a bank, a hospital, or the private dwelling of a man of wealth and opulence!

"The expression of the purpose for which every building is erected," says the writer before quoted, "is the first and most essential beauty, and should be obvious from its architecture, although independent of any particular style; in the same manner as the reasons for things are altogether independent of the language in which they are conveyed. As in literary composition, no beauty of language can ever compensate for poverty of sense, so in architectural composition, no beauty of style can ever compensate for want of expression of purpose." Applying this excellent principle to our own country houses and their offices or out-buildings, we think every reasonable person will, at the first glance, see how lamentably deficient are many of the productions of our architects and builders, in one of the leading principles of the art. The most common form for an American country villa is the pseudo-Greek Temple; that is, a rectangular oblong building, with the chimney-tops concealed, if possible, and instead of a pretty and comfortable porch, veranda, or piazza, four, six, or eight lofty wooden columns are seen supporting a portico, so high as neither to afford an agreeable promenade, nor a sufficient shelter from the sun and rain.

There are two features, which it is now generally admitted contribute strongly to the expression of purpose in a dwelling-house, and especially in a country residence. These are the chimney-tops and the entrance porch. Chimney-tops, with us, are generally square masses of brick, rising above the roof, and presenting certainly no very elegant appearance, which may perhaps serve as the apology of those who studiously conceal them. But in a climate where fires are requisite during a large portion of the year, chimney-tops are expressive of a certain comfort resulting from the use of them, which characterizes a building intended for a dwelling in that climate. Chimney-tops being never, or rarely, placed on those buildings intended for the inferior animals, are also undoubtedly strongly indicative of human habitations. Instead, therefore, of hiding or concealing them, they should be in all dwellings not only boldly avowed, but rendered ornamental; for whatever is a characteristic and necessary feature, should undoubtedly, if possible, be rendered elegant, or at least prevented from being ugly.

Much of the picturesque effect of the old English and Italian houses, undoubtedly arises from the handsome and curious stacks of chimneys which spring out of their roofs. These, while they break and diversify the sky outline of the building, enrich and give variety to its most bare and unornamented part. Examples are not wanting, in all the different styles of architecture, of handsome and characteristic chimneys, which may be adopted in any of our dwellings of a similar style. The Gothic, or old English chimney, with octagonal or cylindrical flues or shafts united in clusters, is made in a great variety of forms, either of bricks or artificial stone. The former materials, moulded in the required shape, are highly taxed in England, while they may be very cheaply made here.

A Porch strengthens or conveys expression of purpose, because, instead of leaving the entrance door bare, as in manufactories and buildings of an inferior description, it serves both as a note of preparation, and an effectual shelter and protection to the entrance. Besides this, it gives a dignity and importance to that entrance, pointing it out to the stranger as the place of approach. A fine country house, without a porch or covered shelter to the doorway of some description, is therefore as incomplete, to the correct eye, as a well printed book without a title page, leaving the stranger to plunge at once *in medias res*, without the friendly preparation of a single word of introduction. Porches are susceptible of every variety of form and decoration, from the embattled and buttressed portal of the Gothic castle, to the latticed arbor porch of the cottage, around which the festoons of luxuriant climbing plants cluster, giving an effect not less beautiful than the richly carved capitals of the classic portico.

In this country no architectural feature is more plainly expressive of purpose in our dwelling-houses than the *veranda,* or piazza. The unclouded splendor and fierce heat of our summer sun, render this very general appendage a source of real comfort and enjoyment; and the long veranda round many of our country residences stands instead of the paved terraces of the English mansions as the place for promenade; while during the warmer portions of the season, half of the days or evenings are there passed in the enjoyment of the cool breezes, secure under low roofs supported by the open colonnade, from the solar rays, or the dews of night. The obvious utility of the veranda in this climate (especially in the middle and southern states) will, therefore, excuse its adoption into any style of architecture that may be selected for our domestic uses, although abroad, buildings in the style in question, as the Gothic, for example, are not usually accompanied by such an appendage. An artist of the least taste or invention will easily compose an addition of this kind, that will be in good keeping with the rest of the edifice.

These various features, or parts of the building, with many others which convey *expression of purpose* in domestic architecture, because they recall to the mind the different uses to which they are applied, and the several enjoyments connected with them, also contribute greatly to the interest of the building itself, and heighten its good effect as part of a harmonious whole, in the landscape. The various projections and irregularities, caused by verandas, porticoes, etc., serve to connect the otherwise square masses of building, by gradual transition with the ground about it.

The reader, who thus recognises features as expressive

of purpose in a dwelling intended for the habitation of man, we think, can be at no great loss to understand what would be characteristic in out-buildings or offices, farm-houses, lodges, stables, and the like, which are necessary structures on a villa or mansion residence of much size or importance. A proper regard to the expression of use or purpose, without interfering with the beauty of style, will confer at all times another, viz. the beauty of truth, without which no building can be completely satisfactory; as deceptions of this kind (buildings appearing to be what they are not) always go far towards destroying in the mind those pleasurable emotions felt on viewing any correct work of art, however simple in character or design.

We have now to consider rural architecture under the guidance of the third leading principle, as *an art of taste*. The expression of architectural *style* in buildings is undoubtedly a matter of the first importance, and proper care being taken not to violate fitness and expression of purpose, it may be considered as appealing most powerfully, at once, to the mind of almost every person. Indeed, with many, it is the only species of beauty which they perceive in buildings, and to it both convenience and the expression of purpose are often ignorantly sacrificed.

A marked style of architecture appears to us to have claims for our admiration or preference for rural residences, for several reasons. As it is intrinsically beautiful in itself; as it interests us by means of the associations connected with it; as it is fitted to the wants and comforts of country life; and as it is adapted to, or harmonizes with, the locality or scenery where it is located.

The harmonious union of buildings and scenery, is a point of taste that appears to be but little understood in any country; and mainly, we believe, because the architect and the landscape painter are seldom combined in the same person, or are seldom consulted together. It is for this reason that we so rarely see a country residence, or cottage and its grounds, making such a composition as a landscape painter would choose for his pencil. But it does not seem difficult, with a slight recurrence to the leading principle of unity of expression, to suggest a mode of immediately deciding which style of building is best adapted to harmonize with a certain kind of scenery.

The reader is, we trust, already familiar with our division of landscapes into two natural classes,—the Beautiful and the Picturesque,—and the two accordant systems of improvement in Landscape Gardening which we have based upon these distinct characters. Now, in order to render our buildings perfectly harmonious, we conceive it only to be necessary to arrange (as we may very properly do) all the styles of domestic architecture in corresponding divisions.

Some ingenious writer has already developed this idea, and, following a hint taken from the two leading schools of literature and art, has divided all architecture into the *Classical* and the *Romantic* schools of design. The Classical comprises the Grecian style, and all its near and direct offspring, as the Roman and Italian modes; the Romantic school, the Gothic style, with its numberless variations of Tudor, Elizabethan, Flemish, and old English modes.

It is easy to see, at a glance, how well these divisions correspond with our Beautiful and Picturesque phases of Landscape Gardening, so that indeed we might call the Grecian or Classical style, Beautiful, and the Gothic or

Romantic style, the Picturesque schools in architecture. In classical buildings, as in beautiful landscape, we are led to admire simplicity of forms and outlines, purity of effect, and grace of composition. In the Romantic or Picturesque buildings, we are struck by the irregularity of forms and outlines, variety of effect, and boldness of composition. What, therefore, can be more evident in seeking to produce unity of effect than the propriety of selecting some variations of the classical style for Beautiful landscape, and some species of romantic irregular building for Picturesque landscape?

In a practical point of view, all buildings which have considerable simplicity of outline, a certain complete and graceful style of ornament, and a polished and refined kind of finish, may be considered as likely to harmonize best with all landscape where the expression is that of simple or graceful beauty—where the lawn or surface is level or gently undulating, the trees rich and full in foliage and form, and the general character of the scenery peaceful and beautiful. Such are the Grecian, Roman, Tuscan, and the chaster Italian styles.

On the other hand, buildings of more irregular outline, in which appear bolder or ruder ornaments, and a certain free and more rustic air in finishing, are those which should be selected to accompany scenery of a wilder or more picturesque character, abounding in striking variations of surface, wood, and water. And these are the Castellated, the Tudor, and the old English in all its forms.

There is still an intermediate kind of architecture, originally a variation of the classical style, but which, in becoming adapted to different and more picturesque situations, has lost much of its graceful character, and has become quite picturesque in its outlines and effects. Of this kind are the *Swiss* and the *bracketed* cottage, and the different highly irregular forms of the *Italian villa*. The more simple and regular variations of these modes of building, may be introduced with good effect in any plain country; while the more irregular and artistical forms have the happiest effect only in more highly varied and suitable localities.

The *Egyptian*, one of the oldest architectural styles, characterized by its heavy colossal forms, and almost sublime expression, is supposed to have had its origin in caverns hewn in the rocks. The *Chinese* style, easily known by its waving lines, probably had its type in the eastern tent. The Saracenic, or Moorish style, rich in fanciful decoration, is striking and picturesque in its details, and is worthy of the attention of the wealthy amateur.

Neither of these styles, however, is, or can well be, thoroughly adapted to our domestic purposes, as they are wanting in fitness, and have comparatively few charms of association for residents of this country.

The only styles at present in common use for domestic architecture, throughout the enlightened portions of Europe and America, are the Grecian and Gothic styles, or some modifications of these two distinct kinds of building. These modifications, which of themselves are now considered styles by most authors, are, the *Roman* and *modern Italian* styles, which have grown out of Greek architecture; the *Castellated*, the *Tudor*, the *Elizabethan*, and the *rural Gothic* or old English cottage styles, all of which are variations of Gothic architecture.

Grecian or classic architecture was exhibited in its purity in those splendid temples of the golden days of

Athens, which still remain in a sufficient degree of preservation to bear ample testimony to the high state of architectural art among the Greeks. The best works of that period are always characterized by *unity* and *simplicity*, and in them an exquisite proportion is united with a chasteness of decoration, which stamps them perfect works of art. Each of the five orders was so nicely determined by their profound knowledge of the harmony of forms, and admirably executed, that all modern attempts at improving them have entirely failed, for they are, individually, complete models.

>   ———" First unadorned
> And nobly plain, the manly Doric rose ;
> The Ionic then with decent matron grace
> Her airy pillar heaved ; luxuriant last
> The rich Corinthian spread her wanton wreath."

A single or double portico of columns supporting a lofty pediment, the latter connected with the main body of the building, which in most cases was a simple parallelogram, were the characteristic features of the pure Grecian architecture. And this very simplicity of form, united with the chasteness of decoration and elegance of proportion, enhanced greatly the beauty of the Grecian temple as a whole.

To the scholar and the man of refined and cultivated mind, the *associations* connected with Grecian architecture are of the most delightful character. They transport him back, in imagination, to the choicest days of classic literature and art, when the disciples of the wisest and best of Athens listened to eloquent discourses that were daily delivered from her grove-embowered porticoes. When her temples were designed by a Phidias, and her architecture encouraged and patronized by a Pericles; when, in short, all the splendor of Pagan mythology, and the wisdom of Greek philosophy, were combined to perfect the arts and sciences of that period, and the temples dedicated to the Olympian Jove or the stately Minerva, were redolent with that *beauty*, which the Greeks worshipped, studied, and so well knew how to embody in material forms.

As it is admitted, then, that Grecian architecture is intrinsically beautiful in itself, and highly interesting in point of associations, it may be asked what are the objections, if any, to its common introduction into domestic Rural Architecture.

To this we answer, that although this form meagrely copied, Fig. 42, is actually in more common use than any other style in the United States, it is greatly inferior to the Gothic and its modifications in fitness, including under that head all the comforts and conveniences of country life.

[Fig. 42. Grecian Residence.]

We have already avowed that we consider fitness and expression of purpose, two leading principles of the first importance in Rural Architecture; and Grecian architecture in its pure form, viz. the *temple*, when applied to the purposes of domestic life, makes a sad blow at both these established rules. As a public building, the Greek temple form is perfect, both as to fitness (having one or more large rooms) and expression of purpose ;—showing a high, broad portico for masses of people, with an ample opening for egress and ingress. Domestic life, on the contrary, requires apartments of various dimensions, some large and others smaller, which, to be conveniently, *must* often be irregularly placed, with perhaps openings or windows of different sizes or dimensions. The comforts of a country residence are so various, that verandas, porches, wings of different sizes, and many other little accommodations expressive of purpose, become necessary, and, therefore, when properly arranged, add to the beauty of Rural Architecture. But the admirer of the true Greek models is obliged to forego the majority of these; and to come within the prescribed form of the rectangular parallelogram, his apartments must be of a given size and a limited number, while many things, both exterior and interior, which convenience might otherwise prompt, have to bow to the despotic sway of the pure Greek model.* In a dwelling of moderate dimensions how great a sacrifice of room is made to enable the architect to display the *portico* alone! We speak now chiefly of houses of the ordinary size, for if one chooses to build a palace, it is evident that ample accommodations may be obtained in any style.

* We are well aware that such is the rage for this style among us just now, and so completely have our builders the idea of its unrivalled supremacy in their heads, that many submit to the most meagre conveniences, under the name of closets, libraries, etc., in our country houses, without a murmur, believing that they are realizing the perfection of domestic comfort.

It has been well observed by modern critics, that there is no reason to believe the temple form was ever, even by the Greeks, used for private dwellings, which easily accounts for our comparative failure in constructing well arranged, small residences in this style.

[Fig. 43. Roman Residence.]

The Romans, either unable to compose in the simple elegance and beauty of the Grecian style, or feeling its want of adaptation to the multifarious usages of a more

[Fig. 44. View at Presque Isle, the residence of Wm. Denning, Esq., Dutchess Co., N. Y.]

luxurious state of society, created for themselves what is generally considered a less beautiful and perfect, yet which is certainly a more rich, varied, and, if we may use the term, *accommodating* style. The *Roman style* is distinguished from its prototype by the introduction of arched openings over the doors and windows, story piled over story,—often with columns of different orders—instead of the simple unbroken line of the Greek edifices. In decoration, the buildings in this style vary from plain, unornamented exteriors, to the most highly decorated façades; and instead of being confined to the few fixed principles of the Greek, the greatest latitude is often observed in the proportions, forms, and decorations of buildings in the Roman style. These very circumstances, while they rendered the style less perfect as a fine art, or for public edifices, gave it a pliability or facility of adaptation, which fits it more completely for domestic purposes. For this reason, a great portion of the finest specimens of the modern domestic architecture of the other continent is to be found in the Roman style.*

The *Italian style* is, we think, decidedly the most beautiful mode for domestic purposes, that has been the direct offspring of Grecian art. It is a style which has evidently grown up under the eyes of the painters of more modern Italy, as it is admirably adapted to harmonize with general nature, and produce a pleasing and picturesque effect in fine landscapes. Retaining more or less of the columns, arches, and other details of the Roman style, it has intrinsically a bold irregularity, and strong contrast of light and shadow, which give it a peculiarly striking and painter-like effect.

* Perhaps the finest façade of a private residence, in America, is that of the "Patroon's house," near Albany, the ancient seat of the Van Rensselaer family, lately remodelled and improved by that skilful architect, Mr. Upjohn, of New York.

386   LANDSCAPE GARDENING.

"The villa architecture of modern Italy," says Mr. Lamb, an able architect,* "is characterized, when on a moderate scale, by scattered irregular masses, great contrasts of light and shade, broken and plain surfaces, and great variety of outline against the sky. The blank wall on which the eye sometimes reposes; the towering campanile, boldly contrasted with the horizontal line of roof only broken by a few straggling chimney-tops: the row of equal sized, closely placed windows, contrasting with the plain space and single window of the projecting balcony; the prominent portico, the continued arcade, the terraces, and the variously formed and disposed out-buildings, all combine to form that picturesque whole, which distinguishes the modern Italian villa from every other."†

A building in the Italian style may readily be known at first sight, by the peculiar appearance of its roofs. These are always projecting at the eaves, and deeply furrowed or

[Fig. 45. A Villa in the Italian style.]

ridged, being formed abroad of semi-cylindrical tiles, which give a distinct and highly marked expression to this

* Loudon's Ency. of Arch. p. 951.

† In this country, owing to the greater number of fires, the effect would be improved by an additional number of chimney-tops.

Fig. 48 Villa of Theodore Lyman Esq. near Boston.

Fig. 49. Residence of Bishop Doane, Burlington, N. J.

portion of the building.* So many appliances of comfort and enjoyment suited to a warm climate appear, too, in the villas of this style, that it has a peculiarly elegant and refined appearance. Among these are *arcades*, with the Roman arched openings, forming sheltered promenades; and beautiful *balconies* projecting from single windows, or sometimes from connected rows of windows, which are charming places for a *coup d'œil*, or to enjoy the cool breeze—as they admit, to shelter one from the sun, of a fanciful awning shade, which may be raised or lowered at pleasure. The windows themselves are bold, and well marked in outline, being either round-arched at the tops, or finished with a heavy architrave.

[Fig. 46. Residence of Gov. Morehead, North Carolina.]

All these balconies, arcades, etc., are sources of real pleasure in the hotter portions of our year, which are quite equal in elevation of temperature to summers of the south of Europe; while by increased thickness of walls and

* In some situations in this country, where it might be difficult to procure tiles made in this form, their effect may be very accurately imitated by deeply ridged zinc or tin coverings. The bold projection of the eaves, in the Italian style, offers great protection to a house against storms and dampness.

closeness of window fixtures, the houses may also be made of the most comfortable description in winter.

The Italian chimney-tops, unlike the Grecian, are always openly shown and rendered ornamental; and as we have already mentioned, the irregularity in the masses of the edifice and shape of the roof, renders the sky outline of a building in this style, extremely picturesque. A villa, however small, in the Italian style, may have an elegant and expressive character, without interfering with convenient internal arrangements, while at the same time this style has the very great merit of allowing additions to be made in almost any direction, without injuring the effect of the original structure; indeed such is the variety of sizes and forms which the different parts of an Italian villa may take, in perfect accordance with architectural propriety, that the original edifice frequently gains in beauty by additions of this description. Those who are aware how many houses are every year erected in the United States by persons of moderate fortune, who would gladly make additions at some subsequent period, could this be done without injuring the effect or beauty of the main building, will, we think, acknowledge how much,

[Fig. 47. The New Haven Suburban Villa.*]

---

* New Haven abounds with tasteful residences. "Hillhouse Avenue," in particular, is remarkable for a neat display of Tuscan or Italian Suburban Villas. Moderate in dimension and economical in construction, these exceedingly neat edifices may be considered as models for this kind of dwelling. *Fig.* 47, without being a precise copy of any one of these buildings, may be taken as a pretty accurate representation of their general appearance.

even were it in this single point alone, the Italian style is superior to the Grecian for rural residences.*

* The villa of Theodore Lyman, Esq., at Brookline, near Boston, *Fig.* 48, is a highly interesting specimen of this style, designed by Mr. Upjohn—beautiful in exterior effect, and replete internally with every comfort and convenience.

Riverside Villa, the residence of Bishop Doane, at Burlington, New Jersey, is one of the best examples of the Italian style in this country. For the drawings from which *Figures* 49 and 50 are engraved, and for the following description, we are indebted to the able architect, John Notman, Esq., of Philadelphia, from whose designs the whole was constructed.

The site of this villa is upon the east bank of the Delaware river, near the town of Burlington, and within a few rods of the margin of this lovely stream.

The Delaware, at this part of its course, takes a direction nearly west; and while the river front (comprising the drawing room, hall, and library), commanding the finest water views, which are enjoyed to the greatest advantage in summer, has a cool aspect: the opposite side of the house, including the dining room, parlor, etc., is the favorite quarter in winter, being fully exposed to the genial influence of the sunbeams during the absence of foliage at that season. From this side of the house, a view is obtained of the pretty suburbs of Burlington, studded with neat cottages and gardens.

In the accompanying plan, *fig.* 50, *a*, is the hall; *b*, the vestibule; *c*, the dining room; *d*, the library; *e*, the drawing room; *f*, the parlor; *g*, Bishop D.'s room; *h*, dressing room; *i*, water closets; *j*, bath room; *k*, store room; *l*, principal stairs; *m*, back stairs; *o*, conservatory; *p*, veranda, etc.

A small terrace with balustrade, which surrounds the hall door, gives importance to this leading feature of the entrance front. The hall, *a*, is 17 feet square; on the right of the arched entrance is a casement window, opening to the floor, occasionally used as a door in winter, when the wind is north. The vestibule, *b*, opens from the hall, 17 by 21 feet. In the ceiling of this central apartment is a circular opening, with railing in the second story, forming a gallery above, which communicates with the different chambers, and affords ventilation to the whole house. Over this circular opening is a sky-light in the roof, which, mellowed and softened by a second colored one below it, serves to light the vestibule. From the vestibule we enter the dining room, *b*, 17 by 25 feet. The fine vista through the hall, vestibule, and dining room, 70 feet in length, is here terminated by the bay-window at the extremity of the dining room, which, through the balcony, opens on the lawn, varied by groups of shrubbery. On the left side of the vestibule, through a wide circular headed opening, we enter upon the principal

390   LANDSCAPE GARDENING.

Pleasing associations are connected with Roman and Italian architecture, especially to those who have studied

stairs, *l*. This opening is balanced by a recess on the opposite side of the vestibule. From the latter, a door also opens into the library, *d*, and another into the drawing room, *e* : offering, by a window in the library, in a line with

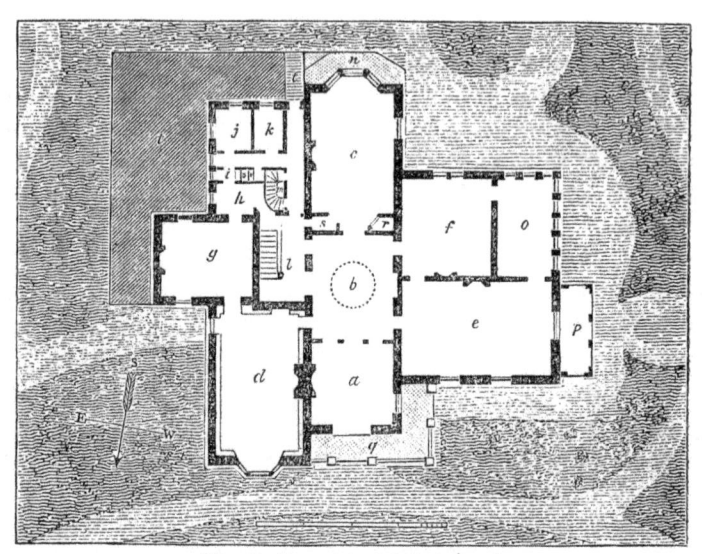

[Fig. 50. Plan of the Principal Floor.]

these doors, another fine vista in this direction. The library, 18 by 30 feet, and 16 feet high, is fitted up in a rich and tasteful manner, and completely filled with choice books. The bay-window, seen on the left in the perspective view, *Fig.* 49, is a prominent feature in this room, admitting, through its colored panes, a pleasing, subdued light, in keeping with the character of the apartment. The drawing room is 19 by 30 feet, with an enriched panelled ceiling, 15 feet high. At the extremity of this apartment, the veranda, *p*, with a charming view, affords an agreeable lounge in summer evenings, cooled by the breeze from the river. From the drawing room, a glazed door opens to the conservatory, *o*, and another door to the parlor, *f*. The latter is 18 by 20 feet, looking across the lawn and into the conservatory. Among the minor details are a china closet, *r*, and a butler's closet, *s*, in the dining room ; through the latter, the dishes are carried to and from the kitchen, larder, etc. The smaller passage leading from the main staircase, opens to the store room, *k*, and other apartments already designated, and communicates their effect in all the richness and beauty with which they are invested in the countries where they originated; and they may be regarded with a degree of classic interest by every cultivated mind. The modern Italian style recalls images of that land of painters and of the fine arts, where the imagination, the fancy, and taste, still revel in a world of beauty and grace. The great number of elegant forms which have grown out of this long cultivated feeling for the beautiful in the fine arts,—in the shape of fine vases, statues, and other ornaments, which harmonize with, and are so well adapted to enrich, this style of architecture,— combine to render it in the fine terraced gardens of Florence and other parts of Italy, one of the richest and most attractive styles in existence. Indeed we can hardly imagine a mode of building, which in the hands of a man of wealth and taste, may, in this country, be made productive of more beauty, convenience, and luxury, than the modern Italian style; so well suited to both our hot summers and cold winters, and which is so easily susceptible of enrichment and decoration, while it is at the same time so well adapted to the material in the most common use at present in most parts of the country,—wood. Vases, and other beautiful architectural ornaments, may now be procured in our cities, or imported direct from the Mediterranean, finely cut in Maltese stone, at very

by the back stairs, *m*, with the servants' chambers, placed over this part of the house, apart from those in the main body of the edifice. The large kitchen area, *t*, is sunk one story, by which the noise and smells of the kitchen, situated under the dining room, are entirely excluded from the principal story. In this sunk story, are also a wash room, scullery, and ample room for cellerage, wine, coals, etc. A forcing-pump supplies the whole house with water from the river; and in the second story are eight principal chambers, averaging 360 square feet each, making in all 25 rooms in the house, of large size.

392　LANDSCAPE GARDENING.

moderate prices, and which serve to decorate both the grounds and buildings in a handsome manner.

From the Italian style it is an easy transition to the Swiss mode, a bold and spirited one, highly picturesque and interesting in certain situations. To build an exact copy of a Swiss cottage in a smooth cultivated country, would, both as regards association and intrinsic want of fitness, be the height of folly. But in a wild and mountainous region, such as the borders of certain deep valleys and rocky glens in the Hudson Highlands, or rich bits of the Alleghanies, positions may be found where the Swiss cottage (Fig. 51), with its low and broad roof, shedding off the heavy snows, its ornamented exterior gallery, its strong and deep brackets, and its rough and rustic exterior, would be in the highest degree appropriate.

[Fig. 51. The Swiss Cottage.]

A modification, partaking somewhat of the Italian and Swiss features, is what we have described more fully in our "Cottage Residences" as the Bracketed mode. It possesses a good deal of character, is capable of considerable picturesque effect, is very easily and cheaply constructed of wood or stone, and is perhaps more entirely adapted to our

[Fig. 52 The Bracketed Mode.]

hot summers and cold winters than any other equally simple mode of building. We hope to see this Bracketed style becoming every day more common in the United States, and especially in our farm and country houses, when wood is the material employed in their construction.

Gothic, or more properly, *pointed architecture*, which sprang up with the Christian religion, reached a point of great perfection about the thirteenth century ; a period when the most magnificent churches and cathedrals of England and Germany were erected. These wonderful structures, reared by an almost magical skill and contrivance. with their richly groined roofs of stone supported in mid-air; their beautiful and elaborate tracery and carving of plants, flowers, and animate objects ; their large windows, through which streamed a rich glow of rainbow light; their various buttresses and pinnacles, all contributing to strengthen, and at the same time give additional beauty to the exterior; their clustered columns, airy-like, yet firm; and, surmounting the whole, the tall spire, piled up to an almost fearful height towards the heavens, are lasting monuments of the genius, scientific skill, and mechanical ingenuity of the artists of those times. That person, who, from ignorance or prejudice, fully supposes there is no architecture but that of the Greeks, would do well to study one of these unrivalled specimens of human skill. In so doing, unless he closes his eyes against the evidences of his senses, he cannot but admit that there is far more genius, and more mathematical skill, evinced in one of these cathedrals, than would have been requisite in the construction of the most celebrated of the Greek temples. Though they may not exhibit that simplicity and harmony of proportion which Grecian buildings display, they abound in much higher proofs of genius, as is abundantly evinced in the conception and execution of Cathedrals so abounding in unrivalled sublimity, variety, and beauty.

Gothic architecture, in its purity, was characterized mainly by the *pointed arch*. This novel feature in architecture, which, probably, in the hands of artists of great mathematical skill, was suggested by the inefficiency of the Roman arch first used, has given rise to all the superior boldness and picturesqueness of this style compared with the Grecian; for while the Greek artist was obliged to cover his narrow openings with architraves, or solid blocks of stone, resting on columns at short intervals, and filling up the open space, the Gothic artist, by a single span of his pointed arch, resting on distant pillars, kept the whole area beneath free and unencumbered. Applied, too, to openings for the admission of light, which were deemed of comparatively little or no importance by the Greeks, the arch was of immense value, making it possible to pierce the solid wall with large and lofty apertures, that diffused a magical brilliancy of light in the otherwise dim and shadowy interior.

We have here adverted to the Gothic cathedral (as we did to the Greek temple) as exhibiting the peculiar style in question in its greatest purity. For domestic purposes, both, for the same reasons, are equally unfitted; as they were never so intended to be used by their original inventors, and being entirely wanting in fitness for the purposes of habitation in domestic life; the Greek temple, as we have already shown, from its massive porticoes and the simple rectangular form of its interior; and the Gothic cathedral, from its high-pointed windows, and immense vaulted apartments. It would scarcely, however, be more absurd to build a miniature cathedral, for a dwelling in the Gothic style, than to make an exact copy of the Temple of Minerva 30 by 50 feet in size, for a country residence, as we often witness in this country.

The *Gothic Style*, as applied to Domestic Architecture, has been varied and adapted in a great diversity of ways, to the wants of society in different periods, from the 12th century to the present time. The baronial castle of feudal days, perched upon its solitary, almost inaccessible height, and built strongly for defence; the Collegiate or monastic abbey of the monks, suited to the rich fertile plains which these jolly ascetics so well knew how to select; the Tudor or Elizabethan mansion, of the English gentleman, surrounded by its beautiful park, filled with old ancestral trees;

and the pretty, rural, gabled cottage, of more humble pretensions; are all varieties of this multiform style, easily adapting itself to the comforts and conveniences of private life.

Contrasted with Classic Architecture and its varieties, in which horizontal lines are most prevalent, all the different Gothic modes or styles exhibit a preponderance of vertical or perpendicular lines. In the purer Gothic Architecture, the style is often determined by the form of the arch predominant in the window and door openings, which in all edifices (except Norman buildings) were lancet-shaped, or high pointed, in the 13th century; four centred, or low arched, in the times of Henry VII. and VIII.; and finally square-headed, as in most domestic buildings of later date.

Castellated Gothic is easily known, at first sight, by the line of battlements cut out of the solid parapet wall, which surmounts the outline of the building in every part. These generally conceal the roof, which is low, and were originally intended as a shelter to those engaged in defending the building against assaults. Modern buildings in the castellated style, without sacrificing almost everything to strength, as was once necessary, preserve the general character of the ancient castle, while

[Fig. 53. The Castellated Mode.]

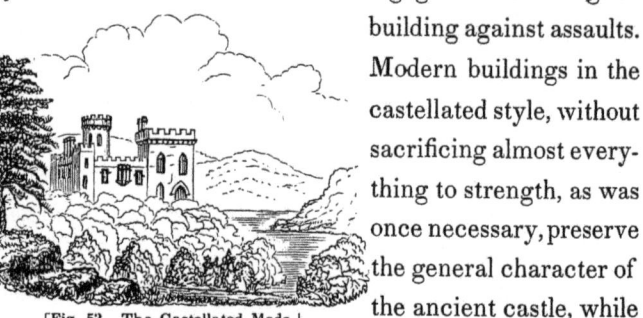

they combine with it almost every modern luxury. In their exteriors, we perceive strong and massive octagonal or circular towers, rising boldly, with corbelled or projecting cornices, above the ordinary level of the building. The windows are either pointed or square-headed, or perhaps a mixture of both. The porch rises into a turreted and embattled gateway, and all the offices and out-buildings connected with the main edifice, are constructed in a style corresponding to that exhibited in the main body of the building. The whole is placed on a distinct and firm terrace of stone, and the expression of the edifice is that of strength and security.

This mode of building is evidently of too ambitious and expensive a kind for a republic, where landed estates are not secured by entail, but divided, according to the dictates of nature, among the different members of a family. It is, perhaps, also rather wanting in appropriateness, castles never having been used for defence in this country. Notwithstanding these objections, there is no very weighty reason why a wealthy proprietor should not erect his mansion in the castellated style, if that style be in unison with his scenery and locality. Few instances, however, of sufficient wealth and taste to produce edifices of this kind, are to be met with among us; and the castellated style is therefore one which we cannot fully recommend for adoption here. Paltry imitations of it, in materials less durable than brick or stone, would be discreditable to any person having the least pretension to correct taste.

The Castellated style never appears completely at home except in wild and romantic scenery, or in situations where the neighboring mountains, or wild passes, are sufficiently near to give that character to the landscape. In such localities the Gothic castle affects us agreeably, because we know that baronial castles were generally built in similar spots, and because the battlements, towers, and other bold features, combine well with the rugged and spirited character of the surrounding objects. To place such a building in this country on a smooth surface in the midst of fertile plains, would immediately be felt to be bad taste by every one, as from the style not having been before our eyes from childhood, as it is in Europe, we immediately refer to its original purposes,—those of security and defence.

A mansion in the *Tudor Style* affords the best example of the excellence of Gothic architecture for domestic purposes. The roof often rises boldly here, instead of being concealed by the parapét wall, and the gables are either plain or ornamented with crockets. The windows are divided by mullions, and are generally enriched with tracery in a style less florid than that employed in churches, but still sufficiently elegant to give an appearance of decoration to these parts of the building. Sometimes the low, or Tudor arch, is displayed in the window-heads, but most commonly the square-headed window with the Gothic label is employed. Great latitude is allowed in this particular, as well as in the size of the window, provided the general details of style are attended to. Indeed, in the domestic architecture of this era, the windows and doors are often sources of great architectural beauty, instead of being left mere bare openings filled with glass, as in the Classic styles. Not only is each individual window divided by mullions into compartments whose tops are encircled by tracery; but in particular apartments, as the dining-hall, library, etc., these are filled with richly stained glass, which gives a mellow, pleasing light to the apartment. Added to this, the windows, in the best Tudor mansions, affect a great variety of forms and sizes. Among these stand conspicuous the *bay* and *oriel* windows. The bay-

Mr. Paulding's Residence, Tarrytown, N. Y.

Residence of the Author, near Newburgh, N. Y.

window, which is introduced on the first or principal floor, in most apartments of much size or importance, is a window of treble or quadruple the common size, projecting from the main body of the room in a semi-octagonal or hexagonal form, thereby affording more space in the apartment, from the floor to the ceiling, as well as giving an abundance of light, and a fine prospect in any favorite direction. This, while it has a grander effect than several windows of moderate size, gives a variety of form and outline to the different apartments, that can never be so well attained when the windows are mere openings cut in the solid walls. The oriel-window is very similar to the bay-window, but projecting in a similar manner from the upper story, supported on corbelled mouldings. These windows are not only elegant in the interior, but by standing out from the face of the walls, they prevent anything like too great a formality externally, and bestow a pleasing variety on the different fronts of the building.

The sky outline of a villa in the Tudor Gothic style, is highly picturesque. It is made up of many fine features. The pointed gables, with their finials, are among the most striking, and the neat parapet wall, either covered with a moulded coping, or, perhaps, diversified with battlements; the latter not so massive as in the castellated style, but evidently intended for ornament only. The roof line is often varied by the ornamented gablet of a dormer window, rising here and there, and adding to the quaintness of the whole. We must not forget, above all, the highly enriched chimney shaft, which, in the English examples, is made of fancifully moulded bricks, and is carried up in clusters some distance above the roof. How much more pleasing for a dwelling must be the outline of such a building, than that of a simple square roof whose summit is one unbroken straight line!*

The inclosed entrance porch, approached by three or four stone steps, with a seat or two for servants waiting, is a distinctive mark of all the old English houses. This projects, in most cases, from the main body of the edifice, and opens directly into the *hall*. The latter apartment is not merely (as in most of our modern houses) an entry, narrow and long, running directly through the house, but has a peculiar character of its own, being rather spacious, the roof or ceiling ribbed or groined, and the floor often inlaid with marble tiles. A corresponding and suitable style of finish, with Gothic details, runs through all the different apartments, each of which, instead of being finished and furnished with the formal sameness here so prevalent, displays, according to its peculiar purposes— as the dining-room, drawing-room, library, etc.—a marked and characteristic air.

We have thus particularized the Tudor mansion, because we believe that for a cold country like England or the United States, it has strong claims upon the attention of large landed proprietors, or those who wish to realize in a country residence the greatest amount of comfort and enjoyment. With the addition, here, of a veranda, which the cool summers of England render needless, we believe the Tudor Gothic to be the most convenient and comfortable, and decidedly the most picturesque and striking

---

* Two miles south of Albany, on a densely wooded hill, is the villa of Joel Rathbone, Esq., Fig. 54, one of the most complete specimens of the Tudor style in the United States. It was built from the designs of Davis, and is, to the amateur, a very instructive example of this mode of domestic architecture.

Fig. 54. Residence of Joel Rathbone, Esq. near Albany, N. Y.

Fig. 55. Cottage of S. E. Lyon, Esq. White Plains, N. Y.

Fig. 56. A Mansion in the Elizabethan style.

Fig. 57. The Residence of the Rev. Robert Bolton, near New Rochelle, N. Y

style, for country residences of a superior class.* The materials generally employed in their construction in England, are stone aud brick; and of late years, brick and stucco has come into very general use.

The *Elizabethan Style*, that mode of building so common in England in the 17th century,—a mixture of Gothic and Grecian in its details—is usually considered as a barbarous kind of architecture, wanting in purity of taste. Be this as it may, it cannot be denied that in the finer specimens of this style, there is a surprising degree of richness and picturesqueness for which we may look in vain elsewhere. In short it seems, in the best examples, admirably fitted for a *bowery*, thickly foliaged country, like England, and for the great variety of domestic enjoyments of its inhabitants. In the most florid examples of this style, of which many specimens yet remain, we often meet with every kind of architectural feature and ornament, oddly, and often grotesquely combined—pointed gables, dormer-windows, steep and low roofs, twisted columns, pierced parapets, and broad windows with small lights. Sometimes the effect of this fantastic combination is excellent, but often bad. The florid Elizabethan style is, therefore, a very dangerous one in the hands of any one but an architect of profound taste; but we think in some of its simpler forms (Fig. 56), it may be adopted for country residences here in picturesque situations with a quaint and happy effect.†

* The residence of Samuel E. Lyon, Esq., at White Plains, N. Y., Fig. 55, is a very pleasing example of the *Tudor Cottage*.

The seat of Robert Gilmor, Esq., near Baltimore, in the Tudor style, is a very extensive pile of building.

† A highly unique residence in the old English syle, is Pelham Priory, the seat of the Rev. Robert Bolton, near New Rochelle, N. Y., Fig. 57. The

The English cottage style, or what we have denominated *Rural Gothic*, contains within itself all the most striking and peculiar elements of the beautiful and picturesque in its exterior, while it admits of the greatest possible variety of accommodation and convenience in internal arrangement.

In its general composition, Rural Gothic really differs from the Tudor style more in that general *simplicity* which serves to distinguish a cottage or villa of moderate size from a mansion, than in any marked character of its own. The square-headed windows preserve the same form, and display the Gothic label and mullions, though the more expensive finish of decorative tracery is frequently omitted. Diagonal or latticed lights are also more commonly seen in the cottage style than in the mansion. The general form and arrangement of the building, though of course much reduced, is not unlike that of the latter edifice. The entrance porch is always preserved, and the bay-window jutting out from the best apartment, gives variety, and an agreeable expression of use and enjoyment, to almost every specimen of the old English cottage.

Perhaps the most striking feature of this charming style as we see it in the best old English cottages, is the *pointed gable*. This feature, which grows out of the high roofs

exterior is massive and picturesque, in the simplest taste of the Elizabethan age, and being built amidst a fine oak wood, of the dark rough stone of the neighborhood, it has at once the appearance of considerable antiquity. The interior is constructed and fitted up throughout in the same feeling,—with harmonious wainscoting, quaint carving, massive chimney pieces, and old furniture and armor. Indeed, we doubt if there is, at the present moment, any recent private residence, even in England, where the spirit of the antique is more entirely carried out, and where one may more easily fancy himself in one of those " mansions builded curiously " of our ancestors in the time of " good Queen Bess."

adopted, not only appears in the two ends of the main building, but terminates every wing or projection of almost any size that joins to the principal body of the house. The gables are either of stone or brick, with a handsome moulded coping, or they are finished with the widely projecting roof of wood, and *verge boards*, carved in a fanciful and highly decorative shape. In either case, the point or apex is crowned by a finial, or ornamental octagonal shaft, rendering the gable one of the greatest sources of interest in these dwellings. The projecting roof renders the walls always dry.

The porch, the labelled windows, the chimney shafts, and the ornamented gables, being the essential features in the composition of the English cottage style, it is evident that this mode of building is highly expressive of purpose, for country residences of almost every description and size, from the humblest peasant's cottage, to the beautiful and picturesque villa of the retired gentleman of fortune. In the simple form of the cottage, the whole may be constructed of wood very cheaply, and in the more elaborate villa residence, stone, or brick and cement, may be preferred, as being more permanent. No style so readily admits of enrichment as that of the old English cottage when on a considerable scale; and by the addition of pointed verandas, bay windows, and dormer-windows, by the introduction of mullions and tracery in the window openings, and indeed, by a multitude of interior and exterior enrichments generally applied to the Tudor mansions, a villa in the rural Gothic style may be made a perfect gem of a country residence. Of all the styles hitherto enumerated, we consider this one of the most suitable for this country, as, while it comes within the reach of all persons of moderate means, it unites, as we before stated, so much of convenience and rural beauty.*

To the man of taste, there is no style which presents greater attractions, being at once rich in picturesque beauty, and harmonious in connexion with the surrounding forms of vegetation. The Grecian villa, with its simple forms and horizontal lines, seems to us only in good keeping when it is in a smooth, highly cultivated, peaceful scene. But the Rural Gothic, the lines of which point upwards, in the pyramidal gables, tall clusters of chimneys, finials, and the several other portions of its varied outline, harmonizes easily with the tall trees, the tapering masses of foliage, or the surrounding hills; and while it is seldom or never misplaced in spirited rural scenery, it gives character and picturesque expression to many landscapes entirely devoid of that quality.

What we have already said in speaking of the Italian style, respecting the facility with which additions may be made to irregular houses, applies with equal, or even greater force, to the varieties of the Gothic style, just described. From the very fact that the highest beauty of these modes of building arises from their irregularity (opposed to Grecian architecture, which, in its chaste simplicity, should be regular), it is evident that additions

---

* The only objection that can be urged against this mode of building, is that which applies to all cottages with a low second story, viz. want of coolness in the sleeping chambers during mid-summer. An evil which may be remedied by constructing a false inner-roof—leaving a vacuity between the two roofs of six or eight inches, which being occupied with air and ventilated at the top, will almost entirely obviate the objection.

In our *Cottage Residences*, Design II., we have shown how the comfort of a full second story, suitable for this climate, may be combined with the expression of the English cottage style.

judiciously made will tend to increase this beauty, or afford more facility for its display; while it is equally evident that in the interior arrangement, including apartments of every description, superior opportunities are afforded for attaining internal comfort and convenience, as well as external effect.

The ideas connected in our minds with Gothic architecture are of a highly *romantic* and *poetical* nature, contrasted with the classical associations which the Greek and Roman styles suggest. Although our own country is nearly destitute of ruins and ancient time-worn edifices, yet the literature of Europe, and particularly of what we term the mother country, is so much our own, that we form a kind of delightful ideal acquaintance with the venerable castles, abbeys, and strongholds of the middle ages. Romantic as is the real history of those times and places, to our minds their charm is greatly enhanced by distance, by the poetry of legendary superstition, and the fascination of fictitious narrative. A castellated residence, therefore, in a wild and picturesque situation, may be interesting, not only from its being perfectly in keeping with surrounding nature, but from the delightful manner in which it awakens associations fraught with the most enticing history of the past.

The older domestic architecture of the English may be viewed in another pleasing light. Their buildings and residences have not only the recommendation of beauty and complete adaptation, but the additional charm of having been the homes of our ancestors, and the dwellings of that bright galaxy of English genius and worth, which illuminates equally the intellectual firmament of both hemispheres. He who has extended his researches, *con amore*, into the history of the domestic life and habits of those illustrious minds, will not, we are sure, forget that lowly cottage by the side of the Avon, where the great English bard was wont to dwell; the tasteful residence of Pope at Twickenham; or the turrets and battlements of the more picturesque Abbotsford; and numberless other examples of the rural buildings of England, once the abodes of renowned genius. In truth, the cottage and villa architecture of the English has grown out of the feelings and habits of a refined and cultivated people, whose devotion to country life, and fondness for all its pleasures, are so finely displayed in the beauty of their dwellings, and the exquisite keeping of their buildings and grounds.

We must be permitted to quote, in further proof of English taste and habits, and their results in their country residences, the testimony of our countryman, Washington Irving, in one of his most elegant essays. "The taste of the English in the cultivation of land, and in what is called Landscape Gardening, is unrivalled. They have studied nature intently, and discovered an exquisite sense of her beautiful forms and harmonious combinations. Those charms which in other countries she lavishes in wild solitudes, are here assembled around the haunts of domestic life. They seem to have caught her coy and furtive graces, and spread them like witchery about their rural abodes. Nothing can be more imposing than the magnificence of English park scenery. Vast lawns that extend like sheets of vivid green, with here and there clumps of gigantic trees heaping up rich piles of foliage The solemn group of groves and woodland glades, with the deer trooping in silent herds across them; the hare bounding away to the covert, or the pheasant bursting suddenly upon the wing. The brook, taught to wind in natural meanderings, or to expand into a glassy lake,—the sequestered pool reflecting the quivering trees, with the yellow leaf sleeping upon its bosom, and the trout roaming fearlessly about its limpid waters ; while some rustic temple or sylvan statue, grown green and dark with age, gives an air of classic sanctity to the seclusion."

"These are but a few of the features of park scenery ; but what most delights me, is the creative talent with which the English decorate the unostentatious abodes of middle life. The rudest habitation, the most unpromising and scanty portion of land, in the hands of an Englishman of taste, becomes a little paradise. With a nicely discriminating eye he seizes at once upon its capabilities, and pictures in his mind the future landscape. The sterile spot grows into loveliness under his hand ; and yet the operations of art which produce the effect are scarcely to be perceived ; the cherishing and training of some trees : the cautious pruning of others ; the nice distribution of flowers and plants of tender and graceful foliage ; the introduction of a green slope of velvet turf ; the partial opening to a peep of blue distance, or silver gleam of water,—all these are managed with a delicate tact, a pervading, yet quiet assiduity, like the magic touchings with which a painter finishes up a favorite picture."

" The residence of people of fortune and refinement in the country, has diffused a degree of taste and elegance that descends to the lowest class. The very laborer, with his thatched cottage and narrow slip of ground, attends to their embellishment. The trim hedge, the grass-plot before the door, the little flower bed bordered with snug box, the woodbine trained up against the wall, and hanging its blossoms about the lattice; the pot of flowers in the window; the holly providentially planted about the house to cheat winter of its dreariness, and to throw in a semblance of green summer to cheat the fireside:—all these bespeak the influence of taste flowing down from high sources, and pervading the lowest levels of the public mind. If ever Love, as the poets sing, delights to visit a cottage, it must be the cottage of an English peasant."

It is this love of rural life and this nice feeling of the harmonious union of nature and art, that reflects so much credit upon the English as a people, and which sooner or later we hope to see completely naturalized in this country. Under its enchanting influence, the too great bustle and excitement of our commercial cities will be happily counterbalanced by the more elegant and quiet enjoyments of country life. Our rural residences, evincing that love of the beautiful and the picturesque, which, combined with solid comfort, is so attractive to the eye of every beholder, will not only become sources of the purest enjoyment to the refined minds of the possessors, but will exert an influence for the improvement in taste of every class in our community. The ambition to build "shingle palaces" in starved and meagre grounds, we are glad to see giving way to that more refined feeling which prefers a neat villa or cottage, tastily constructed, and surrounded by its proper accessories, of greater or less extent, of verdant trees and beautiful shrubbery.

It is gratifying to see the progressive improvement in Rural Architecture, which within a few years past has evinced itself in various parts of the country, and particularly on the banks of the Hudson and Connecticut

Fig. 58. Cottage Residence of Thomas W. Ludlow, Esq. near Yonkers, N. Y

Fig. 59. Residence of Washington Irving, Esq. near Tarrytown, N. Y.

Rivers, as well as in the suburbs of our largest cities. Here and there, beautiful villas and cottages in the Italian or old English styles, are being erected by proprietors who feel the pre-eminent beauty of these modes for domestic architecture. And from the rapidity with which improvements having just claims for public favor advance in our community, we have every reason to hope that our Rural Architecture will soon exhibit itself in a more attractive and agreeable form than it has hitherto generally assumed. We take pleasure in referring to a few of these buildings more in detail.

The cottage of Thomas W. Ludlow, Esq., near Yonkers, on the Hudson (Fig. 58), is one of the most complete examples on this river. The interior is very carefully and harmoniously finished, the apartments are agreeably arranged, and the general effect of the exterior is varied and pleasing.

There is scarcely a building or place more replete with interest in America, than the cottage of Washington Irving, near Tarrytown (Fig. 59). The "Legend of Sleepy Hollow," so delightfully told in the Sketch-Book, has made every one acquainted with this neighborhood, and especially with the site of the present building, there celebrated as the "Van Tassel House," one of the most secluded and delightful nooks on the banks of the Hudson. With characteristic taste, Mr. Irving has chosen this spot, the haunt of his early days, since rendered classic ground by his elegant pen, and made it his permanent residence. The house of "Baltus Van Tassel" has been altered and rebuilt in a quaint style, partaking somewhat of the English cottage mode, but retaining strongly marked symptoms of its Dutch origin. The quaint old weather-cocks and finials, the crow-stepped gables, and the hall paved with Dutch tiles, are among the ancient and venerable ornaments of the houses of the original settlers of Manhattan, now almost extinct among us. There is also a quiet keeping in the cottage and the grounds around it, that assists in making up the charm of the whole; the gently swelling slope reaching down to the water's edge, bordered by prettily wooded ravines through which a brook meanders pleasantly; and threaded by foot-paths ingeniously contrived, so as sometimes to afford secluded walks, and at others to allow fine vistas of the broad expanse of river scenery. The cottage itself is now charmingly covered with ivy and climbing roses, and embosomed in thickets of shrubbery.

Mr. Sheldon's residence (Fig. 60), in the same neighborhood, furnishes us with another example of the Rural Gothic mode, worth the study of the amateur. Captain Perry's spirited cottage, near Sing Sing, partakes of the same features; and we might add numerous other cottages now building, or in contemplation, which show how fast the feeling for something more expressive and picturesque is making progress among us.

Mr. Warren's residence at Troy, N. Y. (Fig. 61), is a very pretty example of the English cottage, elegantly finished internally as well as externally. A situation in a valley, embosomed with luxuriant trees, would have given this building a more appropriate and charming air than its present one, which, however, affords a magnificent prospect of the surrounding country.

It is the common practice here to place a portion of what are called the *domestic offices*, as the kitchen, pantries, etc., in the basement story of the house,

Fig. 60. Residence of H. Sheldon, Esq. near Tarrytown, N. Y.

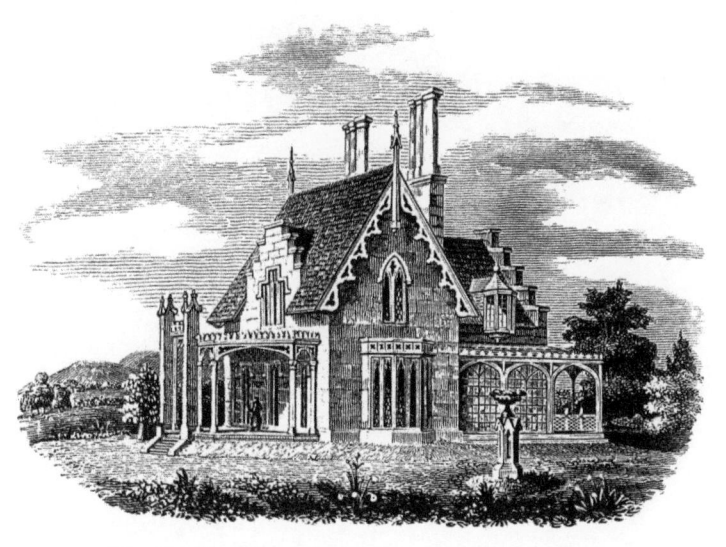

Fig. 61. Mr. Warren,'s Cottage, near Troy, N. Y

directly beneath the living rooms. This has partly arisen from the circumstance of the comparative economy of this method of constructing them under the same roof; and partly from the difficulty of adding wings to the main building for those purposes, which will not mar the simplicity and elegance of a Grecian villa. In the better class of houses in England, the domestic offices, which include the kitchen and its appurtenances, and also the stable, coach-house, harness-room, etc., are in the majority of cases attached to the main body of the building on one side. The great advantage of having all these conveniences on the same floor with the principal rooms, and communicating in such a way as to be easily accessible at all times without going into the open air, is undeniable. It must also be admitted that these domestic offices, extending out from the main building, partly visible and partly concealed by trees and foliage, add much to the extent and importance of a villa or mansion in the country. In the old English style these appendages are made to unite happily with the building, which is in itself irregular. Picturesque effect is certainly increased by thus extending the pile and increasing the variety of its outline.

A blind partiality for any one style in building is detrimental to the progress of improvement, both in taste and comfort. The variety of means, habits, and local feelings, will naturally cause many widely different tastes to arise among us; and it is only by the means of a number of distinct styles, that this diversity of tastes can be accommodated. There will always be a large class of individuals in every country who prefer a plain square house because it is more economical, and because they have little feeling for architectural, or, indeed, any other species of beauty. But besides such, there will always be found some men of finer natures, who have a sympathetic appreciation of the beautiful in nature and art. Among these, the classical scholar and gentleman may, from association and the love of antiquity, prefer a villa in the Grecian or Roman style. He who has a passionate love of pictures and especially fine landscapes, will perhaps, very naturally, prefer the modern Italian style for a country residence. The wealthy proprietor, either from the romantic and chivalrous associations connected with the baronial castle, or from desire to display his own resources, may indulge his fancy in erecting a castellated dwelling. The gentleman who wishes to realize the *beau ideal* of a genuine old English country residence, with its various internal comforts, and its spirited exterior, may establish himself in a Tudor villa or mansion; and the lover of nature and rural life, who, with more limited means, takes equal interest in the beauty of his grounds or garden (however small) and his house—who is both an admirer of that kind of beauty called the picturesque, and has a lively perception of the effect of a happy adaptation of buildings to the landscape,—such a person will very naturally make choice of the rural cottage style.

Entrance Lodges are not only handsome architectural objects in the scenery of country residences of large size, but are in many cases exceedingly convenient, both to the family and the guests or visitors having frequent ingress and egress. The entrance lodge may further be considered a matter strictly useful, in serving as the dwelling of the gardener or farmer and his family. In this point of view, arrangements for the comfort and convenience of the inmates should be regarded as more important than the fanciful decoration of the exterior—as no exterior, however charming, can, to a reflective and well regulated mind, apologize for contracted apartments, and imperfect light and ventilation, in human habitations.

Among the numerous entrance lodges which we remember to have seen in the United States, we scarcely recall a single example where the means, or rather the facility, of opening and shutting the gate itself, has been sufficiently considered. Most generally the lodge is at too great a distance from the gate, consuming too much time in attendance, and exposing the persons attending, generally women or children, to the inclemencies of the weather. Besides this, service of this kind is less cheerfully performed in this country than in Europe, from the very simple reason of the greater equality of conditions here, and therefore everything which tends to lessen labor, is worthy of being taken into account.

For these reasons we would place the gate very near the lodge; it would be preferable if it were part of the same architectural composition: and if possible adopt the contrivance now in use at some places abroad, by which the gate, being hung nearest the building, may be opened by the occupant without the latter being seen, or being scarcely obliged to leave his or her employment.* This

---

* In *Fig.* 62, is shown the section of a gate arranged upon this plan. At the bottom of the hanging post of the gate, is a bevelled iron pinion, that works into another pinion, *b*, at the end of the horizontal shaft, *a*, which shaft is fixed in a square box or tunnel under the road. The part to the right of the partition line, *f*, is the *interior* of the gate-keeper's house ; and by turning the winch, *e*, the upright shaft, *c*, is put in motion, which moves by means of the bevelled pinions, *g*, *d*, the shaft *a*, and therefore, through *d*, the back post of the gate,

is certainly the ultimatum of improvements in gate lodges; and where it cannot be attained, something may still be done towards amelioration, by placing the gate within a convenient distance, instead of half a dozen rods apart from the lodge, as is frequently done.

That the entrance lodge should correspond in style with the mansion, is a maxim insisted upon by all writers on Rural Architecture. Where the latter is built in a mixed style, there is more latitude allowed in the choice of forms for the lodge, which may be considered more as a thing by itself. But where the dwelling is a strictly architectural composition, the lodge should correspond in style, and bear evidence of emanating from the same mind. A variation of the same style may be adopted with pleasing effect, as a

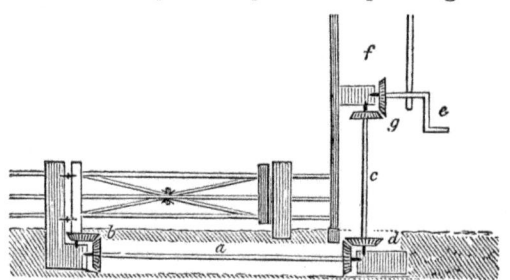

[Fig. 62. Plan for opening the gate from the interior of the Lodge.]

lodge in the form of the old English cottage for a castellated mansion, or a Doric lodge for a Corinthian villa; but never two distinct styles on the same place (a Gothic gate-house and a Grecian residence) without producing in minds imbued with correct principles a feeling of incongruity. A certain correspondence in size is also agreeable; where the dwelling of the proprietor is simply an ornamental

which is opened and shut by the motion of the winch, without obliging the inmates to leave the house.

cottage, the lodge, if introduced, should be more simple and unostentatious; and even where the house is magnificent, the lodge should rather be below the general air of the residence than above it, that the stranger who enters at a showy and striking lodge may not be disappointed in the want of correspondence between it and the remaining portions of the demesne.

[Fig. 63. The New Gate Lodge at Blithewood.]

The gate-lodge at Blithewood, on the Hudson, the seat of R. Donaldson, Esq., is a simple and effective cottage in the bracketed style—octagonal in its form, and very compactly arranged internally.

Nearly all the fine seats on the North river have entrance lodges—often simple and but little ornamented, or only pleasingly embowered in foliage; but, occasionally, highly picturesque and striking in appearance.

A view of the pretty gate lodge at Netherwood, Duchess County N. Y., the seat of Gardi-

[Fig. 64. The Gate Lodge at Netherwood.]

ner Howland, Esq., is shown in Fig. 64. Half a mile north of this seat is an interesting lodge in the Swiss style, at the entrance to the residence of Mrs. Sheafe.

In Fig. 65, is shown an elevation of a lodge in the Italian style, with projecting eaves supported by cantilevers or brackets, round-headed windows with balconies, characteristic porch, and other leading features of this style.

[Fig. 65. Gate Lodge in the Italian style.]

Mr. Repton has stated it as a principle in the composition of residences, that neither the house should be visible from the entrance nor the entrance from the house, if there be sufficient distance between them to make the approach through varied grounds, or a park, and not immediately into a court-yard.

Entrance lodges, and indeed all small ornamental buildings, should be supported, and partially concealed, by trees and foliage ; naked walls, in the country, hardly admitting of an apology in any case, but especially when the building is ornamental, and should be considered part of a whole, grouping with other objects in rural landscape.

RURAL ARCHITECTURE. 417

NOTE.—To readers who desire to cultivate a taste for rural architecture, we take pleasure in recommending the following productions of the English press. Loudon's *Encyclopædia of Cottage, Farm, and Villa Architecture*, a volume replete with information on every branch of the subject; Robinson's *Rural Architecture and Designs for Ornamental Villas;* Lugar's *Villa Architecture;* Goodwin's *Rural Architecture;* Hunt's *Picturesque Domestic Architecture, and Examples of Tudor Architecture;* Pugin's *Examples of Gothic Architecture*, etc. The most successful American architects in this branch of the art, with whom we are acquainted, are Alexander J. Davis, Esq., of New York, and John Notman, Esq., of Philadelphia.

[Fig. 66. The Gardener's House, Blithewood.]

SECTION X.

EMBELLISHMENTS; ARCHITECTURAL, RUSTIC, AND FLORAL.

Value of a proper connexion between the house and grounds. Beauty of the architectural terrace, and its application to villas and cottages. Use of vases of different descriptions. Sun-dials. Architectural flower-garden. Irregular flower-garden. French flower-garden. English flower-garden. General remarks on this subject. Selection of showy plants, flowering in succession. Arrangement of the shrubbery, and selection of choice shrubs. The conservatory or green-house. Open and covered seats. Pavilions. Rustic seats. Prospect tower. Bridges. Rockwork. Fountains of various descriptions. Judicious introduction of decorations.

> Nature, assuming a more lovely face,
> Borrowing a beauty from the works of grace.
> COWPER.

> ———— Each odorous bushy shrub
> Fenced up the verdant wall ; each beauteous flower ;
> Iris all hues, Roses and Jessamine
> Rear'd high their flourished heads between,
> And wrought Mosaic.
> MILTON.

N our finest places, or those country seats where much of the polish of pleasure ground or park scenery is kept up, one of the most striking defects is the want of *"union between the house and the grounds."*

EMBELLISHMENTS. 419

We are well aware that from the comparative rarity of anything like a highly kept place in this country, the want of this, which is indeed like the last finish to the residence, is scarcely felt at all. But this only proves the infant state of Landscape Gardening here, and the little attention that has been paid to the highest details of the art.

If our readers will imagine, with us, a pretty villa, conveniently arranged and well constructed, in short, complete in itself as regards its architecture, and at the same time, properly placed in a smooth well kept lawn, studded with groups and masses of fine trees, they will have an example often to be met with, of a place, in the graceful school of design, about which, however, there is felt to be a certain incongruity between the house, a highly artificial object, and the surrounding grounds, where the prevailing expression in the latter is that of beautiful nature.

Let us suppose, for further illustration, the same house and grounds with a few additions. The house now rising directly out of the green turf which encompasses it, we will surround by a raised platform or terrace, wide enough for a dry, firm walk, at all seasons; on the top of the wall or border of this terrace, we will form a handsome *parapet*, or balustrade, some two or three feet high, the details of which shall be in good keeping with the house, whether Grecian or Gothic. On the coping of this parapet, if the house is in the classical style, we will find suitable places, at proper intervals, for some handsome urns, vases, etc. On the drawing-room side of the house, that is, the side towards which the best room or rooms look, we will place the flower-garden, into which we descend from the terrace by a few steps. This flower-garden may be simply what its name denotes, a place exclusively devoted to the cultivation of flowers, or (if the house is not in a very plain style, admitting of little enrichment) it may be an architectural flower-garden. In the latter case, intermingled with the flowers, are to be seen vases, fountains, and sometimes even statues; the effect of the fine colors and deep foliage of the former, heightened by contrast with the sculptured forms of the latter.

If our readers will now step back a few rods with us and take a second view of our villa residence, with its supposed harmonizing accessories, we think they can hardly fail to be impressed at once with the great improvement of the whole. The eye now, instead of witnessing the sudden termination of the architecture at the base of the house, where the lawn commences as suddenly, will be at once struck with the increased variety and richness imparted to the whole scene, by the addition of the architectural and garden decorations. The mind is led gradually down from the house, with its projecting porch or piazzas, to the surrounding terrace crowned with its beautiful vases, and from thence to the architectural flower-garden, interspersed with similar ornaments. The various play of light afforded by these sculptured forms on the terrace; the projections and recesses of the parapet, with here and there some climbing plants luxuriantly enwreathing it, throwing out the mural objects in stronger relief, and connecting them pleasantly with the verdure of the turf beneath; the still further rambling off of vases, etc., into the brilliant flower-garden, which, through these ornaments, maintains an avowed connexion with the architecture of the house; all this, we think it cannot be denied, forms a rich setting to the architecture, and unites agreeably the forms of surrounding nature with the more regular and uniform outlines of the building.

The effect will not be less pleasing if viewed from another point of view, viz. the terrace, or from the apartments of the house itself. From either of these points, the various objects enumerated, will form a rich *foreground* to the pleasure-grounds or park—a matter which painters well know how to estimate, as a landscape is incomplete and unsatisfactory to them, however beautiful the middle or distant points, unless there are some strongly marked objects in the foreground. In fine, the intervention of these elegant accompaniments to our houses prevents us, as Mr. Hope has observed, " from launching at once from the threshold of the symmetric mansion, in the most abrupt manner, into a scene wholly composed of the most unsymmetric and desultory forms of mere nature, which are totally out of character with the mansion, whatever may be its style of architecture and furnishing "*

The highly decorated terrace, as we have here supposed it, would, it is evident, be in unison with villas of a somewhat superior style; or, in other words, the amount of enrichment bestowed upon exterior decoration near the house, should correspond to the style of art evinced in the exterior of the mansion itself. An humble cottage with sculptured vases on its terrace and parapet, would be in bad taste; but any Grecian, Roman, or Italian villa, where a moderate degree of exterior ornament is visible, or a Gothic villa of the better class, will allow the additional enrichment of the architectural terrace and its ornaments. Indeed the terrace itself, in so far as it denotes a raised dry

* *Essay on Ornamental Gardening,* by Thomas Hope.

platform around the house, is a suitable and appropriate appendage to every dwelling, of whatever class.

The width of a terrace around a house may vary from five to twenty feet, or more, in proportion as the building is of greater or less importance. The surrounding wall, which supports its level, may also vary from one to eight feet. The terrace, in the better class of English residences, is paved with smooth flag stones, or in place of this, a surface of firm well-rolled gravel is substituted. In residences where a parapet or balustrade would be thought too expensive, a square stone or plinth is placed at the angles or four corners of the terrace, which serves as the pedestal for a vase or urn. When a more elegant and finished appearance is desirable, the parapet formed of open work of stone, or wood painted in imitation of stone, rises above the level of the terrace two or three feet with a suitably bold coping. On this vases may be placed, not only at the corners, but at regular intervals of ten, twenty, or more feet. We have alluded to the good effect of climbers, here and there planted, and suffered to intermingle their rich foliage with the open work of the parapet and its crowning ornaments. In the climate of Philadelphia, the Giant Ivy, with its thick sculpturesque looking masses of foliage, would be admirably suited to this purpose. Or the Virginia Creeper (the Ivy of America) may take its place in any other portion of the Union. To these we may add, the Chinese twining Honeysuckle (Lonicera flexuosa) and the Sweet-scented Clematis, both deliciously fragrant in their blossoms, with many other fine climbers which will readily recur to the amateur.

There can be no reason why the smallest cottage, if its occupant be a person of taste, should not have a terrace

EMBELLISHMENTS. 423

decorated in a suitable manner. This is easily and cheaply effected by placing neat flower-pots on the parapet, or border and angles of the terrace, with suitable plants growing in them. For this purpose, the American or Century *Aloe*, a formal architectural-looking plant, is exceedingly well adapted, as it always preserves nearly the same appearance. Or in place of this, the *Yuccas*, or "*Adam's needle* and *thread*," which have something of the same character, while they also produce beautiful heads of flowers, may be chosen. *Yucca flaccida* is a fine hardy species, which would look well in such a situation. An aloe in a common flower pot is shown in Fig. 67; and a Yucca in an ornamental flower-pot in Fig. 68.

[Fig. 67.]

[Fig. 68.]

Where there is a terrace ornamented with urns or vases, and the proprietor wishes to give a corresponding air of elegance to his grounds, vases, sundials, etc., may be placed in various appropriate situations, not only in the architectural flower-garden, but on the lawn, and through the pleasure-grounds in various different points *near the house*. We say near the house, because we think so highly artificial and architectural an object as a sculptured vase, is never correctly introduced unless it appear in some way connected with buildings, or objects of a like architectural character. To place a beautiful vase in a distant part of the grounds, where there is no direct allusion to art, and where it is accompanied only by natural objects, as the overhanging trees and the sloping turf, is in a measure doing violence to our reason or taste, by bringing two objects so strongly contrasted, in direct union. But when we see a statue or a vase placed in any part of the grounds where a near view is obtained of the house (and its accompanying statues or vases), the whole is accounted for, and we feel the distant vase to be only a part of, or rather a repetition of the same idea,—in other words, that it forms part of a whole, harmonious and consistent.

Vases of real stone, as marble or granite, are decorations of too costly a kind ever to come into general use among us. Vases, however, of equally beautiful forms, are manufactured of artificial stone, of fine pottery, or of cast iron, which have the same effect, and are of nearly equal durability, as garden decorations.

A vase should never, in the open air, be set down upon the ground or grass, without being placed upon a firm base of some description, either a *plinth* or a *pedestal*. Without a base of this kind it has a temporary look, as if it had been left there by mere accident, and without any intention of permanence. Placing it upon a pedestal, or square plinth (block of stone), gives it a character of art, at once more dignified and expressive of stability. Besides this, the pedestal in reality serves to preserve the vase in a perpendicular position, as well as to expose it fairly to the eye, which could not be the case were it put down, without any preparation, on the bare turf or gravel.

Figure 69 is a Gothic, and Figures 70, 71, are Grecian vases, commonly manufactured in plaster in our cities, but which are also made of Roman cement. They are here shown upon suitable pedestals—*a* being the vase, and *b* the pedestal. These with many other elegant vases and urns are manufactured in an artificial stone, as durable as marble, by *Austin* of London, and together with a great

[Fig. 69.]

EMBELLISHMENTS. 425

variety of other beautiful sculpturesque decorations, may be imported at very reasonable prices.

Figures 70, 71, are beautiful vases of pottery ware manufactured by Peake, of Staffordshire—and which may be imported cheaply, or will be made to order at the Salamander works, in New York. These vases, when colored to imitate marble or other stone, are extremely durable and very ornamental. As yet, we are unable to refer our readers to any manufactory here, where these articles are made in a manner fully equal to the English ; but we are satisfied, it is only necessary that the taste for such articles should increase, and the consequent demand, to induce our artisans to produce them of equal beauty and of greater cheapness.

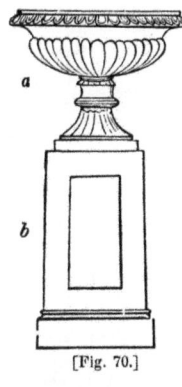

[Fig. 70.]

At Blithewood, the seat of R. Donaldson, Esq., on the Hudson, a number of exquisite vases may be seen in the pleasure-grounds, which are cut in Maltese stone. These were imported by the proprietor, direct from Malta, at very moderate rates, and are not only ornamental, but very durable. Their color is a warm shade of grey which harmonizes agreeably with the surrounding vegetation.

Large vases are sometimes filled with earth and planted with choice flowering plants, and the effect of the blossoms and green leaves growing out of these handsome receptacles, is at least unique and striking

[Fig. 71.]

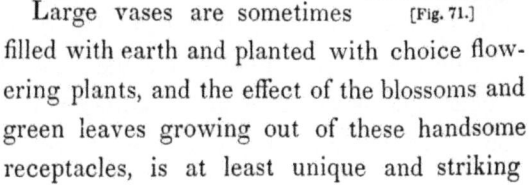

[Fig. 72.]

426   LANDSCAPE GARDENING.

Loudon objects to it in the case of an elegant sculptured vase, "because it is reducing a work of art to the level of a mere garden flower-pot, and dividing the attention between the beauty of the form of the vase and of its sculptured ornaments, and that of the plant which it contains." This criticism is a just one in its general application, especially when vases are considered as architectural decorations. Occasional deviations, however, may be permitted, for the sake of producing variety, especially in the case of vases used as decorations in the flower-garden.

A very pretty and fanciful substitute for the sculptured vase, and which may take its place in the picturesque landscape, may be found in vases or baskets of *rustic work*, constructed of the branches and sections of trees with the bark attached. Figure 74 is a representation of a pleasing rustic vase which we have constructed without difficulty. A tripod of branches of trees forms the pedestal. An octagonal box serves as the body or frame of the vase; on this, pieces of birch and hazel (small split limbs covered with the bark) are nailed closely, so as to form a sort of mosaic covering to the whole exterior. Ornaments of this kind, which may be made by the amateur with the assistance of a common carpenter, are very suitable for the decoration of the grounds and flower-gardens of cottages or picturesque villas. An endless variety of forms will occur to an

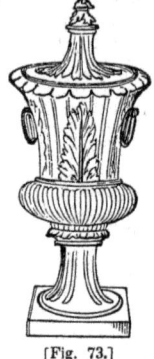

[Fig. 73.]

[Fig. 74.]

ingenious artist in rustic work, which he may call in to the embellishment of rural scenes, without taxing his purse heavily.

Sundials (Fig. 75) are among the oldest decorations for the garden and grounds, and there are scarcely any which we think more suitable. They are not merely decorative, but have also an useful character, and may therefore be occasionally placed in distant parts of the grounds, should a favorite walk terminate there. When we meet daily in our walks for a number of years, with one of these silent monitors of the flight of time, we become in a degree attached to it, and really look upon it as gifted with a species of intelligence, beaming out when the sunbeams smile upon its dial-plate.

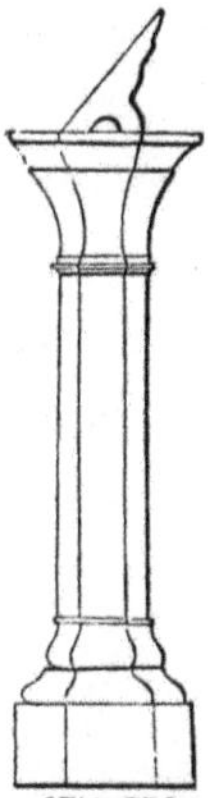

[Fig. 75.]

The *Architectural Flower-garden,* as we have just remarked, has generally a direct connexion with the house, at least on one side by the terrace. It may be of greater or less size, from twenty feet square to half an acre in extent. The leading characteristics of this species of flower-garden, are the regular lines and forms employed in its beds and walks. The flowers are generally planted in beds in the form of circles, octagons, squares, etc., the centre of the garden being occupied by an elegant vase, a sundial, or that still finer ornament, a fountain, or *jet d'eau.* In various parts of the garden, along the principal walks, or in the centre of parterres, pedestals supporting vases, urns, or handsome flower-pots with plants, are placed. When a highly marked character of art is intended, a balustrade or parapet, resembling that of the terrace to which it is connected, is continued round the whole of this garden. Or in other cases the garden is surrounded by a thicket of shrubs and low trees, partly concealing it from the eye on all sides but one.

It is evident that the architectural flower-garden is superior to the general flower-garden, *as an appendage to the house*, on two accounts. First, because, as we have already shown, it serves an admirable purpose in effecting a harmonious union between the house and the grounds. And secondly, because we have both the rich verdure and gay blossoms of the flowering plants, and the more permanent beauty of sculptured forms; the latter heightening the effect of the former by contrast, as well as by the relief they afford the eye in masses of light, amid surrounding verdure.

There are several varieties of general flower-gardens, which may be formed near the house. Among these we will only notice the *irregular* flower-garden, the *old French* flower-garden, and the *modern* or *English* flower-garden.

In almost all the different kinds of flower-gardens, two methods of forming the beds are observed. One is, to cut the beds out of the green turf, which is ever afterwards

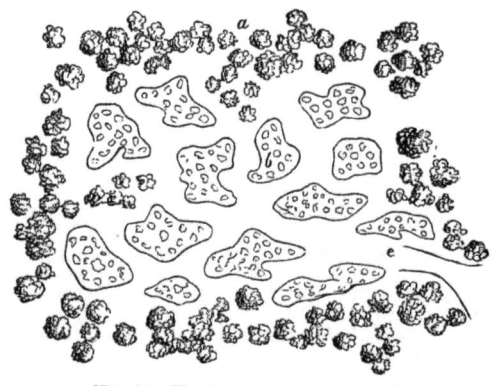

[Fig. 76. The Irregular Flower-garden.]

EMBELLISHMENTS. 429

kept well-mown or cut for the walks, and the edges pared; the other, to surround the beds with edgings of verdure, as box, etc., or some more durable material, as tiles, or cut stone, the walks between being covered with gravel. The turf is certainly the most agreeable for walking upon in the heat of summer, and the dry part of the day; while the gravelled flower-garden affords a dry footing at nearly all hours and seasons.

The *irregular* flower-garden is surrounded by an irregular belt of trees and ornamental shrubs of the choicest species, and the beds are varied in outline, as well as irregularly disposed, sometimes grouping together, sometimes standing singly, but exhibiting no uniformity of arrangement. An idea of its general appearance may be gathered from the accompanying sketch (Fig. 76), which may be varied at pleasure. In it the irregular boundary of shrubs is shown at *a*, the flower-beds *b*, and the walks *e*.

This kind of flower-garden would be a suitable accompaniment to the house and grounds of an enthusiastic lover of the picturesque, whose residence is in the Rural Gothic style, and whose grounds are also eminently varied and picturesque. Or it might form a pretty termination to a distant walk in the pleasure-grounds, where it would be more necessary that the flower-garden should be in keeping with the surrounding plantations and scenery than with the house.

Where the flower-garden is a spot set apart, of any regular outline, not of large size, and especially where it is attached directly to the house, we think the effect is most satisfactory when the beds or walks are laid out in symmetrical forms. Our reasons for this are these: the flower-garden, unlike distant portions of the pleasure-ground scenery, is an appendage to the house, seen in the same view or moment with it, and therefore should exhibit something of the regularity which characterizes, in a greater or less degree, all architectural compositions; and when a given scene is so small as to be embraced in a single glance of the eye, regular forms are found to be more satisfactory than irregular ones, which, on so small a scale, are apt to appear unmeaning.

The *French* flower-garden is the most fanciful of the regular modes of laying out the area devoted to this purpose. The patterns or figures employed are often highly intricate, and require considerable skill in their formation. The walks are either of gravel or smoothly shaven turf, and the beds are filled with choice flowering plants. It is evident that much of the beauty of this kind of flower-garden, or indeed any other where the figures are regular and intricate, must depend on the outlines of the beds, or *parterres of embroidery*, as they are called, being kept distinct and clear. To do this effectually, low growing herbaceous plants or border flowers, perennials and annuals, should be chosen, such as will not exceed on an average, one or two feet in height.

In the English flower-garden, the beds are either in symmetrical forms and figures, or they are characterized by irregular *curved* outlines. The peculiarity of these gardens, at present so fashionable in England, is, that each separate bed is planted with a single variety, or at most two varieties of flowers. Only the most striking and showy varieties are generally chosen, and the effect, when the selection is judicious, is highly brilliant. Each bed, in its season, presents a mass of blossoms, and the contrast of rich colors is much more striking than in any other arrangement. No plants are admitted that are shy bloomers, or which have ugly habits of growth, meagre or starved foliage; the aim being brilliant effect, rather than the display of a great variety of curious or rare plants. To bring this about more perfectly, and to have an elegant show during the whole season of growth, hyacinths and other fine bulbous roots occupy a certain portion of the beds, the intervals being filled with handsome herbaceous plants, permanently planted, or with flowering annuals and green-house plants renewed every season.

To illustrate the mode of arranging the beds and disposing the plants in an English garden, we copy the plan and description of the elegant flower-garden, on the lawn at *Dropmore*, the beds being cut out of the smooth turf.

"The flower-garden at Dropmore is shown in Fig. 77. In this the plants are so disposed, that when in flower the corresponding forms of the figure contain corresponding colored flowers. The following is a list of the plants which occupy this figure during summer, with the order in which they are disposed: and a corresponding enumeration of the bulbs and other plants which occupy the beds during winter and spring.

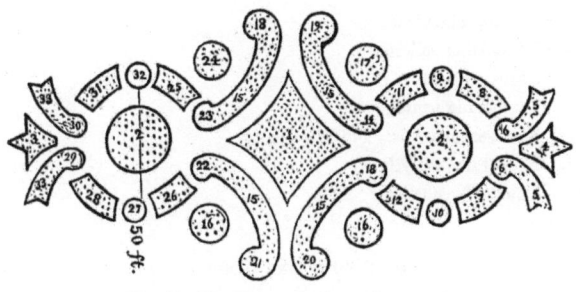

[Fig. 77. The Flower-Garden at Dropmore.]

In Summer.

BEDS.
1. Rosa Indica (blush China), bordered with R. Semperflorens flore pleno, and R. Indica minor.
2. Pelargonium inquinans (Scarlet Geranium).
3. Verbena Lamberti.
4. Senecio elegans, flore pleno. (Double Jacobea.)
5. 5. Alonsoa incisifolia.
6. 6. Agathea excelsis.
7. Fuchsia coccinea (Lady's Eardrop), bordered with Double Primrose.
8. Helitropium peruvianum.
9. Ruellia formosa.
10. Ageratum mexicanum.
11. Dianthus chinensis (Indian Pink), and Mignonette.
12. Lobelia splendens.
13. Dianthus satifolius.
14. Lobelia unidentata.
15. 15. 15. Choice herbaceous plants not exceeding one foot six inches in height.
16. 16. Gladiolus cardinalis.
17. Pelargonium lateripis (pink-flowered variegated Ivy Geranium).
18. Anagallis grandiflora.
19. Anagallis Monelli.
20. Pelargonium coruscans (Fiery-red Geranium.)
21. Prince of Orange Geranium.
22. Œnothera cæspitosa.
23. Œnothera missouriensis (Missouri evening Primrose).
24. Scarlet flowered variegated leaved Geranium.
25. Malope trifida.
26. Lobelia fulgens.
27. Petunia Phœnicea.
28. Commelina cælestis.
29. Cistus guttatus.
30. Campanula pentagona.
31. Four seasons Rose, and Mignonette.
32. Bouvardia triphylla.
33. Double Nasturtium.

In Winter and Spring.

BEDS.
1. Anemone Coronaria.
2. 2. Malcomia maritima (Mediterranean stock).

EMBELLISHMENTS.

Beds.
3 and 4. Fine varieties of Tulips.
5. 5. Double rocket Larkspur (sown in autumn).
6. 6. Agathea cælestis.
7. Scilla nutans (blue harebell).
8. Feathered Hyacinths.
9 and 10. Sweet scented Tulips.
11. Double garden Tulips.
12. Single gesneriana Tulips.
13 and 14. Tritonia crocata, and Tritonia fenestra, kept in frames in mid-winter.
15. 15. 15. 15. Choice herbaceous plants not exceeding one foot six inches in height.
16. 16. Hyacinths, double blue, plunged in pots.
17. Hyacinths, double red, do.
18 and 19. Hyacinths, single blue variety.
20 and 21. Single white Hyacinths.
22 and 23. Crocus vernus and biflorus.
24. Hyacinths, double red.
25 and 26. Tulips, double yellow.
27. Hyacinths, double white.
28. Muscari botryoïdes, (Grape Hyacinth).
29. Oxalis caprina (kept in frames in mid-winter).
30. Scilla verna (Spring Harebell).
31. Muscari racemosum, the border of Viola tricolor in sorts.
32. Hyacinths, double white.
33. Double rose Larkspur.

" As a general principle for regulating the plants in this figure, the winter and spring flowers ought, as much as possible, to be of sorts which admit of being in the ground all the year: and the summer crop should be planted at intervals between the winter plants. Or the summer crop, having been brought forward in pots under glass, or by nightly protection, may be planted out about the middle of June, after the winter plants in pots are removed. A number of hardy bulbs ought to be potted and plunged in the beds in the months of October and November; and when out of bloom, in May or June, removed to the reserve

434   LANDSCAPE GARDENING.

garden and plunged there, in order to perfect their foliage and mature their bulbs for the succeeding season."*

There cannot be a question that this method of planting the flower-garden in groups and masses, is productive of by far the most splendid effect. In England, where flower-gardens are carried to their greatest perfection, the preference in planting is given to exotics which blossom constantly throughout the season, and which are kept in the green-house during winter, and turned out in the beds in the early part of the season, where they flower in the greatest profusion until frost; as Fuchsias, Salvias,

[Fig. 78. English Flower-Garden.]

* Ency. of Gardening, 1000.

EMBELLISHMENTS.   435

Lobelias, Scarlet Geraniums, etc., etc.* This mode can be adopted here where a small green-house or frame is kept. In the absence of these, nearly the same effect may be produced by choosing the most showy herbaceous plants, perennial and biennial, alternating them with hardy bulbs, and the finer species of annuals.

In Fig. 78, we give an example of a small cottage or villa residence of one or two acres, where the flower-beds are disposed around the lawn in the English style: their forms irregular, with curved outlines, affording a great degree of variety in the appearance as viewed from different points on the lawn itself. In this, the central portion is occupied by the lawn; *c, d*, are the flower-beds, planted with showy border-flowers, in separate masses; *b*, the conservatory. Surrounding the whole is a collection of choice shrubs and trees, the lowest near the walk, and those behind increasing in altitude as they approach the boundary wall or fence. In this plan, as there is supposed to be no exterior view worth preserving, the amphitheatre of shrubs and trees completely shuts out all objects but the lawn and its decorations, which are rendered as elegant as possible.

Where the proprietor of a country residence, or the ladies of a family, have a particular taste, it may be indulged at pleasure in other and different varieties of the flower-garden. With some families there is a taste for botany,

* In many English residences, the flower-garden is maintained in never-fading brilliancy by almost daily supplies from what is termed the *reserve garden*. This is a small garden out of sight, in which a great number of duplicates of the species in the flower-garden are grown in pots plunged in beds. As soon as a vacuum is made in the flower-garden by the fading of any flowers, the same are immediately removed and their places supplied by fresh plants just ready to bloom, from the pots in the reserve garden. This, which is the *ultimatum* of refinement in flower-gardening, has never, to our knowledge, been attempted in this country.

when a small botanic flower-garden may be preferred—the herbaceous and other plants being grouped or massed in beds after the *Linnæan,* or the *natural* method. Some persons have an enthusiastic fondness for florist flowers, as Pansies, Carnations, Dahlias, Roses, etc.; others for bulbous roots, all of which may very properly lead to particular modes of laying out flower-gardens.

The desideratum, however, with most persons is, to have a continued display of blossoms in the flower-garden from the opening of the crocus and snowdrop in the spring, until the autumnal frosts cut off the last pale asters, or blacken the stems of the luxuriant dahlias in November. This may be done with a very small catalogue of plants if they are properly selected : such as flower at different seasons, continue long time in bloom, and present fine masses of flowers. On the other hand, a very large number of species may be assembled together; and owing to their being merely botanical rarities, and not bearing fine flowers, or to their blossoming chiefly in a certain portion of the season, or continuing but a short period in bloom, the flower-garden will often have but an insignificant appearance. With a group of Pansies and spring bulbs, a bed of ever-blooming China Roses, including the *Isle de Bourbon* varieties, some few Eschscholtzias, the showy Petunias, Gilias, and other annuals, and a dozen choice double Dahlias, and some trailing Verbenas, a limited spot, of a few yards in diameter, may be made productive of more enjoyment, so far as regards a continued display of flowers, than ten times that space, planted, as we often see flower-gardens here, with a heterogeneous mixture of everything the possesor can lay his hands on, or crowd within the inclosure.

The *mingled* flower-garden, as it is termed, is by far the most common mode of arrangement in this country, though it is seldom well effected. The object in this is to dispose the plants in the beds in such a manner, that while there is no predominance of bloom in any one portion of the beds, there shall be a general admixture of colors and blossoms throughout the entire garden during the whole season of growth.

To promote this, the more showy plants should be often repeated in different parts of the garden, or even the same parterre when large, the less beautiful sorts being suffered to occupy but moderate space. The smallest plants should be nearest the walk, those a little taller behind them, and the largest should be furthest from the eye, at the back of the border, when the latter is seen from one side only, or in the centre, if the bed be viewed from both sides. A neglect of this simple rule will not only give the beds, when the plants are full grown, a confused look, but the beauty of the humbler and more delicate plants will be lost amid the tall thick branches of sturdier plants, or removed so far from the spectator in the walks, as to be overlooked.

Considerable experience is necessary to arrange even a moderate number of plants in accordance with these rules. To perform it successfully, some knowledge of the habits of the plants is an important requisite; their height, time of flowering, and the colors of their blossoms. When a gardener, or an amateur, is perfectly informed on these points, he can take a given number of plants of different species, make a plan of the bed or all the beds of a flower-garden upon paper, and designate the particular situation of each species.

To facilitate the arrangement of plants in this manner,

we here subjoin a short list of the more showy perennial and annual *hardy* border flowers, such as are easily procured here for the use of those who are novices in the art, and who wish to cultivate a taste for the subject.

No. 1, Designates the first class, which grow from six to twelve inches in height.

No. 2, Those which grow from one to two feet.

No. 3, Those which are over two feet in height.

Hardy Perennials.

Flowering in April.

1. *Anemone thalictroides, pl.* Double wood Anemone; white.
1. *Anemone pulsatilla.* Pasque flower; blue.
1. *Anemone hepatica, pl.* Double Hepaticas; blue.
1. *Viola odorata, pl.* Double white and blue European violets.
1. *Omphalodes verna.* Blue Venus Navelwort.
1. *Polemonium reptans.* Greek Valerian; blue.
1. *Phlox stolonifera.* Creeping Phlox; red.
2. *Phlox divaricata.* Early purple Phlox.
1. *Primula veris.* The Cowslip; yellow and red.
1. *Primula polyantha.* The Polyanthus; purple.
1. *Primula auricula.* The Auricula; purple.
1. *Viola tricolor.* Heart's Ease or Pansy; many colors and sorts.
1. *Viola grandiflora.* Purple Pansy.
2. *Saxifraga crassifolia.* Thick-leaved Saxifrage; lilac.
1. *Phlox subulata.* Moss pink Phlox.
1. *Phlox nivea.* White Moss Pink.
1. *Gentiana acaulis.* Dwarf Gentian; purple.
1. *Adonis vernalis.* Spring fl. Adonis; yellow.
2. *Dodecatheon meadia.* American Cowslip; lilac.
2. *Pulmonaria virginica.* Virginian Lungwort; purple.
2. *Alyssum saxatile.* Golden basket; yellow.
2. *Trollius europeus.* European Globe flower; yellow.
1. *Corydalis cucularia.* Breeches-flower; white.

May.

1. *Veronica gentianoides.* Gentian leaved Speedwell; blue.

EMBELLISHMENTS. 439

2. *Veronica spicata.* Blue spiked Speedwell.
2. *Pentstemon ovata.* Oval leaved Pentstemon; blue.
2. *Pentstemon atropurpureus.* Dark purple Pentstemon.
2. *Orobus niger.* Dark purple Vetch.
1. *Jeffersonia diphylla.* Five-leaved Jeffersonia; white.
1. *Lysimachia nummularia.* Trailing Loose-strife; yellow.
1. *Convallaria majalis.* Lily of the Valley; white.
1. *Saponaria ocymoides.* Basil-like Soapwort; red.
1. *Phlox pilosa.* Hairy Phlox; red.
2. *Anchusa Italica.* Italian Bugloss; blue.
2. *Ranunculus acris, pl.* Double Buttercups; yellow.
2. *Tradescantia virginica.* Blue and white Spiderwort.
2. *Lupinus polyphyllus.* Purple Lupin.
2. *Iris siberiaca.* Siberian Iris; blue.
3. *Iris florentina.* Florentine Iris; white.
3. *Pæonia tenuifolia.* Small leaved Pæony; red.
3. *Pæonia albiflora.* Single white Pæony.
2. *Lupinus nootkaensis.* Nootka Sound Lupin; blue.
2. *Hesperis matronalis, alba, pl.* The double white Rocket.
2. *Phlox suaveolens.* The white Phlox or Lychnidea.
2. *Phlox maculata.* The purple spotted Phlox.
3. *Hemerocallis flava.* The yellow Day-Lily.
2. *Lupinus perennis* and *rivularis.* Perennial Lupins; blue.
2. *Lychnis flos cuculi, pl.* Double ragged-Robin; red.
2. *Papaver orientalis.* Oriental scarlet Poppy.
2. *Aquilegia canadensis.* Wild Columbine; scarlet.
1. *Houstonia cærulea.* Blue Houstonia.

JUNE.

1. *Spiræa filipendula, pl.* Double Pride of the Meadow; white.
2. *Spiræa lobata.* Siberian Spirea; red.
2. *Spiræa Ulmaria, pl.* Double Meadow-sweet; white.
2. *Delphinium grandiflorum, pl.* Double dark blue Larkspur.
2. *Delphinium chinense, pl.* Double Chinese Larkspur; blue.
2. *Dianthus hortensis.* Garden Pinks, many double sorts and colors.
2. *Caltha palustris, pl.* Double marsh Marygold; yellow.
1. *Cypripedium pubescens.* Yellow Indian moccasin.
2. *Polemonium cæruleum,* and *album.* Common white and blue Greek Valerian.
2. *Campanula persicifolia, pl.* Double peach-leaved Campanula; white.
2. *Antirrhinum majus.* Red and white Snapdragons.

2. *Geranium sanguineum.* Bloody Geranium; red.
1. *Viscaria vulgaris, pl.* White and red Viscaria.
2. *Œnothera fruticosa.* Shrubby Evening Primrose; yellow.
1. *Eschscholtzia californica.* Golden Eschscholtzia; yellow.
1. *Lychnis fulgens.* Fulgent Lychnis; red.
1. *Dianthus chinensis.* Indian Pinks; variegated.
2. *Dianthus caryophyllus.* Carnation; variegated.
1. *Verbena multifida.* Cut-leaved Verbena; purple.
1. *Verbena Lamberti.* Lambert's Verbena; purple.
2. *Campanula grandiflora.* Large blue Bell-flower.
3. *Aconitum Napellus.* Monkshood; purple.
3. *Aconitum Napellus, variegated.* Purple and white Monkshood.
3. *Campanula ranunculoides.* Nodding Bell-flower; blue.
2. *Clematis integrifolia.* Austrian blue Clematis.
3. *Verbascum phœniceum.* Purple Mullein.
3. *Clematis erecta.* Upright Clematis; white.
3. *Linum perenne.* Perennial Flax; blue.
3. *Pæonia Humei.* Double blush Pæony.
3. *Pæonia fragrans.* Double fragrant Pæony; rose.
3. *Pæonia whitleji.* Double white Pæony.
3. *Gaillardia aristata.* Bristly Gaillardia; yellow.
2. *Asphodelus ramosus.* Branchy Asphodel; white.
2. *Pentstemon speciosa.* Showy Pentstemon; blue.
1. *Iris Susana.* Chalcedonian Iris; mottled.

July.

2. *Dictamnus Fraxinella.* Purple Fraxinella.
2. *Dictamnus alba.* White Fraxinella.
1. *Pentstemon Richardsonii.* Richardson's Pentstemon; purple.
1. *Pentstemon pubescens.* Downy Pentstemon; lilac.
2. *Anchusa officinalis.* Common Bugloss; blue.
1. *Campanula carpathica.* Carpathian Bell-flower; blue.
2. *Monarda didyma.* Scarlet Balm.
2. *Œnothera Fraseri.* Fraser's Evening Primrose; yellow.
2. *Œnothera macrocarpa.* Large podded Evening Primrose; yellow.
1. *Sedum populifolium.* Poplar leaved Sedum; white.
2. *Campanula Trachelium, pl.* Double white and blue Bell-flowers.
1. *Potentilla Russelliana.* Russell's Cinquefoil; red.
1. *Dianthus deltoides.* Mountain Pink; red.
1. *Veronica maritima.* Maritime Speedwell; blue.
2. *Delphinium speciosum.* Showy Larkspur; blue.
2. *Campanula macrantha.* Large blue Bell-flower.

EMBELLISHMENTS.

3. *Pentstemon Digitalis.* Missouri Pentstemon; white.
3. *Hibiscus palustris.* Swamp Hibiscus; red.
3. *Lychnis Chalcedonica.* Single and double scarlet Lychnis.
2. *Chelone Lyoni.* Purple Chelone.
2. *Chelone barbata.* Bearded Chelone; orange.
2. *Dracocephalum grandiflorum.* Dragon's Head; purple.
3. *Lythrum latifolium.* Perennial Pea; purple.

August.

2. *Catananche cærulea.* Blue Catananche.
1. *Corydalis formosa.* Red Fumitory.
1. *Phlox carnea.* Flesh colored Phlox.
2. *Asclepias tuberosa.* Orange Swallowwort.
2. *Veronica carnea.* Flesh colored Speedwell.
2. *Gaillardia bicolor.* Orange Gaillardia.
2. *Hemerocallis japonica.* Japan Day-Lily; white.
2. *Dianthus superbus.* Superb fringed Pink; white.
2. *Lobelia cardinalis.* Cardinal-flower; red.
1. *Lychnis coronata.* Chinese orange Lychnis.
2. *Lythrum salicaria.* Willow Herb; purple.
3. *Yucca filamentosa.* Adam's Thread; white.
2. *Yucca flaccida.* Flaccid Yucca; white.
3. *Phlox paniculata.* Panicled Phlox; purple and white.
3. *Campanula pyramidalis.* Pyramidal Bell-flower; blue and white.
2. *Liatris squarrosa.* Blazing Star; blue.
2. *Epilobium spicatum.* Purple spiked Epilobium.
2. *Coreopsis tenuifolia.* Fine-leaved Coreopsis; yellow.
3. *Cassia Marylandica.* Maryland Cassia; yellow.

September and October.

2. *Achillea Ptarmica, pl.* Double Milfoil; white.
2. *Coreopsis grandiflora.* Large yellow Coreopsis.
1. *Aster linifolius.* Fine-leaved Aster; white.
2. *Eupatorium cælestinum.* Azure blue Eupatorium.
2. *Phlox Wheeleriana.* Wheeler's Phlox; red.
3. *Aster macrophyllus.* Broad-leaved Aster; white.
3. *Eupatorium aromaticum.* Fragrant Eupatorium; white.
3. *Liatris elegans.* Elegant Blazing Star; purple.
3. *Liatris spicata* and *scariosa.* Blue Blazing Stars.
1. *Gentiana saponaria.* Soapwort Gentian; blue.
3. *Aster novæ-angliæ.* New England Aster; purple.

3. *Echinops retro.* Globe Thistle.
3. *Chrysanthemum indicum.* Artemisias, many sorts and colors.

The *shrubbery* is so generally situated in the neighborhood of the flower-garden and the house, that we shall here offer a few remarks on its arrangement and distribution.

A collection of flowering shrubs is so ornamental, that to a greater or less extent it is to be found in almost every residence of the most moderate size: the manner in which the shrubs are disposed, must necessarily depend in a great degree upon the size of the grounds, the use or enjoyment to be derived from them, and the prevailing character of the scenery.

It is evident, on a moment's reflection, that shrubs being intrinsically more ornamental than trees, on account of the beauty and abundance of their flowers, they will generally be placed near and about the house, in order that their gay blossoms and fine fragrance may be more constantly enjoyed, than if they were scattered indiscriminately over the grounds.

Where a place is limited in size, and the whole lawn and plantations partake of the *pleasure-ground* character, shrubs of all descriptions may be grouped with good effect, in the same manner as trees, throughout the grounds; the finer and rarer species being disposed about the dwelling, and the more hardy and common sorts along the walks, and in groups, in different situations near the eye.

When, however, the residence is of larger size, and the grounds have a park-like extent and character, the introduction of shrubs might interfere with the noble and dignified expression of lofty full grown trees, except perhaps they were planted here and there, among large groups, as *underwood;* or if cattle or sheep were allowed to graze in the park, it would of course be impossible to preserve plantations of shrubs there. When this is the case, however, a portion near the house is divided from the park (by a wire fence or some inconspicuous barrier) for the pleasure-ground, where the shrubs are disposed in belts, groups, etc., as in the first case alluded to.

There are two methods of grouping shrubs upon lawns which may separately be considered, in combination with *beautiful* and with *picturesque* scenery.

In the first case, where the character of the scene, of the plantations of trees, etc., is that of polished beauty, the belts of shrubs may be arranged similar to herbaceous flowering plants, in arabesque beds, along the walks, as in Fig. 76, page 428. In this case, the shrubs alone, arranged with relation to their height, may occupy the beds; or if preferred, shrubs and flowers may be intermingled. Those who have seen the shrubbery at *Hyde Park*, the residence of the late Dr. Hosack, which borders the walk leading from the mansion to the hot-houses, will be able to recall a fine example of this mode of mingling woody and herbaceous plants. The belts or borders occupied by the shrubbery and flower-garden there, are perhaps from 25 to 35 feet in width, completely filled with a collection of shrubs and herbaceous plants; the smallest of the latter being quite near the walk; these succeeded by taller species receding from the front of the border, then follow shrubs of moderate size, advancing in height until the background of the whole is a rich mass of tall shrubs and trees of moderate size. The effect of this belt on so large a scale, in high keeping, is remarkably striking and elegant.

Where *picturesque effect* is the object aimed at in the

pleasure-grounds, it may be attained in another way; that is, by planting irregular groups of the most vigorous and thrifty growing shrubs in lawn, without placing them in regular dug beds or belts; but instead of this, keeping the grass from growing and the soil somewhat loose, for a few inches round their stems (which will not be apparent at a short distance). In the case of many of the hardier shrubs, after they become well established, even this care will not be requisite, and the grass only will require to be kept short by clipping it when the lawn is mown.

As in picturesque scenes everything depends upon *grouping well*, it will be found that shrubs may be employed with excellent effect in connecting single trees, or finishing a group composed of large trees, or giving fulness to groups of tall trees newly planted on a lawn, or effecting a union between buildings and ground. It is true that it requires something of an artist's feeling and perception of the picturesque to do these successfully, but the result is so much the more pleasing and satisfactory when it is well executed.

When walks are continued from the house through distant parts of the pleasure-grounds, groups of shrubs may be planted along their margins, here and there, with excellent effect. They do not shut out or obstruct the view like large trees, while they impart an interest to an otherwise tame and spiritless walk. Placed in the projecting bay, round which the walk curves so as to appear to be a reason for its taking that direction, they conceal also the portion of the walk in advance, and thus enhance the interest doubly. The neighborhood of rustic seats, or resting points, are also fit places for the assemblage of a group or groups of shrubs.

For the use of those who require some guide in the

selection of species, we subjoin the accompanying list of hardy and showy shrubs, which are at the same time easily procured in the United States. A great number of additional species and varieties, and many more rare, might be enumerated, but such will be sufficiently familiar to the connoisseur already; and what we have said respecting botanical rarities in flowering plants may be applied with equal force to shrubs, viz. that in order to produce a brilliant effect, a few well chosen species, often repeated, are more effective than a great and ill-assorted *mélange*.

In the following list, the shrubs are divided into two classes—No. 1 designating those of medium size, or low *growth*, and No. 2, those which are of the largest size.

Flowering in April.

1. *Daphne mezereum*, the Pink Mezereum, *D. M. album*, the white Mezereum.
2. *Shepherdia argentea*, the Buffalo berry; yellow.
1. *Xanthorhiza apiifolia*, the parsley-leaved Yellow-root; brown.
1. *Cydonia japonica*, the Japan Quince; scarlet.
1. *Cydonia japonica alba*, the Japan Quince; white.
2. *Amelanchier Botryapium*, the snowy Medlar.
1. *Ribes aureum*, the Missouri Currant; yellow.
1. *Coronilla Emerus*, the Scorpion Senna; yellow.
2. *Magnolia conspicua*, the Chinese chandelier Magnolia; white.

May.

2. *Crategus oxycantha*, the scarlet Hawthorn.
2. *Crategus oxycantha, fl. pleno*, the double white Hawthorn.
2. *Chionanthus virginica*, the white Fringe tree.
1. *Chionanthus latifolius*, the broad-leaved Fringe tree; white.
1. *Azalea*, many fine varieties; red, white, and yellow.
1. *Calycanthus florida*, the Sweet-scented-shrub; brown.
1. *Magnolia purpurea*, the Chinese purple Magnolia.
2. *Halesia tetraptera*, the silver Bell tree; white.
2. *Syringa vulgaris*, the common white and red Lilacs.
1. *Syringa persica*, the Persian Lilac: white and purple.

1. *Syringa persica laciniata*, the Persian cut-leaved Lilac ; purple.
1. *Kerria* or *Corchorus japonica*, the Japan Globe flower ; yellow.
1. *Lonicera tartarica*, the Tartarian upright Honeysuckles ; red and white.
1. *Philadelphus coronarius*, the common Syringo, and the double Syringo ; white.
1. *Spiræa hypericifolia*, the St. Stephen's wreath ; white.
1. *Spiræa corymbosa*, the cluster flowering Spirea ; white.
1. *Ribes sanguineum*, the scarlet flowering Currant.
1. *Amygdalus pumila, pl.*, the double dwarf Almond ; pink.
1. *Caragana Chamlagu*, the Siberian Pea tree ; yellow.
2. *Magnolia soulangeana*, the Soulange Magnolia ; purple.
1. *Pæonia Moutan banksia*, and *rosea*, the Chinese tree Pæonias ; purple.
1. *Benthamia frugifera*, the red berried Benthamia ; yellow.

JUNE.

1. *Amorpha fruticosa*, the Indigo Shrub ; purple.
2. *Colutea arborescens*, the yellow Bladder-senna.
1. *Colutea cruenta*, the red Bladder-senna.
1. *Cytisus capitatus*, the cluster-flowered Cytisus ; yellow.
1. *Stuartia virginica*, the white Stuartia.
1. *Cornus sanguinea*, the bloody twig Dogwood ; white.
1. *Hydrangea quercifolia*, the oak-leaved Hydrangea ; white.
2. *Philadelphus grandiflorus*, the large flowering Syringo ; white.
2. *Viburnum Opulus*, the Snow-ball ; white.
2. *Magnolia glauca*, the swamp Magnolia ; white.
1. *Robinia hispida*, the Rose-acacia

JULY.

1. *Spiræa bella*, the beautiful Spirea ; red.
2. *Sophora japonica*, the Japan Sophora ; white.
2. *Sophora japonica pendula*, the weeping Sophora ; white.
2. *Rhus Cotinus*, the Venetian Fringe tree ; yellow. (Brown tufts.)
1. *Ligustrum vulgare*, the common Privet ; white.
2. *Cytisus Laburnum*, the Laburnum ; yellow.
2. *Cytisus l. quercifolia*, the oaked-leaved Laburnum ; white.
1. *Cytisus purpureus*, the purple Laburnum.
1. *Cytisus argenteus*, the silvery Cytisus ; yellow.
1. *Cytisus nigricans*, the black rooted Cytisus ; yellow.
2. *Kolreuteria paniculata*, the Japan Kolreuteria ; yellow.

EMBELLISHMENTS. 447

AUGUST AND SEPTEMBER.

1. *Clethra alnifolia,* the alder-leaved Clethra ; white.
1. *Symphoria racemosa,* the Snowberry ; (in fruit) white.
2. *Hibiscus syriacus,* the double purple, double white, double striped, double blue, and variegated leaved Altheas.
1. *Spiræa tomentosa,* the tomentose Spirea ; red.
2. *Magnolia glauba thompsoniana,* the late flowering Magnolia ; white.
1. *Baccharis halimifolia,* the Groundsel tree ; white tufts.
2. *Euonymus europæus,* the European Strawberry tree (in fruit), red.
2. *Euonymus europæus alba,* the European Strawberry tree ; the fruit white.
2. *Euonymus latifolius,* the broad-leaved Strawberry tree ; red.
1. *Daphne mezereum autumnalis,* the autumnal Mezereum.

Besides the above, there are a great number of charming varieties of hardy roses, some of which may be grown in the common way on their own roots, and others grafted on stocks, two, three, or four feet high, as standards or tree-roses. The effect of the latter, if such varieties as *George the Fourth, La Cerisette, Pallagi,* or any of the new hybrid roses are grown as standards, is wonderfully brilliant when they are in full bloom. Perhaps the situation where they are displayed to the greatest advantage is, in the centre of small round, oval, or square beds in the flower-garden, where the remainder of the plants composing the bed are of dwarfish growth, so as not to hide the stem and head of the tree-roses.

There are, unfortunately, but few evergreen shrubs that will endure the protracted cold of the winters of the northern states. The fine Hollies, Portugal Laurels, Laurustinuses, etc., which are the glory of English gardens in autumn and winter, are not hardy enough to endure the depressed temperature of ten degrees below zero. South of Philadelphia, these beautiful exotic evergreens may be acclimated with good success, and will add greatly to the interest of the shrubbery and grounds in winter.

Besides the Balsam firs and the Spruce firs, the Arbor Vitæ, and other evergreen trees which we have described in the previous pages of this volume, the following hardy species of evergreen shrubs may be introduced with advantage in the pleasure-ground groups, viz :—

> *Rhododendron maximum*, the American rose bay or big Laurel ; white and pink, several varieties (in shaded places).
> *Kalmia latifolia*, the common Laurel ; several colors.
> *Juniperus suecia*, the Swedish Juniper.
> *Juniperus communis*, the Irish Juniper.
> *Buxus arborescens*, the common Tree-box, the Gold striped Tree-box, and the Silver striped Tree-box.
> *Ilex opaca*, the American Holly.
> *Crategus pyracantha*, the Evergreen Thorn.
> *Mahonia aquifolium*, the Holly leaved Berberry.

The *Conservatory* or the *Green-House* is an elegant and delightful appendage to the villa or mansion, when there is a taste for plants among the different members of a family. Those who have not enjoyed it, can hardly imagine the pleasure afforded by a well-chosen collection of exotic plants, which, amid the genial warmth of an artificial climate, continue to put forth their lovely blossoms, and exhale their delicious perfumes, when all out-of-door nature is chill and desolate. The many hours of pleasant and healthy exercise and recreation afforded to the ladies of a family, where they take an interest themselves in the growth and vigor of the plants, are certainly no trifling considerations where the country residence is the place of habitation throughout the whole year. Often during the inclemency of our winter and spring months, there are days when either the excessive cold, or the disagreeable state of the weather, prevents in a great measure many persons, and especially females, from taking exercise in the open air. To such, the conservatory would be an almost endless source of enjoyment and amusement; and if they are true amateurs, of active exertion also. The constant changes which daily growth and development bring about in vegetable forms, the interest we feel in the opening of a favorite cluster of buds, or the progress of the thrifty and luxuriant shoots of a rare plant, are such as serve most effectually to prevent an occupation of this nature from ever becoming monotonous or *ennuyant*.

The difference between the *green-house* and conservatory is, that in the former, the plants are all kept in *pots* and arranged on stages, both to meet the eye agreeably, and for more convenient growth; while in the *conservatory*, the plants are grown in a *bed* or border of soil precisely as in the open air.

When either of these plant habitations is to be attached to the house, the preference is greatly in favor of the conservatory. The plants being allowed more room, have richer and more luxuriant foliage, and grow and flower in a manner altogether superior to those in pots. The allusion to nature is also more complete in the case of plants growing in the ground; and from the objects all being on the same level, and easily accessible, they are with more facility kept in that perfect nicety and order which an elegant plant-house should always exhibit.

On the other hand, the green-house will contain by far the largest number of plants, and the same may be more easily changed or renewed at any time; so that for a particular taste, as that of a botanical amateur, who wishes to grow a great number of species in a small space, the green-house will be found preferable. Whenever either the conservatory or green-house is of moderate size, and intended solely for private recreation, we would in every case, when such a thing is not impossible, have it attached to the house; communicating by a glass door with the drawing-room, or one of the living rooms. Nothing can be more gratifying than a vista in winter through a glass door down the walk of a conservatory, bordered and overhung with the fine forms of tropical vegetation, golden oranges glowing through the dark green foliage, and gay corollas lighting up the branches of Camellias, and other floral favorites. Let us add the exulting song of a few Canaries, and the enchantment is complete. How much more refined and elevated is the taste which prefers such accessories to a dwelling, rather than costly furniture, or an extravagant display of plate!

The best and most economical form for a conservatory is a parallelogram—the deviation from a square being greater or less according to circumstances. When it is joined to the dwelling by one of its *sides* (in the case of the parallelogram form), the roof need only slope in one way, that is from the house. When one of the *ends* of the conservatory joins the dwelling, the roof should slope both ways from the centre. The advantage of the junction in the former case, is, that less outer surface of the conservatory being exposed to the cold, viz. only a side and two ends, less fuel will be required; the advantage in the latter case is, that the main walk leading down the conservatory will be exactly in the line of the vista from the drawing-room of the dwelling.

It is, we hope, almost unnecessary to state, that the roof of a conservatory, or indeed any other house where plants are to be well-grown, *must* be glazed. Opake roofs prevent the admission of perpendicular light, without which the stems of vegetation are drawn up weak and feeble, and are attracted in an unsightly manner towards the glass in front. When the conservatory joins the house by one of its ends, and extends out from the building to a considerable length, the effect will be much more elegant; and the plants will thrive more perfectly, when it is glazed on all of the three sides, so as to admit light in every direction.

The best aspect for a conservatory is directly south; southeast and southwest are scarcely inferior. Even east and west exposures will do very well, where there is plenty of glass to admit light; for though our winters are cold, yet there is a great abundance of sun, and bright clear atmosphere, both far more beneficial to plants than the moist, foggy vapor of an English winter, which, though mild, is comparatively sunless. When the conservatory adjoins and looks into the flower-garden, the effect will be appropriate and pleasing.

Some few hints respecting the construction of a conservatory may not be unacceptable to some of our readers. In the first place, the roof should have a sufficient slope to carry off the rain rapidly, to prevent leakage; from 40 to 45 degrees is found to be the best inclination in our climate. The *roof* should by no means be glazed with large panes, because small ones have much greater strength, which is requisite to withstand the heavy weight of snow that often falls during winter, as well as to resist breakage by hail storms in summer. Four or eight inches by six, is the best size for roof-glass, and with this size the lap of the panes need not be greater than one-eighth of an inch, while it would require to be one-fourth of an inch, were the panes of the usual size. On the front and sides, the sashes may be handsome, and filled in with the best glass; even plate glass has been used in many cases to our knowledge here.

In the second place, some thorough provision must be made for warming the conservatory; and it is by far the best mode to have the apparatus for this purpose entirely independent of the dwelling house; that is (though the furnace may be in the basement), the flues and fire should be intended to heat the conservatory alone; for although a conservatory may, if small, be heated by the same fire which heats the kitchen or one of the living rooms, it is a much less efficient mode of attaining this object, and renders the conservatory more or less liable at all times to be too hot or too cold.

The common square flue, the sides built of bricks, and the top and bottom of tiles manufactured for that purpose, is one of the oldest, most simple, and least expensive methods of heating in use. Latterly, its place has been supplied by hot water circulated in large tubes of three or four inches in diameter from an open boiler, and by Perkins's mode as it is called, which employs small pipes of an inch in diameter, hermetically sealed. Economy of fuel and in the time requisite in attendance, are the chief merits of the hot water systems, which, however, have the great additional advantage of affording a more moist and genial temperature.

In a green-house, the flues, or hot water pipes, may be concealed under the stage. In conservatories they should by all means be placed out of sight also. To effect this, they are generally conducted into a narrow, hollow

Fig. 80. The Conservatory and Flower Garden, at Montgomery Place

chamber, under the walk, which has perforated sides or a grated top, to permit the escape of heated air.*

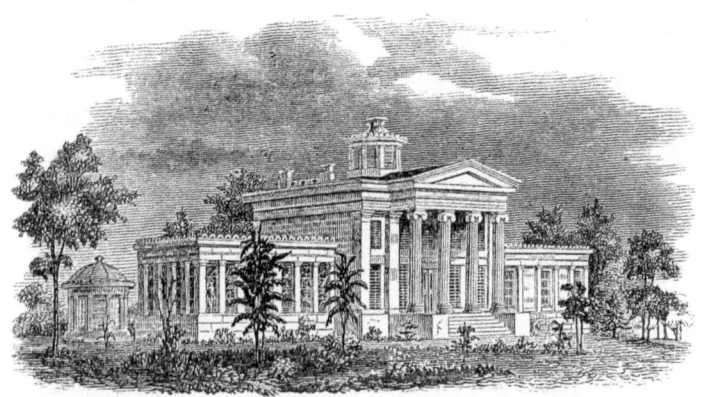

[Fig. 79. Villa at Brooklyn, N. Y., with the Conservatory attached.]

One of the most beautiful conservatories attached to the dwelling, to which we can refer our readers, for an example, is one built by J. W. Perry, Esq., Brooklyn, near New York (Fig. 79), forming the left wing of this elegant villa. Among the most magnificent detached conservatories are those of J. P. Cushing, Esq., at his elegant seat, *Belmont Place*, Watertown, near Boston; and that at Montgomery Place, the seat of Mrs. Edward Livingston, on the Hudson, Fig. 80.

A conservatory is frequently made an addition to a rectangular Grecian villa, as one of its wings—the other being a living or bed-room. The more varied and irregular outline of Gothic buildings enables them to receive an appendage of this nature with more facility in almost any direction, where the aspect is suitable.

* The circulation of warm air is greatly accelerated when an opening through the outer air is permitted to enter the hot air passage, thus becoming heated and passing into the conservatory.

Whatever be the style of the architecture of the house, that of the conservatory should in every case conform to it, and evince a degree of enrichment according with that of the main building.

Though a conservatory is often made an expensive luxury, attached only to the better class of residences, there is no reason why cottages of more humble character should not have the same source of enjoyment on a more moderate scale. A small green-house, or *plant cabinet*, as it is sometimes called, eight or ten feet square, communicating with the parlor, and constructed in a simple style, may be erected and kept up in such a manner, as to be a source of much pleasure, for a comparatively trifling sum; and we hope soon to see in this country, where the comforts of life are more equally distributed than in any other, the taste for enjoyments of this kind extending itself with the means for realizing them, into every portion of the northern and middle States.

Open and covered seats, of various descriptions, are among the most convenient and useful decorations for the pleasure-grounds of a country residence. Situated in portions of the lawn or park, somewhat distant from the house, they offer an agreeable place for rest or repose. If there are certain points from which are obtained agreeable prospects or extensive views of the surrounding country, a seat, by designating those points, and by affording us a convenient mode of enjoying them, has a double recommendation to our minds.

Open and covered seats are of two distinct kinds; one *architectural*, or formed after artist-like designs, of stone or wood, in Grecian, Gothic, or other forms; which may, if they are intended to produce an elegant effect, have vases on pedestals as accompaniments; the other, *rustic*, as they are called, which are formed out of trunks and branches of trees, roots, etc., in their natural forms.

There are particular sites where each of these kinds of seats, or structures, is, in good taste, alone admissible. In the proximity of elegant and decorated buildings where all around has a polished air, it would evidently be doing violence to our feelings and sense of propriety to admit many rustic seats and structures of any kind; but architectural decorations and architectural seats are there correctly introduced. For the same reason, also, as we have already suggested, that the sculptured forms of vases, etc., would be out of keeping in scenes where nature is predominant (as the distant wooded parts or walks of a residence), architectural, or, in other words, highly artificial seats, would not be in character: but rustic seats and structures, which, from the nature of the materials employed and the simple manner of their construction, appear but one remove from natural forms, are felt at once to be in unison with the surrounding objects. Again, the mural and highly artistical vase and statue, most properly accompany the beautiful landscape garden; while rustic baskets, or vases, are the most fitting decorations of the Picturesque Landscape Garden.

The simplest variety of covered architectural seat is the latticed arbor for vines of various descriptions, with the seat underneath the canopy of foliage; this may with more propriety be introduced in various parts of the grounds than any other of its class, as the luxuriance and natural gracefulness of the foliage which covers the arbor, in a great measure destroys or overpowers the expression of its original form. Lattice arbors, however, neatly formed of rough poles and posts, are much more picturesque and suitable for wilder portions of the scenery.

[Fig. 81.] The temple and the pavilion are highly finished forms of covered seats, which are occasionally introduced in splendid places, where classic architecture prevails. There is a circular pavilion of this kind at the termination of one of the walks at Mr. Langdon's residence, Hyde Park. Fig. 81.

We consider rustic seats and structures as likely to be much preferred in the villa and cottage residences of the country. They have the merit of being tasteful and picturesque in their appearance, and are easily constructed by the amateur, at comparatively little or no expense.

[Fig. 82.] There is scarcely a prettier or more pleasant object for the termination of a long walk in the pleasure-grounds or park, than a neatly thatched structure of rustic work, with its seat for repose, and a view of the landscape beyond. On finding such an object, we are never tempted to think that there has been a lavish expenditure to serve a trifling purpose, but are gratified to see the exercise of taste and ingenuity, which completely answers the end in view.

[Fig. 83.] Figure 82 is an example of a simple rustic seat formed of the crooked and curved branches of the oak, elm, or any other of our forest trees Fig. 83 is a seat of the same character, made at the foot of a tree, whose overhanging branches afford a fine shade.

Figure 84 is a covered seat or rustic arbor, with a thatched roof of straw. Twelve posts are set securely in the ground, which make the frame of this structure, the

EMBELLISHMENTS.   457

[Fig. 84.]

openings between being filled in with branches (about three inches in diameter) of different trees—the more irregular the better, so that the perpendicular surface of the exterior and interior is kept nearly equal. In lieu of thatch, the roof may be first tightly boarded, and then a covering of bark or the slabs of trees with the bark on, overlaid and nailed on. The figure represents the structure as formed round a tree. For the sake of variety this might be omitted, the roof formed of an open lattice work of branches like the sides, and the whole covered by a grape, bignonia, or some other vine or creeper of luxuriant growth. The seats are in the interior.

Figure 85 represents a covered seat of another kind. The central structure, which is circular, is intended for a collection of minerals, shells, or any other curious objects for which an amateur might have a *penchant*. Geological or mineralogical specimens of the adjacent neighborhood, would be very proper for such a cabinet. The seat surrounds it on the outside, over which is a thatched roof or veranda, supported on rustic pillars formed of the trunks of saplings, with the bark attached.

[Fig. 85.]

458   LANDSCAPE GARDENING.

Many of the English country places abound with admirable specimens of rustic work in their parks and pleasure-grounds. White Knight's, in particular, a residence of the Duke of Marlborough, has a number of beautiful structures of this kind. Figure 86 is a view of a

[Fig. 86. Rustic Covered Seat.]

round seat with thatched roof, in that demesne. Three or four rustic pillars support the architrave, and the whole of the exterior and interior (being first formed of framework) is covered with straight branches of the maple and larch. The seat on the interior looks upon a fine prospect; and the seat on the back of the exterior fronts the park.

There is no limit to the variety of forms and patterns in which these rustic seats, arbors, summer-houses, etc., can be constructed by an artist of some fancy and ingenuity. After the frame-work of the structure is formed of posts and rough boards, if small straight rods about an inch in diameter, of hazel, white birch, maple, etc., are selected in sufficient quantity, they may be nailed on in squares, diamonds, medallions, or other patterns, and have the effect of a *mosaic* of wood.

Among the curious results of this fancy for rustic work, we may mention the *moss-house*—erected in several places abroad. The skeleton or frame-work of the arbor or house is formed as we have just stated; over this small rods half an inch in diameter are nailed, about an inch from centre to centre; after the whole surface is covered with this sort of rustic lathing, a quantity of the softer wood-moss of different colors is collected; and taking small parcels in the hand at a time, the tops being evenly arranged, the bottoms or roots are crowded closely between the rods with a small wooden wedge. When this is done with some little skill, the tufted ends spread out and cover the rods entirely, showing a smooth surface of mosses of different colors, which has an effect not unlike that of a thick Brussels carpet.

The mosses retain their color for a great length of time, and when properly rammed in with the wedge, they cannot be pulled out again without breaking their tops. The prettiest example which we have seen of a handsome moss-house in this country, is at the residence of Wm. H. Aspinwall, Esq., on Staten Island.

A *prospect tower* is a most desirable and pleasant structure in certain residences. Where the view is comparatively limited from the grounds, on account of their surface being level, or nearly so, it often happens that the spectator, by being raised some twenty-five or thirty feet above the surface, finds himself in a totally different position, whence a charming *coup d'œil* or bird's-eye view of the surrounding country is obtained.

Those of our readers who may have visited the delightful garden and grounds of M. Parmentier, near Brooklyn, some half a dozen years since, during the lifetime of that amiable and zealous amateur of horticulture, will readily remember the rustic prospect-arbor, or tower,

460   LANDSCAPE GARDENING.

[Fig. 87.] Fig. 87, which was situated at the extremity of his place. It was one of the first pieces of rustic work of any size, and displaying any ingenuity, that we remember to have seen here; and from its summit, though the garden walks afforded no prospect, a beautiful reach of the neighborhood for many miles was enjoyed.

Figure 88 is a design for a rustic prospect tower of three stories in height, with a double thatched roof. It is formed of rustic pillars or columns, which are well fixed in the ground, and which are filled in with a fanciful lattice of rustic branches. A spiral staircase winds round [Fig. 88.] the interior of the platform of the second and upper stories, where there are seats under the open thatched roof.

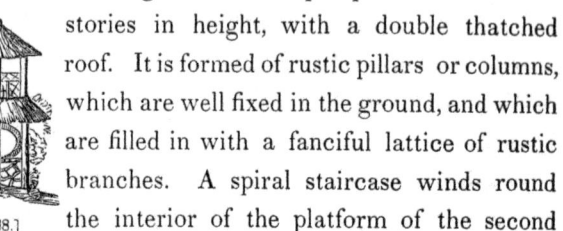

On a *ferme ornée*, where the proprietor desires to give a picturesque appearance to the different appendages of the place, rustic work offers an easy and convenient method of attaining this end. The *dairy* is sometimes made a detached building, and in this country it may be built of logs in a tasteful manner with a thatched roof; the interior being studded, lathed, and plastered in the usual way. Or the ice-house, which generally shows but a rough gable and ridge roof rising out of the ground, might be covered with a neat structure in rustic work, overgrown with vines, which would give it a pleasing or picturesque air, instead of leaving it, as at present, an unsightly object which we are anxious to conceal.

A species of useful decoration, which is perhaps more naturally suggested than any other, is the *bridge*. Where a constant stream, of greater or less size, runs through the grounds, and divides the banks on opposite sides, a bridge of some description, if it is only a narrow plank over a rivulet, is highly necessary. In pieces of artificial water that are irregular in outline, a narrow strait is often purposely made, with the view of introducing a bridge for effect.

When the stream is large and bold, a handsome architectural bridge of stone or timber is by far the most suitable; especially if the stream is near the house, or if it is crossed on the Approach road to the mansion; because a character of permanence and solidity is requisite in such cases. But when it is only a winding rivulet or crystal brook, which meanders along beneath the shadow of tufts of clustering foliage of the pleasure-ground or park, a rustic bridge may be brought in with the happiest effect. Fig. 89 is a rustic bridge erected under our direction. The foundation is made by laying down a few large square stones beneath the surface on both sides of the stream to be spanned; upon these are stretched two round posts or sleepers with the bark on, about eight or ten inches in diameter. The rustic hand-rail is framed into these two sleepers. The floor of the bridge is made by laying down small posts of equal size, about four or six inches in diameter, crosswise upon the sleepers, and nailing them down securely. The bark is allowed to remain on in every piece of wood employed in the construction of this little bridge; and when the wood is cut at the proper season (durable kinds being chosen), such a bridge, well made, will remain in excellent order for many years.

Rockwork is another kind of decoration sometimes introduced in particular portions of the scenery of a residence, Fig. 90. When well executed, that is, so as to have a natural and harmonious expression, the effect is highly pleasing. We have seen, however, in places where a high

[Fig. 90. Rockwork.]

keeping and good taste otherwise prevailed, such a barbarous *mélange*, or confused pile of stones mingled with soil, and planted over with dwarfish plants dignified with the name of rockwork, that we have been led to believe that it is much better to attempt nothing of the kind, unless there is a suitable place for its display, and at the same time, the person attempting it is sufficiently an artist, imbued with the spirit of nature in her various compositions and combinations, to be able to produce something higher than a caricature of her works.

The object of *rockwork* is to produce in scenery or portions of a scene, naturally in a great measure destitute of groups of rocks and their accompanying drapery of plants and foliage, something of the picturesque effect which such natural assemblages confer. To succeed in this, it is evident that we must not heap up little hillocks of mould and smooth stones, in the midst of an open lawn, or the centre of a flower-garden. But if we can make choice of a situation where a rocky bank or knoll already partially exists, or would be in keeping with the form of the ground and the character of the scene, then we may introduce such accompaniments with the best possible hope of success.

It often happens in a place of considerable extent, that somewhere in conducting the walks through the grounds, we meet with a ridge with a small rocky face, or perhaps with a large rugged single rock, or a bank where rocky summits just protrude themselves through the surface. The common feeling against such uncouth objects, would direct them to be cleared away at once out of sight. But let us take the case of the large rugged rock, and commence our picturesque operations upon it. We will begin by collecting from some rocky hill or valley in the neighborhood of the estate, a sufficient quantity of rugged rocks, in size from a few pounds to half a ton or more, if necessary, preferring always such as are already coated with mosses and lichens. These we will assemble around the base of a large rock, in an irregular somewhat pyramidal group, bedding them sometimes partially, sometimes almost entirely in soil heaped in irregular piles around the rock. The rocks must be arranged in a natural manner, avoiding all regularity and appearance of formal art, but placing them sometimes in groups of half a dozen together, overhanging each other, and sometimes half bedded in the soil, and a little distance apart. There are no *rules* to be given for such operations, but the study of natural groups, of a character similar to that which we wish to produce, will afford sufficient hints if the artist is

"Prodigue de génie," and has a perception of the natural beauty which he desires to imitate.

The rockwork once formed, choice trailing, creeping, and alpine plants, such as delight naturally in similar situations, may be planted in the soil which fills the interstices between the rocks: when these grow to fill their proper places, partly concealing and adorning the rocks with their neat green foliage and pretty blossoms, the effect of the whole, if properly done, will be like some exquisite portion of a rocky bank in wild scenery, and will be found to give an air at once striking and picturesque to the little scene where it is situated.

In small places where the grounds are extremely limited, and the owner wishes to form a rockwork for the growth of alpine and other similar plants, if there are no natural indications of a rocky surface, a rockwork may sometimes be introduced without violating good taste by *preparing natural indications* artificially, if we may use such a term. If a few of the rocks to be employed in the rockwork are sunk half or three-fourths their depth in the soil near the site of the proposed rockwork, so as to have the appearance of a rocky ridge just *cropping out*, as the geologists say, then the rockwork will, to the eye of a spectator, seem to be connected with, and growing out of this rocky spur or ridge below: or, in other words, there will be an obvious reason for its being situated there, instead of its presenting a wholly artificial appearance.

In a previous page, when treating of the banks of pieces of water formed by art, we endeavored to show how the natural appearance of such banks would be improved by the judicious introduction of rocks partially imbedded into and holding them up. Such situations, in the case of a small lake or pond, or a brook, are admirable sites for rockwork. Where the materials of a suitable kind are abundant, and tasteful ingenuity is not wanting, surprising effects may be produced in a small space. Caves and grottoes, where ferns and mosses would thrive admirably with the gentle drip from the roof, might be made of the overarching rocks arranged so as to appear like small natural caverns. Let the exterior be partially planted with low shrubs and climbing plants, as the wild Clematis, and the effect of such bits of landscape could not but be agreeable in secluded portions of the grounds.

In many parts of the country, the secondary blue limestone abounds, which, in the small masses found loose in the woods, covered with mosses and ferns, affords the very finest material for artificial rockwork.*

After all, much the safest way is never to introduce rockwork of any description, unless we feel certain that it will have a good effect. When a place is naturally picturesque, and abounds here and there with rocky banks, etc., little should be done but to heighten and aid the expressions of these, if they are wanting in spirit, by adding something more; or softening and giving elegance to the expression, if too wild, by planting the same with

---

* Our readers may see an engraving and description of a superb *extravaganza* in rockwork in a late number of Loudon's Gardener's Magazine. Lady Broughton, of Hoole House, Chester, England, has succeeded in forming, round a natural valley, an imitation of the hills, glaciers, and scenery of a *passage* in Switzerland. The whole is done in rockwork, the snow-covered summits being represented in white spar. The appropriate plants, trees, and shrubs on a small scale, are introduced, and the illusion, to a spectator standing in the valley surrounded by these glaciers, is said to be wonderfully striking and complete.

beautiful shrubs and climbers. On a tame sandy level, where rocks of any kind are unknown, their introduction in rockworks, nine times in ten, is more likely to give rise to emotions of the ridiculous, than those of the sublime or picturesque.

Fountains are highly elegant garden decorations, rarely seen in this country; which is owing, not so much, we apprehend, to any great cost incurred in putting them up, or any want of appreciation of their sparkling and enlivening effect in garden scenery, as to the fact that there are few artisans here, as abroad, whose business it is to construct and fit up architectural, and other *jets d'eau.*

The first requisite, where a fountain is a desideratum, is a constant supply of water, either from a natural source or an artificial reservoir, some distance higher than the level of the surface whence the jet or fountain is to rise.

[Fig. 91. Design for a Fountain.]

Where there is a pond, or other body of water, on a higher level than the proposed fountain, it is only necessary to lay pipes under the surface to conduct the supply of water to the required spot; but where there is no such head of water, the latter must be provided from a reservoir artificially prepared, and kept constantly full.

There are two very simple and cheap modes of effecting this, which we shall lay before our readers, and one or the other of which may be adopted in almost every locality. The first is to provide a large flat cistern of sufficient size, which is to be placed under the roof in the upper story of one of the outbuildings, the carriage-house for example, and receive its supplies from the water collected on the roof of the building; the amount of water collected in this way from a roof of moderate size being much more than is generally supposed. The second is to sink a well of capacious size (where such is not already at command) in some part of the grounds where it will not be conspicuous, and over it to erect a small tower, the top of which shall contain a cistern and a small horizontal windmill; which being kept in motion by the wind more or less almost every day in summer, will raise a sufficient quantity of water to keep the reservoir supplied from the well below. In either of these cases, it is only necessary to carry leaden pipes from the cistern (under the surface, below the reach of frost) to the place where the jet is to issue; the supply in both these cases will, if properly arranged, be more than enough for the consumption of the fountain during the hours when it will be necessary for it to play, viz. from sunrise to evening.

The steam-engine is often employed to force up water for the supply of fountains in many of the large public and royal gardens; but there are few cases in this country where private expenditures of this kind would be justifiable.

But where a small stream, or even the overflow of a

perpetual spring, can be commanded, the *Hydraulic Ram* is the most perfect as well as the simplest and cheapest of all modes of raising water. A supply pipe of an inch in diameter is in many cases sufficient to work the Ram and force water to a great distance; and where sufficient to fill a "driving pipe" of two inches diameter can be commanded, a large reservoir may be kept constantly filled. As the Hydraulic Ram is now for sale in all our cities we need not explain its action.

"In conducting the water from the cistern or reservoir to the jet or fountain, the following particulars require to be attended to: In the first place, all the pipes must be laid sufficiently deep in the earth, or otherwise placed and protected so as to prevent the possibility of their being reached by frost; next, as a general rule, the diameter of the orifice from which the jet of water proceeds, technically called the bore of the quill, ought to be four times less than the bore of the conduit pipe; that is, the quill and the pipe ought to be in a quadruple proportion to each other. There are several sorts of quills or spouts, which throw the water up or down, into a variety of forms: such as fans, parasols, sheaves, showers, mushrooms, inverted bells, etc. The larger the conduit pipes are, the more freely will the jets display their different forms; and the fewer the holes in the quill or jet (for sometimes this is pierced like the rose of a watering pot) the greater certainty there will be of the form continuing the same; because the risk of any of the holes choking up will be less. The diameter of a conduit pipe ought in no case to be less than one inch; but for jets of very large size, the diameter ought to be two inches. Where the conduit pipes are of great length, say upwards of 1000 feet, it is found advantageous to begin, at the reservoir or cistern, with pipes of a diameter somewhat greater than those which deliver the water to the quills, because the water, in a pipe of uniform diameter of so great a length, is found to lose much of its strength, and become what is technically called sleepy: while the different sizes quicken it, and redouble its force. For example, in a conduit pipe of 1800 feet in length, the first six hundred feet may be laid with pipes of eight inches in diameter, the next 600 feet with pipes of six inches in diameter, and the last 600 feet with pipes of four inches in diameter. In conduits not exceeding 900 feet, the same diameter may be continued throughout. When several jets are to play in several fountains, or in the same, it is not necessary to lay a fresh pipe from each jet to the reservoir; a main of sufficient size, with branch pipes to each jet, being all that is required. Where the conduit pipe enters the reservoir or cistern, it ought to be of increased diameter, and the grating placed over it to keep out leaves and other matters which might choke it up, ought to be semi-globular or conical; so that the area of the number of holes in it may exceed the area of the orifice of the conduit pipe. The object is to prevent any diminution of pressure from the body of water in the cistern, and to facilitate the flow of the water Where the conduit pipe joins the fountain, there, of course, ought to be a cock for turning the water off and on; and particular care must be taken that as much water may pass through the oval hole of this cock as passes through the circular hole of the pipe. In conduit pipes, all elbows, bendings, and right angles should be avoided as much as possible, since they diminish the force of the water. In very long conduit pipes, air-holes formed by soldering on upright pieces of pipe, terminating in inverted valves or suckers, should be made at convenient distances, and protected by shafts built of stone or brick, and covered with movable gratings, in order to let out the air. Where pipes ascend and descend on very irregular surfaces, the strain on the lowest parts of the pipe is always the greatest; unless care is taken to relieve this by the judicious disposition of cocks and air-holes. Without this precaution, pipes conducted over irregular surfaces will not last nearly so long as those conducted over a level."— *Encycl. of Cottage, Farm, and Villa Architecture*, page 989.

Where the reservoir is but a short distance, as from a dozen to fifty yards, all that is necessary is to lay the conduit pipes on a regular uniform slope, to secure a steady uninterrupted flow of water. Owing to the friction in the pipes, and pressure of the atmosphere, the water in the fountain will of course, in no case, rise quite as high as the level of the water in the reservoir; but it will nearly as high. For example, if the reservoir is ten feet four inches high, the water in the jet will only rise ten feet, and in like proportion for the different heights. The following table*

Height of the Reservoir.		Diameter of the Conduit pipes.		Diameters of the Orifices.		Height the water will rise to.	
Feet.	Inches.	Inches.	Lines.	Lines.	Parts.	Feet.	Inches.
5	1	0	22	4	0	5	0
10	4	0	25	5	0	10	0
15	9	2¼	0	6	0	15	0
21	4	2½	0	6½	0	20	0
33	0	3	0	7	0	30	0
45	4	4½	0	7	8	40	0
58	4	5	0	8	10	50	0
72	0	5½	0	10	12	60	0
86	4	6	0	12	14	70	0
100	0	7	0	12	15	80	0

* Switzer's *Introduction to a General System of Hydrostatics.*

shows with a given height of reservoirs and diameter of conduit pipes and orifices, the height to which the water will rise in the fountain.

A simple jet (Fig. 92) issuing from a circular basin of water, or a cluster of perpendicular jets (candelabra jets), is at once the simplest and most pleasing of fountains. Such are almost the only kinds of fountains which can be introduced with propriety in simple scenes where the predominant objects are sylvan, not architectural.

[Fig. 92.]

Weeping, or *Tazza Fountains*, as they are called, are simple and highly pleasing objects, which require only a very moderate supply of water compared with that demanded by a constant and powerful jet. The conduit pipe rises through and fills the vase, which is so formed as to overflow round its entire margin. Figure 93 represents a beautiful Grecian vase for tazza fountains. The ordinary jet and the tazza fountain may be combined in one, when the supply of water is sufficient, by carrying the conduit pipe to the level of the top of the vase, from which the water rises perpendicularly, then falls back into the vase and overflows as before.

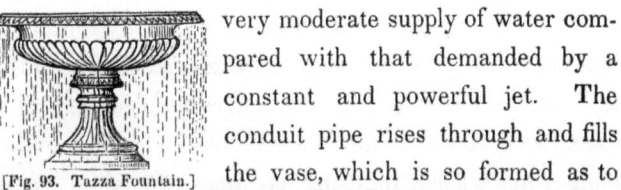

[Fig. 93. Tazza Fountain.]

We might enumerate and figure a great many other designs for fountains; but the connoisseur will receive more ample information on this head than we are able to afford, from the numerous French works devoted to this branch of Rural Embellishment.

A species of rustic fountain which has a good effect, is made by introducing the conduit pipe or pipes among the groups of *rockwork* alluded to, from whence (the orifice of the pipe being concealed or disguised) the water issues among the rocks either in the form of a cascade, a weeping fountain, or a perpendicular jet. A little basin of water is formed at the foot or in the midst of the rockwork; and the cool moist atmosphere afforded by the trickling streams, would offer a most congenial site for aquatic plants, ferns, and mosses.

Fountains of a highly artificial character are happily situated only when they are placed in the neighborhood of buildings and architectural forms. When only a single fountain can be maintained in a residence, the centre of the flower-garden, or the neighborhood of the piazza or terrace-walk, is, we think, much the most appropriate situation for it. There the liquid element, dancing and sparkling in the sunshine, is an agreeable feature in the scene, as viewed from the windows of the rooms; and the falling watery spray diffusing coolness around is no less delightful in the surrounding stillness of a summer evening.

After all that we have said respecting architectural and rustic decorations of the grounds, we must admit that it requires a great deal of good taste and judgment, to introduce and distribute them so as to be in good keeping with the scenery of country residences. A country residence, where the house with a few tasteful groups of flowers and shrubs, and a pretty lawn, with clusters and groups of luxuriant trees, are all in high keeping and evincing high order, is far more beautiful and pleasing than the same place, or even one of much larger extent, where a profusion of statues, vases, and fountains, or rockwork and rustic seats, are distributed throughout the garden and grounds, while the latter, in themselves, show slovenly keeping, and a crude and meagre knowledge of design in Landscape Gardening.

Unity of expression is the maxim and guide in this department of the art, as in every other. Decorations can never be introduced with good effect, when they are at variance with the character of surrounding objects. A beautiful and highly architectural villa may, with the greatest propriety, receive the decorative accompaniments of elegant vases, sundials, or statues, should the proprietor choose to display his wealth and taste in this manner; but these decorations would be totally misapplied in the case of a plain square edifice, evincing no architectural style in itself.

In addition to this, there is great danger that a mere lover of fine vases may run into the error of assembling these objects indiscriminately in different parts of his grounds, where they have really no place, but interfere with the quiet character of surrounding nature. He may overload the grounds with an unmeaning distribution of sculpturesque or artificial forms, instead of working up those parts where art predominates in such a manner, by means of appropriate decorations, as to heighten by contrast the beauty of the whole adjacent landscape.

With regard to pavilions, summer-houses, rustic seats, and garden edifices of like character, they should, if possible, in all cases be introduced where they are manifestly appropriate or in harmony with the scene. Thus a grotto should not be formed in the side of an open bank, but in a deep shadowy recess; a classic temple or pavilion may crown a beautiful and prominent knoll, and a rustic covered seat may occupy a secluded, quiet portion of the grounds, where undisturbed meditation may be enjoyed. As our favorite Delille says:

> " Sachez ce qui convient ou nuit au caractère.
> Un réduit écarté, dans un lieu solitaire,
> Peint mieux la solitude encore et l'abandon.
> Montrez-vous donc fidèle à chaque expression ;
> N'allez pas au grand jour offrir un ermitage :
> Ne cachez point un temple au fond d'un bois sauvage."
>
> LES JARDINS.

Or if certain objects are unavoidably placed in situations of inimical expression, the artist should labor to alter the character of the locality. How much this can be done by the proper choice of trees and shrubs, and the proper arrangement of plantations, those who have seen the difference in aspect of certain favorite localities of wild nature, as covered with wood, or as denuded by the axe, can well judge. And we hope the amateur, who has made himself familiar with the habits and peculiar expressions of different trees, as pointed out in this work, will not find himself at a loss to effect such changes, by the aid of time, with ease and facility.

APPENDIX.

I.

Notes on transplanting trees. Reasons for frequent failures in removing large trees. Directions for performing this operation. Selection of subjects. Preparing trees for removal. Transplanting evergreens.

THERE is no subject on which the professional horticulturist is more frequently consulted in America, than transplanting trees. And, as it is an essential branch of Landscape Gardening—indeed, perhaps, the most important and necessary one to be practically understood in the improvement or embellishment of new country residences—we shall offer a few remarks here, with the hope of rendering it a more easy and successful practice in the hands of amateurs.

Although there are great numbers of acres of beautiful woods and groves, the natural growth of the soil, in most of the older states, yet a considerable portion of our ordinary country seats are meagrely clothed with trees, while many beautiful sites for residences have, in past years, been so denuded that the nakedness of their appearance constitutes a serious objection to them as places of residence. To be able, therefore, to transplant, from natural copses, trees of ten or twenty years' growth, is so universally a desideratum, that great numbers of experiments are made annually with this view; though few persons succeed in obtaining what they desire, viz. the immediate effect of wood; partly from a want of knowledge of the nature of vegetable physiology, and partly from malpractice in the operation of removal itself.

When the admirably written "Planter's Guide," by Sir Henry

Steuart, made its appearance some ten years ago, not only describing minutely the whole theory of transplanting nearly full grown trees, but placing before its readers a report of a committee of the Highland Society of Edinburgh attesting the complete success of the practice, as exemplified in the woods, copses, and groups, which, removed by the transplanting machine, beautified with their verdure and luxuriance the baronet's own park, the whole matter of transplanting was apparently cleared up, and numbers of individuals in this country, with sanguine hopes of success, set about the removal of large forest trees.

Of the numerous trials made upon this method, *with trees of extra size*, we have known but a very few instances of even tolerable success. This is no doubt owing partly to the want of care and skill in the practical part of the process, but mainly to the ungenial nature of our climate.

The climate of Scotland during four-fifths of the year is, in some respects, the exact opposite of that of the United States. An atmosphere which, for full nine months of the twelve, is copiously charged with fogs, mist, and dampness, may undoubtedly be considered as the most favorable in the world for restoring the weakened or impaired vital action of large transplanted trees. In this country, on the contrary, the dry atmosphere and constant evaporation under the brilliant sun of our summers, are most important obstacles with which the transplanter has to contend, and which render complete success so much more difficult here than in Scotland. And we would therefore rarely attempt in this country the extensive removal of trees larger than twenty feet in height. When of the size of fifteen feet they are sufficiently large to produce very considerable immediate effect, while they are not so large as to be costly or very difficult to remove, or to suffer greatly by the change of position, like older ones.

The great want of success in transplanting trees of moderate size in this country arises, as we conceive, mainly from two causes; the first, a want of skill in performing the operation, arising chiefly from ignorance of the nature of the vital action of plants, in roots, branches, etc., and the second, a bad or improper selection of subjects on which the operation is to be performed. Either of these causes would account for bad success in removals; and where, as is frequently the case, both are combined, total failure can scarcely be a matter of surprise to those really familiar with the matter.

An uninformed spectator, who should witness for the first time the removal of a forest tree, as ordinarily performed by many persons, would scarcely suppose that anything beyond mere *physical strength* was required. Commencing as near the tree as possible, cutting off many of the roots, with the very smallest degree of reluctance, wrenching the remaining mass out of their bed as speedily and almost as roughly as possible, the operator hastens to complete his destructive process, by cutting off the best part of the head of the tree, to make it correspond with the reduced state of the roots. Arrived at the hole prepared for its reception, his replanting consists in shovelling in, while the tree is held upright, the surrounding soil, paying little or no regard to filling up all the small interstices among the roots; and finally, after treading the earth as hard as possible, completing the whole by pouring two or three pails of water upon the top of the ground. How any reflecting person, who looks upon a plant as a delicately organized individual, can reasonably expect or hope for success after such treatment in transplanting, is what we never could fully understand. And it has always, therefore, appeared pretty evident that all such operators must have very crude and imperfect notions of vegetable physiology, or the structure and functions of plants.

The first and most important consideration in transplanting should be the *preservation of the roots*. By this we do not mean a certain bulk of the larger and more important ones only, but as far as possible all the numerous small fibres and rootlets so indispensably necessary in assisting the tree to recover from the shock of removal. The coarser and larger roots serve to secure the tree in its position, and convey the fluids; but it is by means of the small fibrous roots, or the delicate and numerous *points* of these fibres called *spongioles*, that the food of plants is imbibed, and the destruction of such is manifestly in the highest degree fatal to the success of the transplanted tree. To avoid this as far as practicable, we should, in removing a tree, commence at such a distance as to include a circumference large enough to comprise the great majority of the roots. At that distance from the trunk we shall find most of the smaller roots, which should be carefully loosened from the soil, with as little injury as possible; the earth should be gently and gradually removed from the larger roots, as we proceed onward from the extremity of the circle to the centre, and when we reach the nucleus of roots surrounding the trunk, and fairly undermine the whole, we shall find ourselves in possession of a tree in such a perfect condition, that even when of considerable size, we may confidently hope for a speedy recovery of its former luxuriance after being replanted.

Now to remove a tree in this manner, requires not only a considerable degree of experience, which is only to be acquired by practice, but also much *patience and perseverance* while engaged in the work. It is not a difficult task to remove, in a careless manner, four or five trees in a day, of fifteen feet in height, by the assistance of three or four men, and proper implements of removal, while one or two trees only can be removed if the roots and branches are preserved entire or nearly so. Yet in the latter case, if the work be well performed, we shall have the satisfaction of beholding the subjects, when removed, soon taking fresh root, and becoming vigorous healthy trees, with fine luxuriant heads, while three-fourths of the former will most probably perish, and the remainder struggle for several years, under the loss of so large a portion of their roots and branches, before they entirely recover, and put on the appearance of handsome trees.

When a tree is carelessly transplanted, and the roots much mutilated, the operator feels obliged to reduce the top accordingly; as experience teaches him, that although the leaves may expand, yet they will soon perish without a fresh supply of food from the roots. But when the largest portion of the roots are carefully taken up with the tree, pruning should be less resorted to, and thus the original symmetry and beauty of the head retained. When this is the case, the leaves contribute as much, by their peculiar action in elaborating the sap, towards re-establishing the tree, as the roots; and indeed the two act so reciprocally with each other, that any considerable injury to the one always affects the other. "The functions of respiration, perspiration, and digestion," says Professor Lindley, "which are the particular offices of leaves, are essential to the health of a plant; its healthiness being in proportion to the degree in which these functions are duly performed. The leaf is in reality a natural contrivance for exposing a large surface to the influence of external agents, by whose assistance the crude sap contained in the stem is altered, and rendered suitable to the particular wants of the species, and for returning into the general circulation, the fluids in their matured condition. In a word, the leaf of a plant is its lungs and stomach traversed by a system of veins."* All the pruning, therefore, that is necessary, when a tree is properly transplanted, will be comprised in paring smooth all bruises or accidental injuries, received by the roots or branches during the operation, or the removal of a few that may interfere with elegance of form in the head.

Next in importance to the requisite care in performing the operation of transplanting, is *the proper choice of individual trees to be transplanted.* In making selections for removal among our fine forest trees, it should never be forgotten that there are two distinct kinds of subjects, even of the same species of every tree, viz. those that grow among and surrounded by other trees or woods, and those which grow alone, in free open exposures, where they are acted upon by the winds, storms, and sunshine, at all times and seasons. The former class it will always be exceedingly difficult to transplant successfully even with the greatest care, while the latter may always be removed with comparatively little risk of failure.

Any one who is at all familiar with the growth of trees in woods or groves somewhat dense, is also aware of the great difference in the external appearance between such trees and those which stand singly in open spaces. In thick woods, trees are found to have tall, slender trunks, with comparatively few branches except at the top, smooth and thin bark, and they are scantily provided with roots, but especially with the small fibres so essentially necessary to insure the growth of the tree when transplanted. Those, on the other hand, which stand isolated, have short thick stems, numerous branches, thick bark, and great abundance of root and small fibres. The latter, accustomed to the full influence of the weather, to cold winds as well as open sunshine, have what Sir Henry Steuart has aptly denominated the "*protecting properties*," well developed; being robust and hardy, they are well cal-

* Theory of Horticulture.

culated to endure the violence of the removal, while trees growing in the midst of a wood sheltered from the tempests by their fellows, and scarcely ever receiving the sun and air freely except at their topmost branches, are too feeble to withstand the change of situation, when removed to an open lawn, even when they are carefully transplanted.

"Of trees in open exposures," says Sir Henry, "we find that their peculiar properties contribute, in a remarkable manner, to their health and prosperity. In the first place, their shortness and greater girth of stem, in contradistinction to others in the interior of woods, are obviously intended to give to the former greater strength to resist the winds, and a shorter lever to act upon the roots. Secondly, their larger heads, with spreading branches, in consequence of the free access of light, are as plainly formed for the nourishment as well as the balancing of so large a trunk, and also for furnishing a cover to shield it from the elements. Thirdly, their superior thickness and induration of bark is, in like manner, bestowed for the protection of the sap-vessels, that lie immediately under it, and which, without such defence from cold, could not perform their functions. Fourthly, their greater number and variety of roots are for the double purpose of nourishment and strength; nourishment to support a mass of such magnitude, and strength to contend with the fury of the blast. Such are the obvious purposes for which the unvarying characteristics of trees in open exposures are conferred upon them. Nor are they conferred equally and indiscriminately upon all trees so situated. They seem, by the economy of nature, to be *peculiar adaptations* to the circumstances and wants of each individual, *uniformly bestowed in the ratio of exposure*, greater where that is more conspicuous, and uniformly decreasing, as it becomes less."*

Trees in which the protecting properties are well developed are frequently to be met with on the skirts of woods; but those standing singly here and there, through the cultivated fields and meadows of our farm lands, where the roots have extended themselves freely in the mellow soil, are the finest subjects for removal into the lawn, park, or pleasure ground.

* The Planter's Guide, p. 105.

APPENDIX. 481

The machine used in removing trees of moderate size is of simple construction, consisting of a pair of strong wheels about five feet high, a stout axle, and a pole about twelve feet long. In transplanting, the wheels and axle are brought close to the trunk of the tree, the pole is firmly lashed to the stem, and when the soil is sufficiently removed and loosened about the roots, the pole, with the tree attached, is drawn down to a horizontal position by the aid of men and a pair of horses. When the tree is thus drawn out of the hole, it is well secured and properly balanced upon the machine, the horses are fastened in front of the mass of roots by gearings attached to the axle, and the whole is transported to the destined location.

In order more effectually to insure the growth of large specimens when transplanted, a mode of *preparing beforehand* a supply of young roots, is practised by skilful operators. This consists in removing the top soil, partially undermining the tree, and shortening back many of the roots; and afterwards replacing the former soil by rich mould, or soil well manured. This is suffered to remain at least one year, and often three or four years; the tree, stimulated by the fresh supply of food, throws out an abundance of small fibres, which render success, when the time for removal arrives, comparatively certain.

It may be well to remark here, that before large trees are transplanted into their final situations, the latter should be well prepared by trenching, or *digging the soil two or three feet deep*, intermingling throughout the whole a liberal portion of well decomposed manure, or rich compost. To those who are in the habit of planting trees of any size in unprepared grounds, or that merely prepared by digging *one spit deep*, and turning in a little surface manure, it is inconceivable how much more rapid is the growth, and how astonishingly luxuriant the appearance of trees when removed into ground properly prepared. It is not too much to affirm, that young trees under favorable circumstances —in soil so prepared—will advance more rapidly, and attain a larger stature in eight years, than those planted in the ordinary way, without deepening the soil, will in twenty—and trees of larger size in proportion; a gain of growth surely worth the trifling expense incurred in the first instance. And the same observation will apply to all planting. A little extra labor and cost expended in *preparing the soil* will, for a long time, secure a surprising rapidity of growth.

In the actual planting of the tree, the chief point lies in bringing every small fibre in contact with the soil, so that no hollows or interstices are left, which may produce mouldiness and decay of the roots. To avoid this, the soil must be pulverized with the spade before filling in, and one of the workmen, with his hands and a flat dibble of wood, should fill up all cavities, and lay out the small roots before covering them in their natural position. When *watering* is thought advisable (and we practise it almost invariably), it should always be done while the planting is going forward. Poured in the hole when the roots are just covered with the soil, it serves to settle the loose earth compactly around the various roots, and thus both furnishes a supply of moisture, and brings the pulverized mould in proper contact for growth. Trees well watered when planted in this way, will rarely require it afterwards; and should they do so, the better way is to remove two or three inches of the top soil, and give the lower stratum a copious supply; when the water having been absorbed, the surface should again be replaced. There is no practice more mischievous to newly moved trees, than that of pouring water, during hot weather, upon the surface of the ground above the roots. Acted upon by the sun and wind, this surface becomes baked, and but little water reaches the roots; or just sufficient, perhaps, to afford a momentary stimulus, to be followed by increased sensibility to the parching drought.

With respect to the proper seasons for transplanting, we may remark that, except in extreme northern latitude, autumn planting is generally preferred for large, hardy, deciduous trees. It may commence as soon as the leaves fall, and may be continued until winter. In planting large trees in spring, we should commence as early as possible, to give them the benefit of the April rains; if it should be deferred to a later period, the trees will be likely to suffer greatly by the hot summer sun before they are well established.

The transplanting of *evergreens* is generally considered so much more difficult than that of deciduous trees, and so many persons who have tolerable success in the latter, fail in the former, that we may perhaps be expected to point out the reason of these frequent failures.

APPENDIX. 483

Most of our horticultural maxims are derived from English authors and among them, that of always planting evergreens either in August or late in autumn. At both these seasons, it is nearly impossible to succeed in the temperate portions of the United States, from the different character of our climate at these seasons. The genial moisture of the English climate renders transplanting comparatively easy at all seasons, but especially in winter, while in this country, our Augusts are dry and hot, and our winters generally dry and cold. If planted in the latter part of summer, evergreens become parched in their foliage, and soon perish. If planted in autumn or early winter, the severe cold that ensues, to which the newly disturbed plant is peculiarly alive, paralyses vital action, and the tree is so much enfeebled that, when spring arrives, it survives but a short period. The only period, therefore, that remains for the successful removal of evergreens here, is the spring. When planted as early as practicable in the spring, so as to have the full benefit of the abundant rains so beneficial to vegetation at that season, they will almost immediately protrude new shoots, and regain their former vigor.

Evergreens are, in their roots, much more delicate and impatient of dryness than deciduous trees; and this should be borne in mind while transplanting them. For this reason, experienced planters always choose a wet or misty day for their removal; and, in dry weather, we would always recommend the roots to be kept watered and covered from the air by mats during transportation. When proper regard is paid to this point, and to judicious selection of the season, evergreens will not be found more difficult of removal than other trees.

Another mode of transplanting large evergreens, which is very successfully practised among us, is that of removing them with frozen balls of earth in mid-winter. When skilfully performed, it is perhaps the most complete of all modes, and is so different from the common method, that the objection we have just made to winter planting does not apply to this case. The trees to be removed are selected, the situations chosen, and the holes dug, while the ground is yet open in autumn. When the ground is somewhat frozen, the operator proceeds to dig a trench around the tree at some distance, gradually undermining it, and leaving all the principal mass of roots embodied in the ball of earth.

484   APPENDIX.

The whole ball is then left to freeze pretty thoroughly (generally till snow covers the ground), when a large sled drawn by oxen is brought as near as possible, the ball of earth containing the tree rolled upon it, and the whole is easily transported to the hole previously prepared, where it is placed in the proper position, and as soon as the weather becomes mild, the earth is properly filled in around the ball. A tree, either evergreen or deciduous, may be transplanted in this way, so as scarcely to show, at the return of growth, any ill effects from its change of location.

II.

Description of an English Suburban residence, CHESHUNT COTTAGE. With views and plans showing the arrangement of the house and grounds. And the mode of managing the whole premises.

[The following description of an interesting suburban residence near London, with the numerous engravings illustrating it, has been kindly furnished us for this work, by J. C. Loudon, Esq. It was originally published in his "Gardener's Magazine," and affords an admirable illustration of this class of residences, showing what may be done, and how much beauty and enjoyment realized, on a comparatively limited space of ground.]

CHESHUNT COTTAGE, THE RESIDENCE OF WM. HARRISON, ESQ.,
F. L. S., ETC.

"All that can render a country seat delightful, and a well furnished library in the house."
(*Evelyn's Memoirs, by Bray*, vol. i., p. 432.)

THE sides of the road from London to Cheshunt, by Stoke Newington, Edmonton, and Enfield Wash, are thickly studded with suburban houses and gardens the whole distance; but, by going straight on through the Ball's Pond Turnpike, and taking the country road leading out of Newington Green, called the Green Lanes, between the Tottenham and Edmonton road, and the Barnet Road, and threading our way through numerous interesting lanes, we may pass through very rural and umbrageous scenery, with the appearance of but few houses of any kind. Indeed, it may be mentioned as one of the most remarka-

APPENDIX. 485

ble circumstances in the state of the country in the neighborhood of London, that, while all the main roads are bordered by houses for some miles from town, so as almost to resemble streets; there are tracts which lie between the main roads, and quite near town, which have undergone little or no change in the nature of their occupation for

[Fig. 1. Cheshunt Cottage, from the Road.]

several, and apparently many, generations; at all events, not since the days of Queen Elizabeth. The tracts of country to which we allude are in pasture or meadow, with crooked irregular hedges, numerous stiles and footpaths, and occasional houses by the roadsides; the farms characterized by large hay barns. Scenery of this kind is never seen by the citizen who goes to his country seat along the public road, in his family carriage, or in a stage-coach; and it is accordingly only known to pedestrians, and such as are not afraid of driving their horses over rough roads, or meeting wagons or hay-carts in narrow lanes. The road through the Green Lanes to Enfield is an excellent turnpike road, always in a good state, with occasional villas near Bour Farm and Palmer's Green; and near Enfield, at Forty Hill, there is a handsome church, built and endowed by Mr. Myers, opposite to his park, which is filled with large and handsome trees. Afterwards it passes the celebrated park of Theobalds, near where formerly stood a royal palace, the favorite residence of James I., and winds in the most agreeable and picturesque manner, under the shade of overhanging trees. Having made several turns, it leads to a lane with a brook which runs parallel to the road, a foot-bridge across which forms the entrance to Mr. Harrison's cottage, as exhibited in the view Fig. 1.

The ground occupied by Mr. Harrison's cottage and gardens is about seven acres, exclusive of two adjoining grass fields. The grounds lie entirely on one side of the house, as shown in the plan, Fig. 13, in pp. 510, 511. The surface of the whole is flat, and nothing is seen in the horizon in any direction but distant trees. The beauties of the place, to a stranger at his first glance, appear of the quiet and melancholy kind, as shown in the Figs. 2, 3; the one looking to the right from the drawing-room window and the other to the left: but, upon a nearer examination by a person conversant with the subjects of botany and gardening, and knowing in what rural comfort consists, these views will be found to be full of intense interest, and to afford many instructive hints to the possessors of suburban villas or cottages.

In building the house and laying out the grounds, Mr. Harrison was his own architect and Landscape Gardener; not only devising the general design, but furnishing working-drawings of all the details of the interior of the cottage. His reason for fixing on the present situation for the house was, the vicinity (the grounds joining) of a house and walk belonging to a relation of his late wife. The circumstance is mentioned as accounting in one so fond of a garden, for fixing on a spot which had neither tree nor shrub in it when he first inhabited it. Mr. Harrison informs us, and we record it for the use of amateurs commencing, or extending, or improving gardens, that he commenced his operations about thirty years ago, by purchasing, at a large nursery sale, large lots of evergreens, not six inches high, in beds of one hundred each, such as laurels, Portugal laurels, laurustinuses, bays, hollies, &c.; with many lots of deciduous trees, in smaller numbers, which he planted in a nursery on his own ground; and at intervals, as he from time to time extended his garden, he took out every second plant, which, with occasional particular trees and shrubs from nursery grounds, constituted a continual supply for improvement and extension. This, with the hospital ground mentioned hereafter, furnished the

APPENDIX. 487

[Fig. 2. View from the Drawing-room Window at Cheshunt Cottage, looking to the Left.]

means of extensions and improvements at no other expense than labor, which, when completed, gave the place the appearance of an old garden; the plants being larger than could be obtained, or, if obtained, safely transplanted, from nurseries. This is an important consideration, in addition to that of economy, well worth the attention of amateur improvers of grounds or gardens.

By inspecting the plan, Fig 4, it will be found that the house contains, on the ground floor, three good living rooms, and two other rooms (n and g) particularly appropriate to the residence of an amateur fond of botany and gardening; and that it is replete with every description of accommodation and convenience requisite for the enjoyment of all the comforts and luxuries that a man of taste can desire for himself or his friends.

In laying out the grounds, the first object was to insure agricultural and gardening comforts; and hence the completeness of the farm-yard, and of the hot-house and frame departments, as exhibited in the plan, Fig. 6. On the side of the grounds opposite to the hot-houses and flower-garden are the kitchen-garden and orchard; and though in most situations it would have been more convenient to have had the farm buildings, and kitchen garden, and hot-houses on the same side as the kitchen offices, yet in this case no inconvenience results from their separation; because the public road, as will be seen by the plan, Fig. 13, forms a ready medium of communication between them, in cases in which the communication through the ornamented ground would be unsightly or inconvenient. In arranging the pleasure-ground, the great object, as in all similar cases, was to introduce as much variety as could be conveniently done in a comparatively limited space. This has been effected chiefly by distributing over the lawn a collection of trees and shrubs; by forming a small piece of water, and disposing of the earth excavated into hilly inequalities; and by walks leading to different points of view, indicated by different kinds of covered seats or garden structures. In conducting the walks, and distributing the trees and shrubs, considerable skill and taste have been displayed in concealing the distant walks, and those which cross the lawn in different directions, from the windows of the living-rooms; and also in never showing any walk but the one which is being walked on, to a spectator making the circuit of the grounds.

APPENDIX. 489

[Fig. 3. View from the Drawing-room Window at Chesnut Cottage, looking to the Right.]

490 APPENDIX.

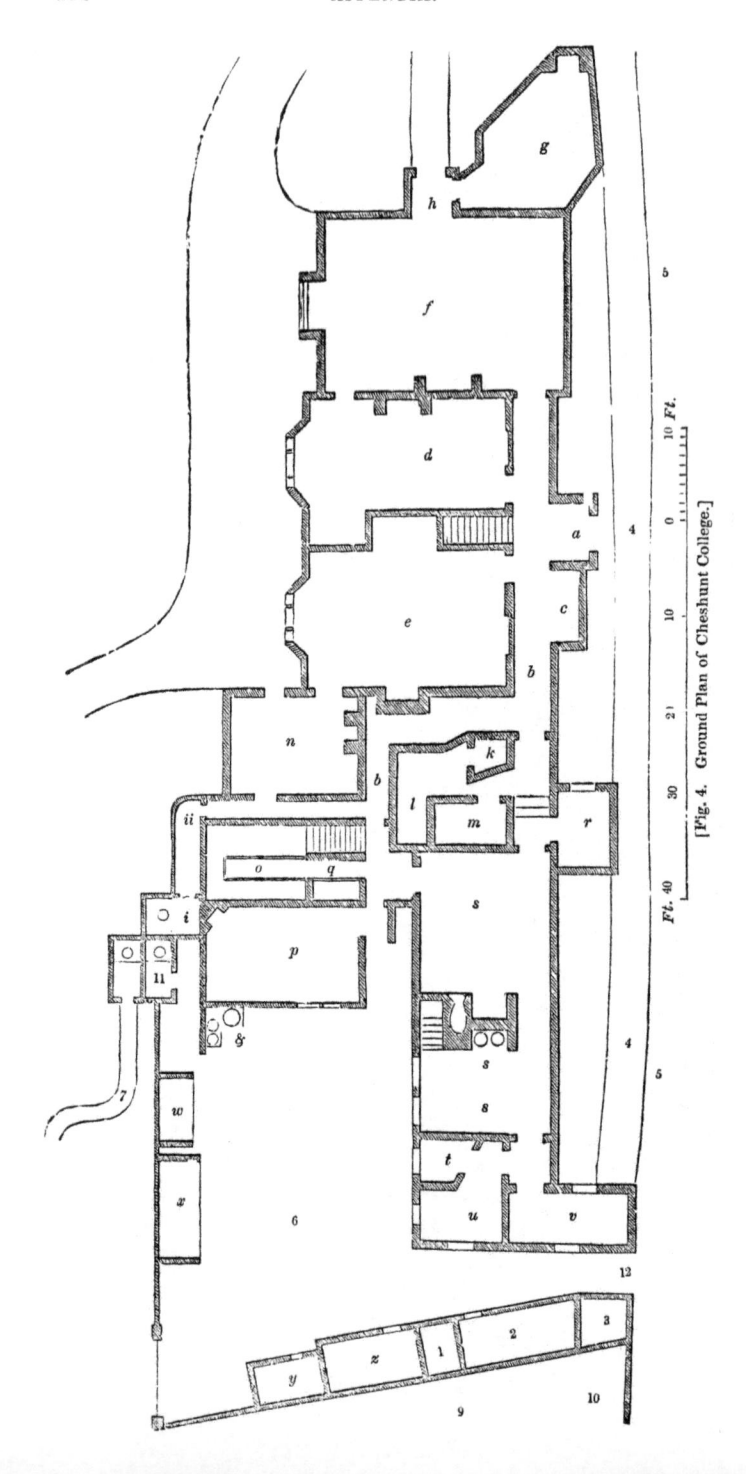

[Fig. 4. Ground Plan of Cheshunt College.]

Before we enter into further details, we shall describe, first, the plan of the house; secondly, that of the farm and garden offices and the hothouses; and, thirdly, the general plan of the grounds.

The house, in its external form and interior arrangement, is to be considered as a cottage, or rather as a villa assuming a cottage character. Hence, the centre part of the house, over the dining and drawing-rooms, appears from the elevation of the entrance front to be only two stories high. There is, however, a concealed story over part of the offices, for servants' bedrooms.

The house, of which Fig. 4 is an enlarged plan, consists of:

a, The porch, entered from a bridge thrown across the brook, 4, as shown in Fig. 4.

b b, Passage from which are seen the stairs to the bedrooms; and in which, at *ii*, there is a jib-door and a ventilating window, to prevent the possibility of the smell from the kitchen or offices, or water-closet, penetrating to the other parts of the passage.

c, Recess for coats, hats, &c., fitted up with a hat and umbrella-stand, tables, &c.

d, Drawing-room, with a recess at the further end, fitted up with a sofa and a writing-table.

e, Dining-room, with a recess for the largest sideboard, and another for a smaller sideboard and cellarets.

f, Library, chiefly lighted from the roof, but having one window to the garden, and a glass door to the porch, *h*, also looking into the garden, and from which the view, Fig. 5, is obtained. This room is fitted up with book-cases all round; those on each side of the fire-place being over large cabinets, about 4 ft. 6 in. high, filled with a collection of shells, minerals, and organic remains, &c.; and, to save the space that would otherwise be lost at the angles, pentagonal closets are formed there, in which maps, and various articles that cannot be conveniently put on the regular book-shelves, are kept. The doors to these corner closets are not more than 9 in. in width, and they are of panelled wainscot. The shelves are fitted in front with mahogany double reeds, fixing the cloth which protects the tops of the books, thus giving the appearance of mahogany.

g, Museum for specimens of minerals and other curiosities, entered

[Fig. 5 View from the Library Porch.]

from the porch, *h,* and lighted from that porch and from a window in the roof.

h, Porch leading to the garden from the library and museum.

i, Ladies' water-closet kept warm by the heat from the back of the servants' hall fire; the back of the fire-place being a cast iron plate. *ii,* Jib-door. *k,* Plate-closet.

l, Butler's pantry, lighted from the roof.

m, China-closet, lighted from the roof.

n, Room serving as a passage between the dining-room and the garden and also between the dining-room and the water-closet *i,* containing a turning-lathe, a carpenter's work bench, a complete set of carpenters' tools, garden tools for pruning, &c., of all sorts; spuds with handles, graduated with feet and inches, fishing tackle, archery articles, &c.

o, Inner wine-cellar, where the principal stock of wine is kept. There is a ventilating opening from this cellar into the passage *b.*

p, Servants' hall.

q, Outer wine-cellar, where the wine given out weekly for use is placed, and entered in the butler's book. Between *q* and the passage *b,* are seen the stairs leading to the servants' bedrooms. *r,* Beer-cellar.

APPENDIX. 493

s, Kitchen, lighted from the roof, and from a window on one side.
ss, Scullery, lighted from one side. *t*, Housekeeper's closet. *u*, Coal-cellar. *v*, Larder. *w*, Bottle rack. *x*, Safe for cold meat. *y*, Wash-house.
z, Knife-house. *&*, Filtering apparatus. 1, Ash-pit. 2, Coal-house.
3, Fire-place to the vinery at 10, in the kitchen-garden 9.
4 4, Brook. 5 5, Public road. 6, Kitchen-court.
7, Concealed path to gentlemen's water-closet.
8, Plantation of evergreens. 9, Kitchen-garden.
10, Vinery. 11, House servants' water-closet.
12, Servants' entrance.

Though it cannot be said that the arrangement of the offices of this house is so good as it would be if they were placed on each side of a straight passage; yet it will not be denied that these offices include everything that is desirable for comfort and even luxury. The chief difficulty which occurs to a stranger, in looking at the plan, is, to discover how several of the rooms which compose the offices are lighted; and this, it may be necessary to state, is chiefly effected from the roof; a mode which, in the case of some rooms, such as a butler's pantry, china-closet, plate-room, &c., is to be preferred; but which in most cases it is desirable to avoid.

The three windows to the three principal rooms being on the same side of the house, and adjoining each other, must necessarily have a sameness of view; but the quiet character intended to be produced by the idea of a cottage by a road side, may be supposed to account for circumstances of this kind, and for various others.

The following are the details of the farmyard, garden offices, and hot-houses, as exhibited in Fig. 6:—

1, Rustic alcove, forming a recess under a thatched roof, which covers the space from the green-house, 3, to the houses or yards, 70, 71, and 72. This rustic alcove has the floor paved with small pebbles, and the sides and ceiling lined with young fir-wood, with the bark on. There is a disguised door on the right, which leads to 69, a house for grinding-mills and other machines; and on the left, which leads to 2, the ship-room. In the upper part of the central compartment, in a square recess fronting the entrance, is a white marble statue of

494   APPENDIX.

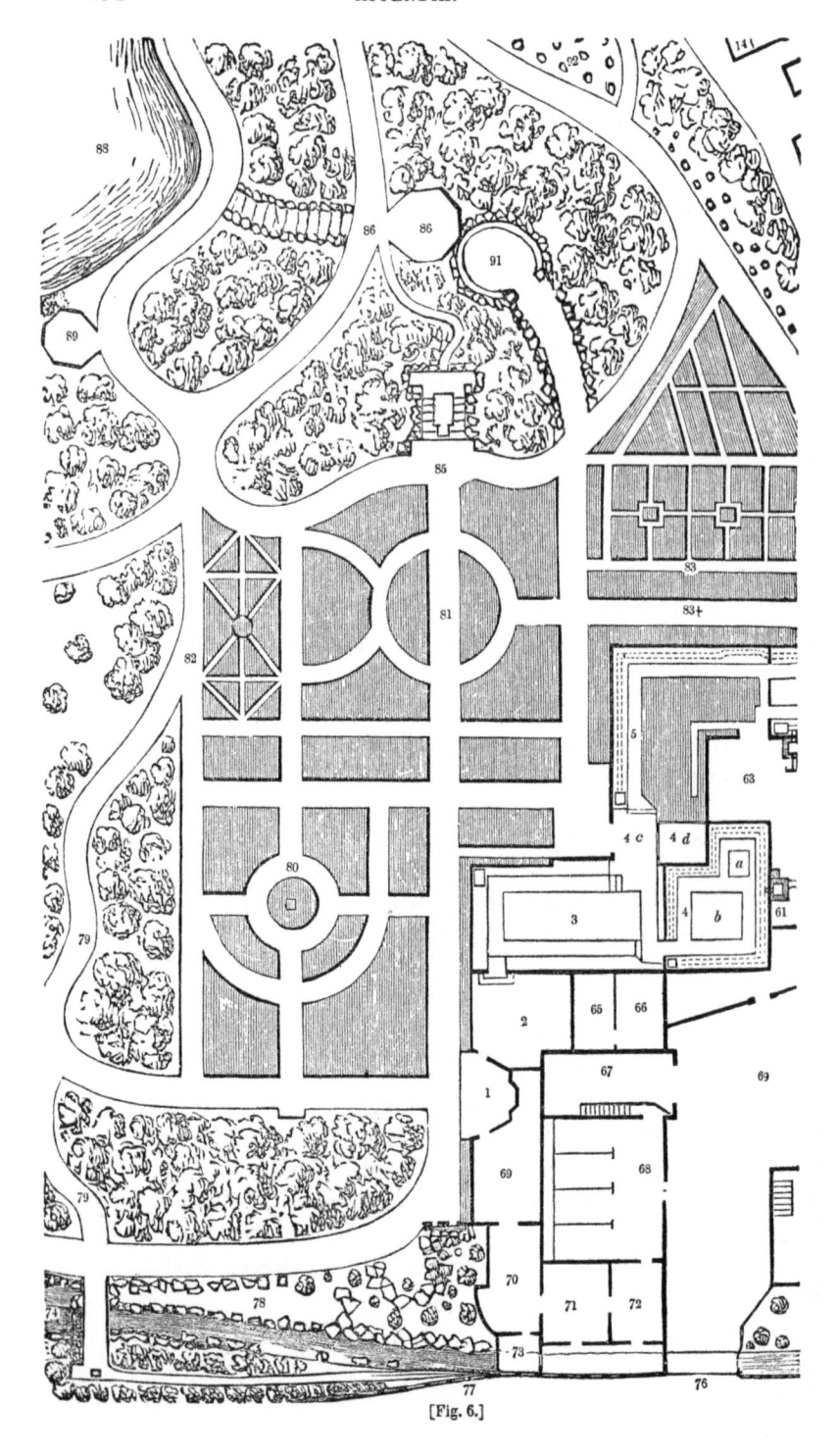

[Fig. 6.]

APPENDIX. 495

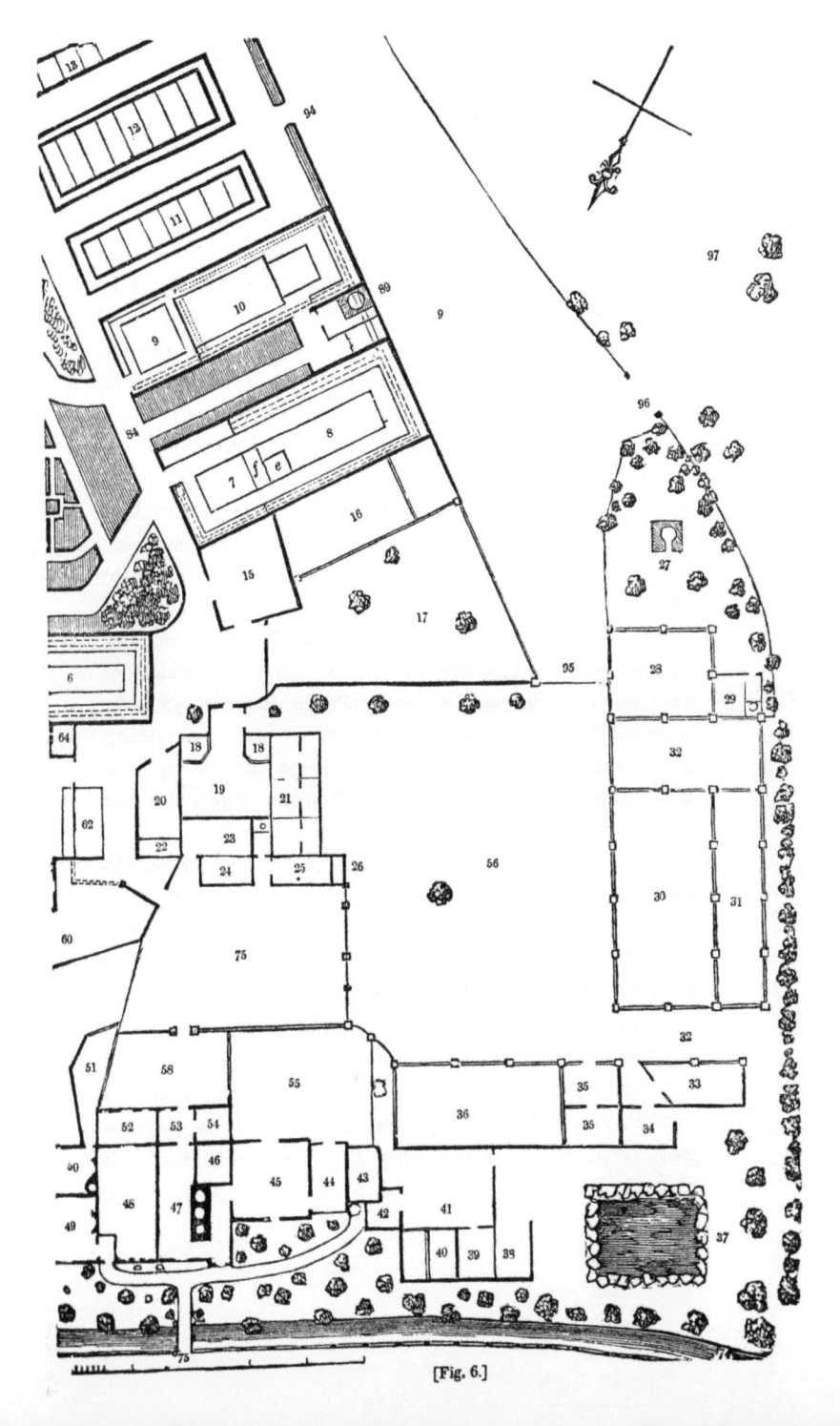

[Fig. 6.]

the Indian god Gaudama, or Gaudmia. Three Elizabethan benches, each as long as one of the sides of the alcove, are placed so as to disguise the doors. The external appearance of this alcove is shown in Fig. 7.

2, Ship-room, paved with slate, and with the walls finished in stucco, and ceiling with beams painted like oak, to which are hung Indian spears, and other curiosities, and serving to contain models of ships and vessels of various sorts during winter. These are placed on the pond in the summer season; square-rigged vessels at fixed anchorage, and the fore-and-aft-rigged ones, whose sails traverse, such as schooners, cutters, and coasting vessels, with cables of lengths to allow of their sailing without touching the edge of the pond; and these continue constantly traversing the pond when there is any wind. This room also contains a variety of the warlike instruments of the savages of different countries, a bust of Lord Nelson, one of the Duke of Wellington, some pictures in mosaic, and a number of East Indian curiosities. It serves also as a lobby to the orangery.

3, The orangery. The paths are of slate, and the centre bed, or pit, for the orange trees, is covered with an open wooden grating, on which are placed the smaller pots; while the larger ones, and the boxes and tubs, are let down through openings made in the grating, as deep as it may be necessary for the proper effect of the heads of the trees. This house, and that for Orchidàceæ, are heated from the boiler indicated at 61.

4, Orchidaceous and fern house, in which *a* is the stage for Orchidàceæ, and *b* a cone of rockwork, chiefly of vitrified bricks, for ferns. These ferns, amounting to above two dozen species, all sprang up accidentally from the soil attached to some plants which were sent to Mr. Harrison from Rio Janeiro and other parts of South America. The shelves round the house are also occupied with Orchidàceæ, all of which are in pots, in order that, when they come into flower, they may be removed to the green-house; as, when thus treated, as practised by the Duke of Devonshire at Chatsworth, they continue much longer in bloom than when kept in the degree of heat necessary for their growth.

APPENDIX. 497

4 *c*, Lobby between the orangery (3) and the conservatory (5).

4 *d*, An aviary for canaries, separated from the conservatory and the lobby by a wire grating, and from the orchidaceous house by a wall. Both the aviary and the lobby have a glass roof in the same plane as that of the conservatory, as may be seen in Fig. 8, in p. 499. In the winter season the temperature of the aviary being the same as that of the conservatory, the birds require little or no care, except giving them food; while they sing freely at that season, and greatly enliven this part of the garden scenery.

5, Conservatory, with vines under the rafters. The walks are slate, the shrubs are planted in a bed of free soil edged with slate, and the back wall is covered with different species of Passiflòra, and with the Tacsònia pinnatistípula.

6, Camellia-house. The camellias kept in pots; the rafters covered with vines, and the back wall with passiflòras and other climbers. This house, and also 5, are heated from one boiler, as indicated at 64.

7, Geranium-house. The roof is in the ridge and furrow manner of Mr. Paxton. This house, and also 8, 9, and 10, are heated from the boiler indicated at 89.

8, Botanic stove. The roof is in the ridge and furrow manner of Paxton. The sides of the pit are formed of slabs of slate; and there is a slate box at *e*, containing a plant of Mùsa Cavendíshii with a spike of fruit, two or three of which ripen off weekly. *F*. is a cistern for stove aquatics. There is a plant of Brugmánsia suavòlens (*Datùra* arbòrea *L.*) 15 ft. high, with a head 13 ft. in diameter. When we saw it, Aug. 10th, 277 blossoms were expanded at once, producing an effect upon the spectator under the tree, when looking up, which no language can describe. Last year it produced successions of blossoms, in one of which 600 were fully expanded at one time. This year it has had five successions of blossoms, and another is now coming out as the plant expands in growth. There is a large Brugmánsia coccínea in this house. Both these plants are in the free soil.

9, House for Cape heaths.

10, Pinery. The roof of this house is in the ridge and furrow manner,

32

498   APPENDIX.

in imitation of Mr. Paxton's mode; from which it differs, in having the ridge about one-third higher in proportion to the breadth, in having the sash-bar deeper, and placed at right angles to the crown of the ridge and to the furrow, and in having the panes of twice the size which they are in Mr. Paxton's roof. This house was built by Mr. Harrison's carpenter, from the general idea given to him; and before he had been to Chatsworth to examine the original house with this kind of roof, built there by Mr. Paxton.

[Fig. 7. Rustic Alcove.]

11, Cucumber-pit, on M'Phail's plan.
12, Succession pine-pit, also on M'Phail's plan, in order to be heated with dung linings.
13, Melon-pit.
14, Dutch cold-pit, for preserving lettuces, cauliflowers, etc., during winter.
15, Tool-house and potting-shed; the tools regularly hung on irons fixed to the ceiling, or set against the wall, or laid on shelves, the place for each sort of tool or implement, ropes, etc., being painted in large white letters on black boards. The following rules are painted on a board which is hung up in the tool-house:—

APPENDIX. 499

[Fig. 8. General View of the Hot-houses, as seen across the American Garden.]

"*Rules to be observed by all persons working on these Premises, Master and Man.*

"I. For every tool or implement of any description not returned to the usual place at night, or returned to a wrong place not appointed for it, or returned or hung up in a dirty or unfit state for work, the forfeit is 3d.

"II. For every heap of sweepings or rakings left at night uncleared, forfeit 3d.

"III. Every person making use of bad language to any person on these premises shall forfeit, for each and every such offence, 6d.

"IV. Every person found drunk on these premises shall forfeit one shilling; and, if he be in regular employment on the premises, he shall be suspended from his employment one day for every hour he loses through drunkenness.

"V. Every person who shall knowingly conceal or screen any person offending, shall be fined double the amount of the fine for the offence he so conceals, in addition to the fine of the offending party.

"VI. All forfeits to be paid to the gardener, on or before the Saturday night following. If any person working regularly on the premises fail to conform to the above rules and regulations, the gardener shall be at liberty to stop his fines from his wages. Further, should any foreman or journeyman fail to comply with the above rules and regulations (with a knowledge of them), the gardener shall be at liberty to seize and sell his tools or part of them, to pay such fines, in one month from the time the offence was committed.

"VII. All fines to be expended in a supper, yearly, to all the parties who have been fined."

When these rules were first adopted, the fines were sufficient to afford an annual supper with beer, &c.; but of late the amount has been so small, that Mr. Harrison has found it necessary to add to it to supply beer, &c., for the supper; a proof of the excellent working of the rules. Mr. Harrison remarks that these rules were established about eleven years ago, and that they have been most effective in preventing all slovenly practices; an advantage which he considers as thus purchased at a very cheap rate.

APPENDIX. 501

16, Mushroom-shed, in which the mushrooms are grown in Oldacre's manner.
17, Wood-yard, shaded by three elm trees.
18 18, Calf-pens.   19, Cow-house.   20, Tool-house.
21, Piggeries.
22, 23, 24, places for fattening poultry, on Mowbray's plan, not, as usual, in coops. Between this and 25, is a privy for the head gardener.
25, Place for meat for the pigs, which is passed through a shoot to 26.
26, Two tanks sunk in the ground, covered with hinged flaps, the upper edges of which lap under the plate above, so as to shoot off the rain, for souring the food intended for the pigs. One tank, which is much smaller than the other, is used chiefly for milk and meal for the fattening pigs, and sows with pigs; and the other for the wash and other refuse from the house, for the store pigs, which, with the refuse from the garden, apple-loft, &c., amply supplies the store pigs and sows, without any purchased food, except when they have pigs sucking. The good effect of the fermentation or souring is accounted for by chemists, who have found that it ruptures the ultimate particles of the meal or other food; a subject treated in detail in the *Quarterly Journal of Agriculture*, vol. vii. p. 445. According to the doctrine there laid down, the globules of meal, or farinaceous matter of the roots and seeds of plants, lie closely compacted together, within membranes so exquisitely thin and transparent that their texture is scarcely to be discerned with the most powerful microscope. Each farinaceous particle is, therefore, considered as enveloped in a vesicle, which it is necessary to burst, in order to allow the soluble or nutritious part to escape. This bursting is effected by boiling, or other modes of cookery; and also, to a certain extent, by the stomach, when too much food is not taken at a time; but it is also effected by the heat and decomposition produced by fermentation; and hence, fermented food, like food which has been cooked, is more easily digested than uncooked or unfermented food. Plants are nourished by the ultimate particles of manure in the same way that animals are nourished by the ultimate particles of food; and hence fermentation is as essential

to the dunghill as cookery is to food. The young gardener, as well as the young farmer, may learn from this the vast importance of fermentation, in preparing the food both for plants and animals.

27, Furnace and boiler, for boiling dogs' meat, heating pitch, &c.; placed in this distant and concealed spot, to prevent risk from fire when pitch or tar is boiled; and, when meat is boiled for dogs, to prevent the smell from reaching the garden. The reason why it is found necessary to have a boiler for tar is, that, most of the farm-buildings and garden-offices being of wood, it is found conducive to their preservation occasionally to coat them with tar heated to its boiling point.

28, Open shed for lumber.

29, Dog-kennel; adjoining which is a privy for the under gardeners.

30, Hay-barn. 31, Lean-to for straw.

32 32, Places for loaded hay-carts to unload, or to remain in when loaded during the night, in order to be ready to cart to town or to market early in the morning.

33, House for lumber, wood, &c. 34, Duck-house.

35 35, Houses for geese and turkeys.

36, Open shed for carts and farm implements.

37, Pond surrounded by rockwork and quince trees.

38, House for a spring-cart. 39, Coal-house for Mr. Pratt.

40 40, Places for young chickens. 41, Yard to chicken-houses.

42, Hatching-house for hens, containing boxes, each 1 ft. square within, with an opening in front 7 in. wide and 7 in. high, the top being arched, so that the sides of the opening are only 5 in. high.

43, Lobby to Mr. Pratt's house. 44, His kitchen.

45, Living-room.

46, Oven opening to 47.

47, Brewhouse, bakehouse, and scullery, containing a copper for brewing, another for the dairy utensils, and a third for washing, besides the oven already mentioned.

48, Dairy. The milk dishes are of white earthenware; zinc having been tried, but having been found not to throw up the cream so speedily and effectively as had been promised. One zinc dish, with handles, is used for clotted cream, which is regularly made during

APPENDIX. 503

the whole of the fruit season, and occasionally for dinner parties, for preserved tarts, &c. We observed here small tin cases for sending eggs and butter to town. The butter, wrapped in leaves, or a butter cloth, is placed in the bottom of a tin box about a foot square, so as to fill the box completely; and another tin box is placed over it, the inner box resting on a rebate, to prevent its crushing the butter below it. In this latter box, the eggs are packed in bran, after which the cover of the outer one is put on, and the whole may then be sent to any distance by coach. The dairy is supplied with water from a pump in the scullery; the water being conveniently distributed in both places by open tubes and pipes.

49, Coachman's living-room.

50, Coachman's kitchen, and stairs to two bedrooms over.

51, Court for inclosing the coachman's children.

52, Lobby to the dairy. 53, Lobby to Mr. Pratt's brew-house.

54, Cellar. 55, Chicken-yard.

56, Farmer's yard.

57, A gravelled court separating the court-yard, 59, from the stable-yard, 56.

58, Place for slaughtering in. 59, Stable-yard.

60, Shed for compost, and various other garden materials; such as a tub for liquid manure, in which it ferments and forms a scum on the top, while the liquid is drawn off below by a faucêt with a screw spigot, such as is common in Derbyshire and other parts of the north, which admits the water to come out through the under side of the faucet. Here are also kept paint pots, oil cans, boxes, baskets, and a variety of other matters. The whole of this shed is kept warm by the heat which escapes from the fire-place in 61, and from the back of the orchidaceous house, 4.

61, Fire-place and boiler for heating the orchidaceous house.

62, Place for arranging garden pots.

63, Shed, with roof of patent slates, which becomes a cheap mode of roofing in consequence of requiring so few rafters, amply lighted from the roof, and kept warm in the winter time by the heat proceeding from the boilers at 61 and 64. This shed contains a potting-bench, cistern of water, and compartments for mould; and, being lofty, it

504   APPENDIX.

contains in the upper part two apartments inclosed by wirework, for curious foreign pigeons or other birds. On the ground are set, during the winter season, the large agaves and other succulent plants which are then in a dormant state, and which are kept in the open garden during summer. On the whole, this is an exceedingly convenient working shed; being central to the houses 3, 4, 5, and 6; being kept comfortably warm by the boilers; being well lighted from the roof; and having the two windows indicated at 62, before which is the potting-bench.

64, Fire-place to the conservatory and camellia-house.

65, Place for keeping food for the rabbits and pigeons, with stairs to the pigeon-house, which is placed over it.

[Fig. 9. View from the Chinese Temple.]

66, Rabbit-house containing twenty-one hutches, each of which is a cubic box of 20 in. on the side. Each box is in two divisions, an eating-place and a sleeping-place; the sleeping-place is 8 in. wide, and is entered by an opening in the back part of the partition. Both divisions have an outer door in front; and, in order that the door of the sleeping-place may not be opened by any stranger, it is fastened by an iron pin, which cannot be seen or touched till the door of the eating-place is opened. Mr. Pratt pointed this out to us as an improvement in the construction of rabbit-hutches, well deserving of imitation wherever there is any chance of boys or idle persons getting into the rabbit-house. The rabbits are fed on garden vegetables and bran, barley, oatmeal, and hay, making frequent changes; the vegetables being gathered three or four days before being used, and laid in a heap to sweat, in order to deprive them of a portion of their moisture. Salt is also given occasionally with the bran. Cleanliness, and frequent change of food, have now, for five years, kept the rabbits in constant health. It ought never to be forgotten, that attention to the above rules, in partially drying green succulent vegetables, is essential to the thriving of rabbits kept in hutches; and, hence, in London and other large towns, instead of fresh vegetables, they are fed with clover hay. One of the kinds of rabbit bred at Mr. Harrison's is the hare rabbit, mentioned in the *Encyclopædia of Agriculture*, §7355, the flesh of which resembles that of the hare in quantity and flavor. Mr. Pratt has fed rabbits here, which, when killed, weighed 11 lbs. We can testify to their excellence when cooked.

67, Coach-house, with stairs to hay-loft. 68, Stable.

69, Mill-house, containing mills for bruising corn for poultry, a portable flour mill, a lathe, and grinding-machine for sharpening garden instruments and similar articles. In the Angel Inn in Oxford, some years ago, a lathe of this sort was used for cleaning shoes, the brushes being fixed to the circumference of the wheel, and the shoes applied to them, while the wheel was turned round by a tread lever, or treadle.

70, Root-house, containing binns for keeping different kinds of potatoes, carrots, parsnips, Jerusalem artichokes, beets, and yellow, French, and white turnips, with shelves for onions; and a loft over, which is used as a fruit room. The fruit is kept partly on shelves, and partly on cupboard trays.

71, Store place for beer or ale, which is brewed by Mr. Pratt for the use of the family in London, as well as Cheshunt; here is also a regular staircase to the fruit-room.

72, Harness-room, properly fitted up with every convenience, and warmed by a stove.

73, A lobby or court to a door which opens to the brook, for the purpose of clearing out an excavation made in the bottom of the channel, in order to intercept mud, and thus render the water quite clear where it passes along the pleasure-ground, and is seen from the library window and the grand walk (Fig. 5, p. 492). The whole of any mud which may collect in the brook may be wheeled up a plank through this door without dirtying the walk.

74 74, The brook.

75, Foot entrance to Mr. Pratt's house, the coachman's house, the dairy, etc.

76, Carriage entrance to the stable-court, garden offices, farm-yard, etc.

77, Private entrance to the garden, over the rustic bridge shown in Fig. 5.

78, Masses of vitrified bricks and blocks of stone, distributed among lawn and shrubs; among which, large plants of agave, and other rock exotics, are placed in the summer season, the pots and tubs being concealed by covering them with the stones forming the masses of rock-work. Here the semicircular space surrounded by rock contains a collection of Himalayan rhododendrons, etc., in pots, many of them seedlings which have not yet flowered.

79 79, American shrubbery, consisting chiefly of rhododendrons, azaleas, magnolias, etc., growing in the peat earth kept moist by the brook.

80, American garden consisting of choice American shrubs, and American herbaceous plants. In the centre of the circle a handsome tazza vase on a bold pedestal.

81, Two semicircles for dahlias; the surrounding compartments containing a collection of roses.

82. Garden of florist's flowers.

83 83, Garden of herbaceous plants, chiefly annuals. The walks in all these gardens are edged with slate. The bed 83† contains a collection of choice standard roses. 84, Dahlias.

APPENDIX. 507

85, Double ascent of the steps to a mound formed of the earth removed in excavating for the pond. From the platform to which these steps lead, there is a circuitous path to the Chinese temple; and the steps are ornamented with Chinese vases, thus affording a note of preparation for the Chinese temple. The outer sides of the steps are formed of rockwork, and between the two stairs is a pedestal with Chinese ornaments.

86, The Chinese temple, on the highest part of the mount formed of the soil taken from the excavation now constituting the pond. The view from the interior of this temple is shown in Fig. 9, p. 504.

87, Rustic steps descending from the Chinese temple to the walk which borders the pond. 88, The pond.

89, Open tent, with sheet-iron roof supported by iron rods. This structure may be seen in the view Fig. 10.

[Fig. 10. Distant view of the House and Tent across the Pond.]

90 90, Masses of evergreens and deciduous trees and shrubs.

91, Grotto, made late last year, not yet completed. It was formerly an outer ice-house, but it failed as such. The entrance is surrounded by rockwork, and the interior in the form of a horseshoe, furnished with a wooden bench as a seat. Over this grotto, is an umbrella tent, as shown in the view Fig. 11. 92, Dahlias.

[Fig. 11. Grotto, with Umbrella Tent over.]

93, Slip of ground for compost, and various other materials requisite for the garden and farm-yard; communicating with the frame-ground by the door 94, with the farm-yard by the gate 95, and with the farm by the gate 96.

94, Door from the frame-ground to the slip behind.

95, Gate from the slip to the farmyard.

96, A gate from the slip to the fields of the farm.

97, Grass field, forming part of the farm.

Fig. 13, in pp. 510, 511, is a vertical profile of the gardens and pleasure ground, with the farmyard, and a small portion of the farm. This view shows:—

1, The house. 2, The domestic offices and yard. 3, Vinery in a small garden.

4, Back entrance to the domestic offices, and the smaller kitchen garden. On one side of this walk is placed one of Fuller's portable ice-boxes.

5, The smaller kitchen-garden.

6, Broad border for pits; and in which there is a cold pit for protecting vegetables during winter.

7, Boundary plantation.

8, Angular brick wall, for the sake of having different aspects for the fruit trees which are trained against it; and for strength, being only one brick in thickness for lessening the expense.

9, Pond in the largest kitchen garden, supplied from the brook by pipes with waste pipe to the pond on the lawn.

10, Filbert plantation.

11, Orchard and boundary plantation.

[Fig. 12. Covered Seat, of grotesque and rustic Masonry.]

12, Covered seat, of which a view is shown in Fig. 12. In front of this seat there is a mulberry tree of large dimensions, which was transplanted by Mr. Harrison, when it was upwards of 80 years of age. The instruments with which a number of large plants, particularly shrubs, were transplanted under Mr. Harrison's direction, when the grounds were being altered and enlarged, were described for us by Mr. Pratt. (See *Gardener's Magazine*, vol. xi. p. 134.) Mr. Pratt kept for many years large plants which had suffered from many causes, or which were not immediately wanted, in what he called an hospital for these purposes.

13, A flower garden, in which for several years a large Araucària brasiliènsis stood out in the centre bed; but it was killed to the ground in the winter of 1837-8.

14, The rustic covered seat, shown in Fig 14, in p. 513, and of which Fig. 15 is an elevation of the back, showing the manner in which the barked poles are arranged.

510   APPENDIX.

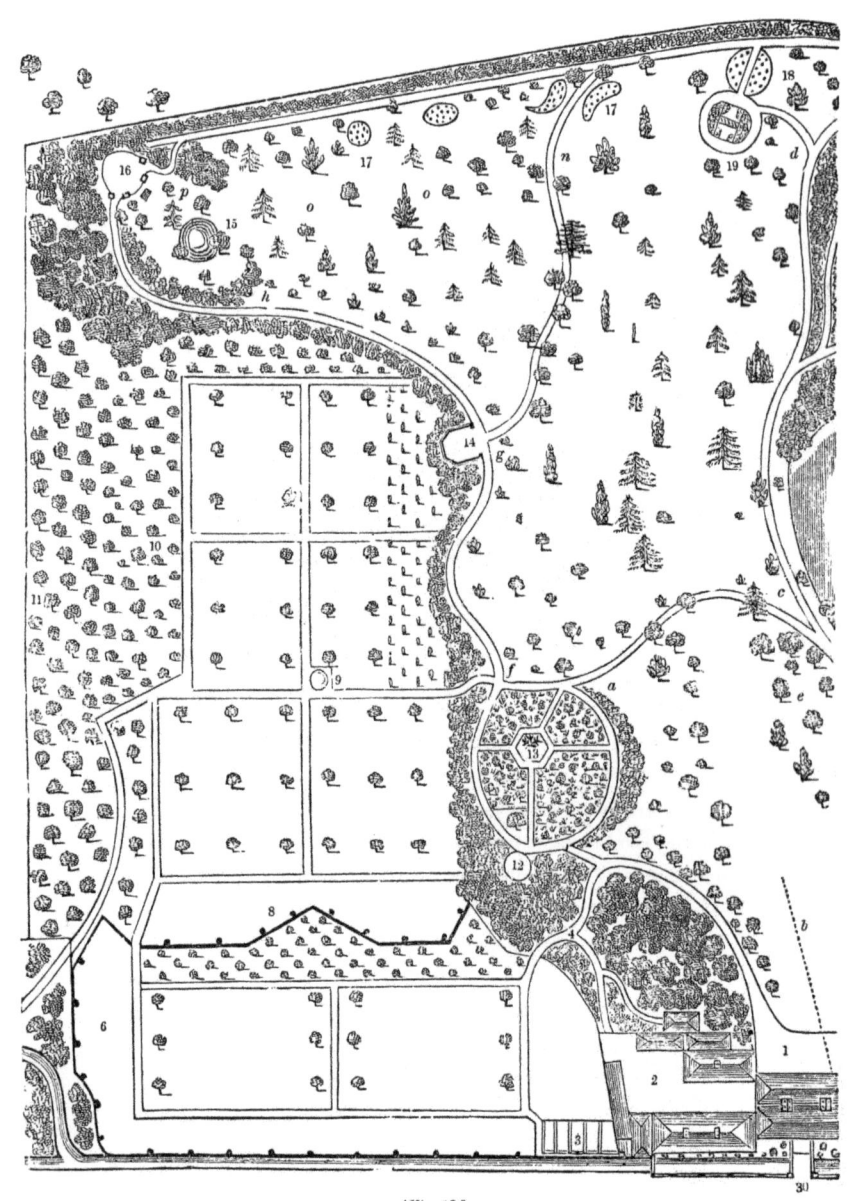

[Fig. 13.]

APPENDIX. 511

[Fig. 13.]

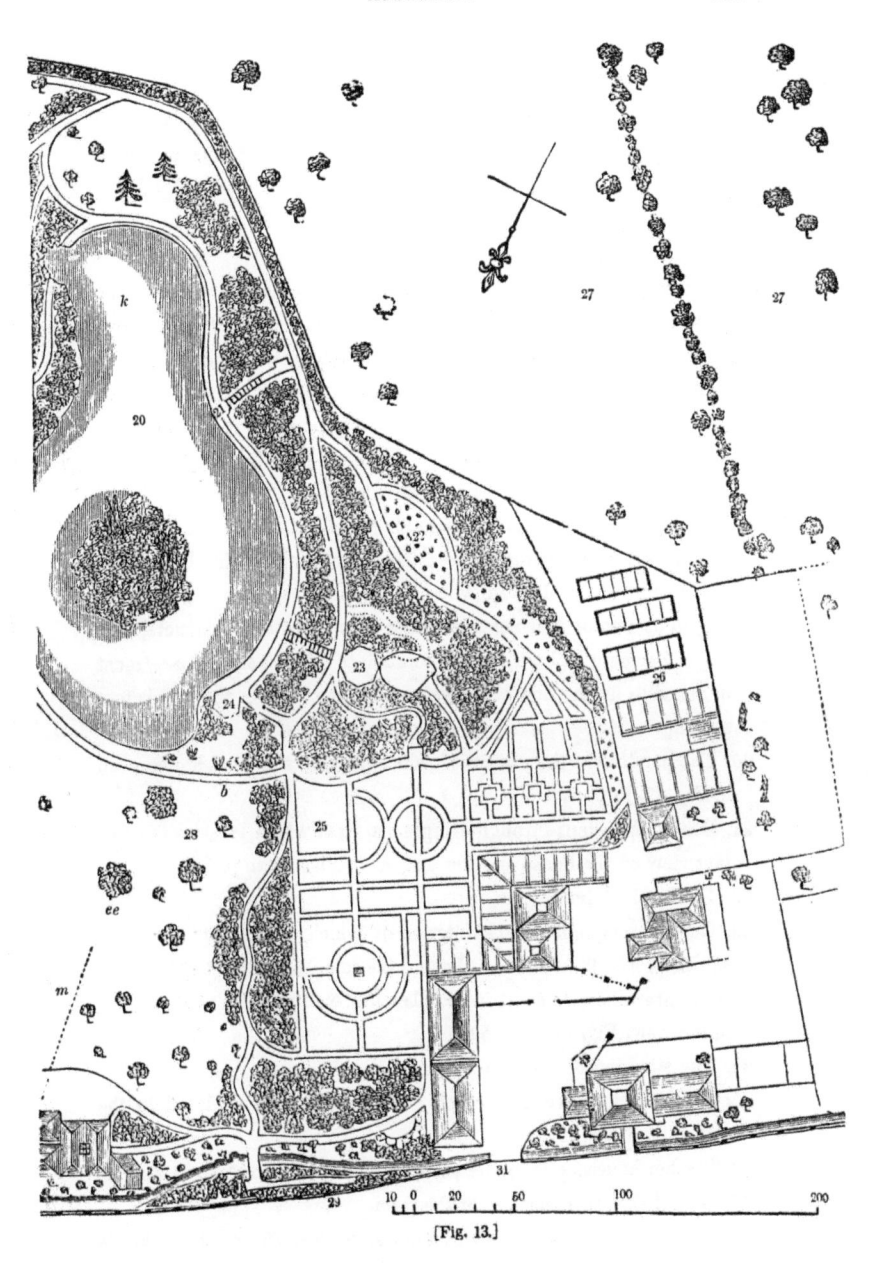

15, Basins of water for aquatics.

16, Rustic building, of which a view is shown in Fig. 16. In the interior is an alto-relievo of statuary marble, representing a female over a funeral vase, surrounded by a sort of broad frame of corals, cornua Ammonis, and large mineral specimens of different kinds.

17, Groups of roses, dahlias, and other ornamental flowers.

18, Two semicircular beds of roses.

19, A covered double seat, one half looking towards the roses and the other in the opposite direction. In the latter are kept the instruments for playing at what is called lawn billiards, which is said to be a game intermediate between bowls and common billiards. This game is little known, but materials for playing at it are sold by Messrs. Cato & Son, wire-workers, Holborn Hill, London, who sent out with them the following printed rules:—

"This game, which differs from all others, should be played on a lawn about twelve yards square; the socket with the ring being fixed in the centre by a block of wood fixed into the earth. It may be played by two or four persons, either separately, or as partners, each player having a ball with a cue pointed to correspond. Care must be taken to fix the ring at the end of the cue close to the ball before striking."

20, The pond. On the margin of which, at k, is the boat-house seen in Fig. 17, in p. 517.

21, Descending steps through evergreens, from which is seen the distant view of the house and the tent, as in Fig. 10, in p. 507.

22, Dahlia plantation.

23, Chinese temple, from the interior of which is obtained the view shown in Fig. 9, in p. 504. Behind the temple, a little to one side, is the grotto shown at 91 in the plan, Fig. 6, in pp. 494, 495, and also in the view, Fig. 11, in p. 508.

24, The situation of the tent shown in Fig. 10.

25, The different flower and shrub gardens described in detail in the plan, Fig. 6, pp. 494, 495.

26, The hot-houses, pits, frames, farm buildings, &c., shown in Fig. 6.

27, Grass fields, forming part of the farm.

28, Point from which the view of the hot-houses, Fig. 8, in p. 499, is

APPENDIX. 513

taken, and also, turning round, the view of the house, Fig. 18, in p. 519.

29, Secret entrance to the grounds. 30, Principal entrance to the house.

31, Entrance to the stable-court and farmyard.

[Fig. 14. Rustic Covered Seat, of Woodwork.]

Remarks.—In pointing out the principal sources of the professional instruction which a young gardener may derive from examining this

[Fig. 15. Elevation of the Back.]

514 APPENDIX.

place, we shall first direct attention to the garden structures. These, whether of the ornamental or useful kind, are executed substantially, and with great care and neatness; while the farm buildings, being chiefly of wood, show how great an extent of accommodation may be obtained without regularity of plan, and without incurring much expense. A good exercise for the young designer would be to distribute the same accommodation, properly classed, along the sides of a square

[Fig. 16. Hermit's Seat, and Classical Vase.]

or squares, or along the sides of a parallelogram or polygon, and either detached from or connected with the horticultural buildings.

The manner in which the working-sheds are heated by the waste heat from the furnaces, in consequence of which, in severe weather, much more work will be done in them, and in a better manner, and in which they are lighted, so as to serve for protecting certain kinds of plants during winter, is worthy of imitation; as is the mode of heating so many different houses from only three boilers. In no garden structures have we seen a more judicious use of the Penrhyn slate;

paths, edgings, shelves, cisterns, boxes for plants, copings, kerbs, partitions, and substitutes for dwarf walls, being all made of it. The order and neatness with which all the different tools, utensils, &c., are kept in the horticultural and farm buildings, are most exemplary, and greatly facilitate the despatch of business.

In the farm buildings, the fittings up of the poultry-houses, the rabbit-house, and the dairy and scullery, well deserve attention; and also the arrangement for fermenting the food of the pigs in underground cisterns, not too warm for summer, nor so cold as to check fermentation in winter. The manure of the horses, of the cows, of the pigs, of the rabbits, of the pigeons, and of the poultry, is kept in separate pits, that it may be used, if desirable, in making up different composts.

There are three liquid-manure tanks, in which the liquid matter, which in most farmyards is wasted, is fermented, and afterwards mixed up with soil for use in the kitchen-garden, or used in forming composts for particular plants. The liquid-manure from the stables is kept apart from that from the cow-house; and the general drainings of the yard, and of the frame-ground in the kitchen-garden, are fermented by themselves. The liquid manure with which Mr. Pratt waters his plants is formed chiefly of the sweepings of the pigeon, rabbit, and cow houses, with lime; and is kept in a cask in a close shed (60 in the plan Fig. 6, in p. 494, 495), so that the temperature admits of its fermenting in winter, as well as in summer: a thick scum rises to the top of the cask, and the liquid is drawn out from the bottom as clear as old ale. The plants which Mr. Pratt waters with this liquid are chiefly those of rapid growth, such as the *Datùra*, Brugmán*sia*, and other soft-wooded tree plants, which, like these, are cut in every year, and appear to profit by the stimulating effect of this manure. He gives it also, occasionally, to various other plants which appear to want vigor; but has not yet had sufficient experience of its effects, to give a list of plants to which it ought to be applied.

In order to produce as much manure as possible, as well for the farm as for the garden, all leaves, haulm, and waste vegetable matters, are carefully collected, and fermented by the addition of fresh stable dung; and heaps of different kinds of soils, procured from different parts of the country, are constantly kept in the slip adjoining the frame-ground, ready for use.

The grounds being nearly level are readily supplied with water from the ponds and from the brook; and there are concealed wells, communicating with these sources by pipes from the brook, in different parts of the grounds, and more especially in the kitchen-garden, from which the plants can be abundantly watered in the growing season with comparatively little labor; there being six different places, including the ponds and brook, from which the gardeners take water, and all the strawberries are planted close to the wells in the inner and outer walled gardens.

The kitchen-gardens, the hot-houses, and the store-houses and some other structures, can be locked up at pleasure, Mr. Harrison and Mr. Pratt being the only persons having complete master keys. Part of the outer kitchen-garden is inclosed with an open iron spike fence, 5 ft. 6 in. high, within which and the inner walled garden are the strawberries and choicest gooseberries, figs, etc., and these inclosures are opened only by the master keys. The whole, therefore, of the wall and best fruit is secured from plunder.

The beauties of this place, as has been already mentioned, depend chiefly on the taste and judgment displayed in laying out the walks, and distributing the trees and shrubs; though the choice of a situation for the pond, and the mount adjoining it, is also a matter of some consequence.

The trees and shrubs, being comparatively limited in number, consist of one of almost every kind that is to be procured in British nurseries, exclusive of those which are common, or not considered ornamental. In selecting these, the more rare kinds have been procured, and planted quite young; Mr. Harrison and Mr. Pratt having found, by experience, that the pines and firs should be planted out when not more than of three or four years' growth. When the plants have been in pots, the balls should be gently broken with the hand, and afterwards all the earth washed away from the roots by the application of water. The plant may then be placed on a hill of prepared mould, and the roots stretched out, so as to radiate from the plant in every direction, and afterwards covered with mould.

APPENDIX. 517

[Fig. 17. Boat House and Agave Mount.]

The masses of trees and shrubs are chiefly on the mount near the lake, and along the margin which shuts out the kitchen-garden; and in these places they are planted in the gardenesque manner, so as to produce irregular groups of trees, with masses of evergreen and deciduous shrubs as undergrowth, intersected by glades of turf. They are scattered over the general surface of the lawn, so as to produce a continually varying effect, as viewed from the walks; and so as to disguise the boundary, and prevent the eye from seeing from one extremity of the grounds to the other, and thus ascertain their extent. The only points at which the lawn is seen directly across from the drawing-room window are in the direction of l and m, Fig. 13, in pp. 510, 511; but, through these openings, the grass field beyond appears united with the lawn; so that the extent thus given to the views from the drawing-room windows is of the greatest assistance to the character of the place, with reference to extent. From every other part of the grounds, the views across the lawn are interrupted by some tree, bush, or object which conceals the boundary; or, if the boundary is seen on one side, as in passing along the walk from 16 by 18 to 22, there is ample space on the lawn side to keep up the idea of extent.

In many situations, this walk, as seen on paper, would be considered to be too near the boundary; but in the grounds the narrow plantation from 22 to 18 is of evergreens, chiefly hollies, which already partially shut out all view of the boundary or the field, and which are ultimately intended to spread their upper branches over the walk, so as to give it a character of shade and gloom, different from any other in these grounds.

In general, it may be laid down as a rule, that the boundary between a lawn and the park or field beyond should not be such as to cut the landscape, as it were, in two; and another rule is, that the walks should never be so near this fence, or should not be so conducted when near it, as to admit of the spectator looking directly across. Indeed, in scenery, no rule is generally more applicable than this, viz. that all straight lines, whether fences, roads, canals, or rivers, and all regular symmetrical objects, such as buildings, should be looked at obliquely. Applying this rule, therefore, to the scenery between the walk and the fence, from 18 to 16, we should say that either the direction of the walk ought to be altered, so as to remove it further from the boundary, or the boundary extended further into the field; and instead of being bordered by a hedge-like fringe of shrubs, it should only be broken here and there by occasional bushes and trees, connected and harmonizing in position with other trees beyond the fence. If it were desirable to avoid altering the boundary, then we should recommend continuing the walk which commences at *d* near 19, by *n* and *o o*, to *p* near 16. If there were nothing to see or be seen beyond the boundary, then, unless the boundary fence were a conservative wall, that is, a wall covered with half-hardy ornamental plants, we should still prefer changing the direction of the walk, so as to take away from the monotonous appearance of continually skirting the boundary. In every place, however small, there ought to be some part left which the visitor has not seen, and which may leave the impression on his mind, that, however much he has been shown, he has not seen everything. We make these observations with great deference to Mr. Harrison, who has paid much attention to the subject of Landscape Gardening, and shown much practical taste and good sense both in that art and in architecture.

APPENDIX. 519

It is, however, right to state that Mr. Harrison accords with our general view of the subject, but " defends the walk in question as an exception founded on his objects in making it; which were, 1st, to have a walk different from any other in the garden; and 2d, a walk sheltertered from the winter southerly gales, and ornamented by the bloom of the laurustinus at that season. It is, therefore, so slightly curved as merely to avoid a straight line, and permits an extent of length, not found in any other part, to be seen on descending the elevation at the east end, or on emerging from wood at the west end, where, when the improvements connected with it are finished, it will enter a dense plantation, the walk going round at the back of the building in that corner. The fence would have been entirely excluded from either near or distant view, and the eye carried so as not to catch a view of the grounds of the field nearer than one hundred yards or more at the least, if the laurustinuses had not suffered so severely in 1837-38; but these will by next year, and by trees already planted along the border, and others to be planted irregularly, at intervals, in the field near the fence in a

[Fig. 18. Garden Front of Cheshunt Cottage.]

great measure, Mr. Harrison thinks, obviate the objection made, or at least lessen the force of it, as future appearances will, he thinks, prove.—W. H."

The trees and shrubs on the lawn are almost all disposed in the gardenesque manner; that is, so that each individual plant may assume its natural shape and habit of growth. The masses are also chiefly planted in the same style; and, as the trees and shrubs advance in growth, they are cut in, or thinned out, so that each individual, if separated from the mass to which it belongs, and considered by itself alone, shall be a handsome plant. At the same time, in order to produce as much variety as possible, the picturesque style of planting, in which trees and shrubs are so closely grouped together as partially to injure each other's growth, occasionally occurs, for the sake of producing variety. With the exception of the pines and firs, the other trees have been selected more for their picturesque effect and variety of foliage, than for their botanical interest. Among these are the Scotch pine for its darkness; the P. pulus angulàta for its large leaves, and for its property of preserving these till destroyed by severe frost, long before which all the other poplars have become naked; the A cer macrophyllum, for its large leaves; the Montpelier maple, for its small ones; the *Negúndo fraxinifòlium*, for its green-barked shoots; the American oaks, for the singular variety in form and color of their foliage; the catalpa, for its broad rich yellowish leaves, and its showy blossoms, which appear late in the season; the deciduous cypress; the bonduc, or Kentucky coffee tree; the cut-leaved alder, the tulip tree, the purple beech, the purple hazel, the Oriental plane, of which there are several fine specimens, the variegated sycamore, and other variegated trees and shrubs, which are always so beautiful in spring; those thorns and crabs which are beautiful or remarkable for their blossoms in the spring, and for their fruit in autumn; the Nepal sorbus, so interesting for its large woolly leaves, which die off of a fine straw color; the magnolias; the rhododendrons, the heaths, the brooms, and the double-blossomed furze, besides various striking or popular plants, such as the variegated hollies, the scarlet arbutus, etc. Among the detached trees and small groups, there is scarcely to be met with a single bush or tree that a general observer will not find noticeable for something in its foliage, general form, flowers, or fruit. The Magnòlia grandiflòra var. exoniénsis flowers freely as a standard without any protection, and was not even injured by the winter of 1837–8; nor was A'rbutus procéra, also un-

APPENDIX. 521

protected. A number of the more rare trees and shrubs, such as Araucària brasiliénsis, which had stood out eight years, A Cunninghàm*ii*, Pìnus insìgnis, P. palústris, P. Girard*iána*, P. canariénsis, etc., were killed during the winter of 1837-8, and a number of others, which were severely injured, are now recovering. Mr. Pratt, the head gardener, did not begin to prune the trees which were injured till the rising of the sap showed the extent of the injury that they had received. After waiting till the middle of summer, it was found that the laurustinus, sweet bay, Chinese privet, and various other shrubs, were alive to the height of from 3 ft. to 5 ft., and after the dead wood was cut out, the plants soon became covered with young shoots and foliage.

The Walks are so laid out and planted as to be sheltered or bordered by evergreens, for the sake of their lively appearance during winter. They are also so contrived as to be shaded from the sun by deciduous trees during summer; while these trees being naked during winter, admit the sun at that season to dry the grounds. The walks are laid out in different directions, in order that, from whatever point the wind may blow, at least one walk will be sheltered from it. The greater number are in the direction of north and south, because walks in that direction are best exposed to the sun in the winter season, which is the period of the year in which the proprietor chiefly resides here. It is always desirable, in a small place, that all the walks should be concealed from the windows, except that immediately under the eye, and that, in walking through the grounds, no path should be seen except the one walked on, and that (except in the case of a straight avenue) only for a moderate distance. These rules (derived from the principle of variety and intricacy) have been carefully attended to by Mr. Harrison, and hence the walk from a to b, in the plan, Fig. 13, in pp. 510, 511, is concealed by raising the turf on the side next the house higher than on the opposite side, while that from c to d is concealed by the bushes and trees at c, and more especially by a large rhododendron at ee. The walk fgh is concealed from the walk i, partly by a swell in the surface of the turf on the side next i, but chiefly by the bushes which are scattered along its margin. At g, there is a clump which prevents any one on the walk i from seeing the line gf, and any one on the walk g

522   APPENDIX.

f from seeing the line *i*. In walking along from *f* to *h*, it is clear that the trees and shrubs on the left hand will always prevent the eye from seeing the walk to any great distance. All the other walks through the lawn are concealed in a similar manner, so that a person walking in the grounds never sees any other walk than that which lies immediately before him, and, therefore, in looking across the lawn, he never can discover the extent either of what he has seen, or of what he has yet to see. To form a great number of walks of this sort, and lead the spectator over them without showing him more than one walk at a time, but taking care, at the same time, to let him have frequent and extensive views across the lawn, and these views always different, constitute the grand secret of making a small place look large.

The walks are filled to the brim with gravel, kept firmly rolled, and their grass margins are clipt, but never cut, because the gravel, being almost as high as the turf, the latter can never sink down, and swell

[Fig. 19. View across the Water, looking towards the House.]

APPENDIX. 523

out over the former. This it invariably does when the turf is a few inches higher than the gravel, and, hence, paring off the part of the turf which had projected was originally, no doubt, adopted only as a remedy for the evil, though it is now erroneously practised by gardeners as an evidence of care and good keeping. As much of the beauty of the walk depends upon the beauty of its boundary, the feeling that this boundary is likely to be disturbed every time the walk is cleaned, or the adjoining turf mown, is extremely disagreeable. The freshly pared turf becomes a spot or scar in the scene, withdrawing the attention from the walk itself, and from the adjoining grounds, to a point, or rather a line, which is in itself of little consequence, but which, by the paring, is obtruded on the eye, so as to destroy all allusion to stability. We are displeased with the paring of the edges, because it conveys the idea that the walks are not finished, or that they are liable to be disturbed in this way from time to time, and nothing either in grounds or in buildings, is more unsatisfactory than an apparent want of stability or fixedness. It is as much the nature of the ground to be fixed and immovable, as it is of trees and shrubs to increase in growth, and hence, any operation, such as clipping, which seems to stop the growth of the one, is as unsatisfactory to the eye as paring, which seems to derange the fixed state of the other. Would that we could impress this on the minds of all gardeners and their employers!

The Pond is of an irregular shape, so arranged as with the assistance of the island to prevent the whole of it, and consequently its limited extent, from being seen from any one point in the garden. For the same reason, the walk only goes along one side, there being but one point on the western side, viz. where the iron seats are close to the agaves, from which any part of the pond can be seen. The pond is so situated as to form the main feature in the right hand view from the drawing-room window, as shown in Fig. 3, in p. 487; the wooded island (which is shown rather too much in the middle in the plan, though, perhaps, not so in reality) disguising the boundary from that and every other point of view. The bank of the pond on one side is rocky, and nearly perpendicular, while on the other it is sloping, and partly covered with shrubs. At *k*, in Fig. 13, in p. 511, there is a boat house, on the top of which are several large agaves, the common, the variegated, and Agàve plicátilis; the tubs containing which are so disguised by rockwork, as to create an allusion to the appearance of these plants in their native habitats. The appearance of these agaves, and also of a large crassula, is indicated in a view of the boat house, Fig. 17, in p. 517, and it is only from a seat among these agaves that any part of the pond can be seen from this side of it. Had a walk been conducted completely round the pond, and near its margin, the charm of partial concealment would have been entirely lost. The high banks have been formed with earth taken out of the pond, and these have given occasion to a considerable variety in the inclination, as well as in the direction, of the walks. The banks are planted on the same principle as the open lawn, that is, with trees and shrubs having striking foliage or showy flowers, and with a judicious mixture of evergreens to give the effect of cheerfulness in winter. In the water are two large plants of Cálla æthiòpica, L*in.*, which cover a space of nearly 5 ft. in diameter; they have lived there through ten winters without any protection, the water being 5 ft. deep, and they flower luxuriantly every year. The views across the water, to the house and to the other parts of the grounds, are singularly varied, owing to the winding direction of the walk, and the consequently changing position of the island, and of the trees in the foreground and middle distance. One of these views may be seen in Fig. 19, and others have been already given in pp. 487, 504, 507, 517.

The *Flower-Garden* (25, in Fig. 13, in pp. 510, 511), is laid out, as the ground plan indicates, in beds, everywhere bordered with slate; a flower-garden of this kind, with the walks gravelled, having the advantage of rendering the flowers accessible to ladies immediately after rain, when they are often in their greatest beauty, and, at all events, in their greatest freshness and vigor, an advantage which is not obtained when the beds are on turf. There are also flower-beds on turf in other parts of the grounds, but these are filled with roses, dahlias, and other large-growing plants in masses, the beauties of which do not require to be closely examined.

APPENDIX. 525

III.

Note on the treatment of Lawns

As a lawn is the *ground-work* of a landscape garden, and as the management of a dressed grass surface is still a somewhat ill-understood subject with us, some of our readers will, perhaps, be glad to receive a very few hints on this subject.

The unrivalled beauty of the "velvet lawns" of England has passed into a proverb. This is undoubtedly owing, in some measure, to their superior care and keeping, but mainly to the highly favorable climate of that moist and sea-girt land. In a very dry climate it is nearly impossible to preserve that emerald freshness in a grass surface, that belongs only to a country of " weeping skies." During all the present season, on the Hudson, where we write, the constant succession of showers has given us, even in the heat of midsummer, a softness and verdure of lawn that can scarcely be surpassed in any climate or country.

Our climate, however, is in the middle states one of too much heat and brilliancy of sun, to allow us to keep our lawns in the best condition without considerable care. Beautifully verdant in spring and autumn, they are often liable to suffer from drought in midsummer. On sandy soils, this is especially the case, while on strong loamy soils, a considerable drought will be endured without injury to the good appearance of the grass. It therefore is a suggestion worthy of the attention of the lover of a fine lawn, who is looking about for a country residence, to carefully avoid one where the soil is *sandy*. The only remedy in such a soil is a tedious and expensive one, that of constant and plentiful topdressing with a compost of manure and heavy soil—marsh mud—swamp muck, or the like. Should it fortunately be the case (which is very rare) that the sub-stratum is loamy, deep ploughing, or trenching, by bringing up and mixing with the light surface soil some of the heavier earth from below, will speedily tend to remedy the evil.

In almost all cases where the soil is of good strength, a permanent lawn may be secured by preparing the soil deeply before finally laying it down. This may be done readily, at but little outlay, by *deep ploughing*—a good and cheap substitute for trenching—that is to say, making the plough follow three times in the same furrow. This, with manure, if necessary, will secure a depth of soil sufficient to allow the roots of plants to strike below the effects of a surface drought.

In sowing a lawn, the best mixture of grasses that we can recommend for this climate, is a mixture of Red-top and white Clover—two natural grasses found by almost every roadside—in the proportion of three fourths of the former, to one of the latter.

There is a common and very absurd notion current (which we have several times practically disproved), that, in order to lay down a lawn well, it is better to sow the seed along with that of some grain; thus, starving the growth of a small plant by forcing it to grow with a larger and coarser one. A whole year is always lost by this process—indeed more frequently two. Many trials have convinced us that the proper mode is to sow a heavy crop of grass at once, and we advise him who desires to have speedily a handsome turf, to follow the English practice, and sow *three to four bushels of seed to the acre.* If this is done early in the spring, he will have a lawn-like surface by mid-summer, and a fine close turf the next season.

After this, the whole beauty of a lawn depends on *frequent mowing.* Once a fortnight at the furthest, is the rule for all portions of the lawn in the neighborhood of the house, or near the principal walks. A longer growth than this will only leave yellow and coarser stubble after mowing, instead of a soft velvet surface. A broad-bladed English scythe (to be had at the shops of the seedsman), set nearly parallel to the surface, is the instrument for the purpose, and with it a clever mower will be able to shave within half an inch of the ground, without leaving any marks. To free the surface from worm casts, etc., it is a common practice to roll the previous evening as much as may be mown the next day.

As the neatness of a well kept lawn depends mainly upon the manner in which it is mown, and as this again can only be well done where there are no inequalities in the ground, it follows that the surface should be kept as smooth as possible. Before sowing a lawn, too much pains cannot be taken to render its surface smooth and even. After this, in the spring, before the grass starts, it should be examined,

APPENDIX. 527

and all little holes and irregularities filled up, and the same should be looked over at any annual top-dressing that may take place. The occasional use of a heavy roller, after rain, will also greatly tend to remedy all defects of this nature.

Where a piece of land is long kept in lawn, it must have an occasional top-dressing every two or three years, if the soil is rich, or every season, if it is poor. As early as possible in the spring is the best time to apply such a top-dressing, which may be a compost of any decayed vegetable or animal matter—heavier and more abounding with marsh mud, etc., just in proportion to the natural lightness of the soil. Indeed almost every season the lawn should be looked over, all weeds taken out, and any poor or impoverished spots plentifully top-dressed, and, if necessary, sprinkled with a little fresh seed. Wood ashes, either fresh or leached, is also one of the most efficient fertilizers of a lawn.

We can already, especially in the finer places on the Hudson, and about Boston, boast of many finely kept lawns, and we hope every day, as the better class of country residences increases, to see this indispensable feature in tasteful grounds becoming better understood and more universal.

IV.

Note on professional quackery.

Landscape Gardening, like all other arts, is not free from ignorant pretenders to knowledge, who, without a spark of appreciation for the beautiful in nature, boldly undertake to remodel, in what they consider a tasteful and fashionable style, every piece of natural landscape, whether of a simple or highly picturesque character. They succeed in leaving behind them, on the places they attempt to improve, indubitable marks of their footsteps, in a sort of labored ease, and stiff striving after grace; but they are pretty certain, also, to mar or obliterate in a great degree, the natural charm of any fine situation. We have seen one or two examples lately where a foreign *soi-disant* landscape gardener has completely spoiled the simple grand beauty of a fine river residence, by cutting up the breadth of a fine lawn with a ridiculous effort at what he considered a very charming arrangement of walks and groups of trees. In this case he only followed a mode sufficiently common and appropriate in a level inland country, like that of Germany, from whence he introduced it, but entirely out of keeping with the bold and lake-like features of the landscape which he thus made discordant.

One of this kind of improvers was, some years ago, very cleverly satirized by Mr. Peacock, an English reviewer of celebrity, in a comic work entitled "Headlong Hall." The latter is the name of the supposed seat of Lord Littlebrain, who has assembled around him during the Christmas feastings an odd party, among whom is Mr. Milestone, the landscape gardener, evidently a portrait of "Capability Brown." Mr. Milestone has been examining the estate, and, full of his projected park, is exhibiting his portfolio of drawings of the proposed improvements to his host and some of the guests.

"Mr. Milestone.—This, you perceive, is the natural state of one part of the grounds. Here is a wood, never yet touched by the finger of taste; thick, intricate, and gloomy. Here is a little stream, dashing from stone to stone, and overshadowed with these untrimmed boughs.

Miss Tenorina.—The sweet romantic spot! How beautifully the birds must sing there on a summer evening.

Miss Graziosa.—Dear sister! how can you endure the horrid thicket?

Mr. Milestone.—You are right, Miss Graziosa; your taste is correct, perfectly *en règle*. Now, here is the same place corrected—trimmed—polished—decorated—adorned. Here sweeps a plantation, in that beautiful regular curve; there winds a gravel walk; here are parts of the old wood, left in these majestic circular clumps disposed at equal distances with wonderful symmetry; there are some single shrubs scattered in elegant profusion; here a Portugal laurel, there a juniper; here a laurustinus, there a spruce fir; here a larch, there a lilac; here a rhododendron, there an arbutus. The stream, you see, is become a canal: the banks are perfectly smooth and green, sloping to the water's edge, and there is Lord Littlebrain, rowing in an elegant boat.

Squire Headlong.—Magical, faith!

Mr. Milestone.—Here is another part of the ground in its natural state. Here is a large rock, with the mountain-ash rooted in its fissures, overgrown, as you see, with ivy and moss, and from this part of it bursts a little fountain, that runs bubbling down its rugged sides.

Miss Tenorina.—O how beautiful! How I should love the melody of that miniature cascade!

Mr. Milestone.—Beautiful, Miss Tenorina! Hideous. Base, common, and popular. Such a thing as you may see anywhere, in wild and mountainous districts. Now, observe the metamorphosis. Here is the same rock, cut into the shape of a giant. In one hand he holds a horn, through which the little fountain is thrown to a prodigious elevation. In the other is a ponderous stone, so exactly balanced as to be apparently ready to fall on the head of any person who may happen to be beneath,* and there is Lord Littlebrain walking under it.

Squire Headlong.—Miraculous, by Mahomet!

Mr. Milestone.—This is the summit of a hill, covered, as you perceive, with wood, and with those mossy stones scattered at random under the trees.

Miss Tenorina.—What a delightful spot to read in, on a summer's day! The air must be so pure, and the wind must sound so divinely in the tops of those old pines!

Mr. Milestone.—Bad taste, Miss Tenorina. Bad taste, I assure you. Here is the spot improved. The trees are cut down; the stones are cleared away; this is an octagonal pavilion, exactly on the centre of the summit, and there you see Lord Littlebrain, on the top of the pavilion, enjoying the prospect with a telescope.

Squire Headlong.—Glorious, egad!

Mr. Milestone.—Here is a rugged, mountainous road, leading through impervious shades; the ass and the four goats characterize a wild uncultured scene. Here, as you perceive, it is totally changed into a beautiful gravel road, gracefully curving through a belt of limes, and there is Lord Littlebrain driving four-in-hand.

Squire Headlong.—Egregious, by Jupiter!

Mr. Milestone.—Here is Littlebrain Castle, a Gothic, moss-grown

* See Knight on Taste.

structure, half-bosomed in trees. Near the casement of that turret is an owl peeping from the ivy.

Squire Headlong.—And devilish wise he looks.

Mr. Milestone.—Here is the new house, without a tree near it, standing in the midst of an undulating lawn; a white, polished angular building, reflected to a nicety in this waveless lake, and there you see Lord Littlebrain looking out of the window."

V.

Note on Walks and Roads.

In our remarks on walks and roads, we omitted to say anything of the best manner of making gravel walks. We may here state that, where it can easily be procured, pure *pit* gravel is preferable to all other materials for this purpose, as it binds almost at once, and becomes a firm and solid mass nearly as hard as a stone floor. Beach gravel, not having any mixture of loamy particles, does not become hard until after a good deal of rolling, and a little loam is often mixed with it to secure its tenacity and firmness. A very thin coat of gravel will render a walk superior to a path which consists only of the natural soil, and such surfacing, in our dry climate (though it frequently requires renewing), is often sufficient for distant walks, or those little used except in fine weather. But the approach road, and all walks immediately about the dwelling, should be laid at least a foot thick with gravel, to insure dryness, and a firm footing at all times and seasons. The lower six inches is better executed when filled with small stones —placing the six inches of gravel on the top of these, and there are few new places where this is not a convenient mode of getting rid of the small stones that require to be taken out of the gardens, and various parts of the premises undergoing improvement.

A word may be said here with regard to the *color* of gravel. Undoubtedly in almost all examples in the natural style of landscape gardening *slate-colored* gravel, the kind common in nearly all parts of the country, is much the most agreeable to the eye, being unobtrusive, just differing sufficiently with the soil to be readily recognised as artistical in its effect, while it harmonizes with the color of the ground, and the soft tints of vegetation. A thirst after something new has induced some persons, even in the interior, to substitute, at considerable cost, the white gravel of the sea-shore for the common pit or beach gravel. The change, we think, is, in point of taste, not a happy one. The strong white of this gravel, as the painters would say, *disturbs the tone* of a simply beautiful landscape, whose prevailing tints are those of the broad lawn and rich overshadowing trees; and the glare of these snowy white pebbles is not, we confess, so pleasing in our eyes as the cooler and more quiet color of the slate or grey gravel. When we add to this, that these sea-side pebbles seldom or never pack or become firm, it would appear very evident that they are far less suitable for walks than the common material. The only situation where this brilliant gravel seems to us perfectly in keeping, is in the highly artificial garden of the ancient or geometric style, or in the symmetrical terrace flower garden adjoining the house. In these instances its striking appearance is in excellent keeping with the expression of all the surrounding objects, and it renders more forcible and striking the highly artificial and artistical character of the scene; and to such situations we would gladly see its use limited.

The labor and expense of keeping the roads and walks clean, and free from weeds, in a place of large extent (and some of our seats have now several miles of private roads and walks within their own limits), is a very considerable item of the annual outlay of a country residence. At a recent visit to Blithewood, we saw in operation there a very simple implement, invented by R. Donaldson, Esq., the intelligent proprietor of that beautiful place, which promises to be of important service as a labor-saving machine in cleaning roads and walks. In Fig. 20 is shown a sketch of this implement, in use. In general appearance it is not unlike the frame of a wheelbarrow, except that instead of the two legs it has two iron bars, reaching down to the earth, and connecting with a transverse blade, about three inches wide, which is set nearly parallel with the ground. The handles of the implement are held by a workman, like those of the common double-tailed plough, while the horse which draws it is led or ridden by a boy. With this implement, which is three and a half feet wide, all the weeds in the space it covers are cleared from a road or walk as rapidly as a horse can walk forward, and it is only necessary to follow with a rake and remove the weeds, and the whole is in good order.

On the lower portion of the upright bars, where they rise from the blade, there is an edge for cutting the turf on the sides of the walk, which performs its work very well and rapidly—the horse being carefully led; and it will, no doubt, answer perfectly for this purpose, in all those walks and roads not directly around the house, or where the greatest nicety is not required.

The simplicity of this machine, the very small cost at which it is made, and the great saving of expense and labor which it secures will, we think, render it a valuable acquisition to all owners of large places, or to those wishing to keep up a long series of private roads and walks in the picturesque manner. For smaller gardens and grounds, where the most scrupulous nicety is observed, there is, of course, nothing that will supersede the common hoe, rake, and roller.

[Fig. 20. Implement in use at Blithewood for cleaning gravel roads.]

THE END.